Virtual Geographic Environments

Edited by

Hui Lin and Michael Batty

Science Press
Beijing, China

Responsible Editors: *Peng Shengchao and Guan Yan*

Published by Science Press
16 Donghuangchenggen North Street
Beijing 100717, P. R. China

Printed in Beijing

ISBN 978-7-03-023467-4

Dedicated to

Professor Chen Shupeng

1920–2008

Senior Fellow of Chinese Academy of Sciences, Fellow of International Eurasian Academy of Sciences and the Third World Academy of Sciences

Chinese Pioneer in Geographic Information Science

About the Editors

Hui Lin is Professor and Director of Institute of Space and Earth Information Science (ISEIS) of The Chinese University of Hong Kong (CUHK). He is Director of Hong Kong Base of National Remote Sensing Center of China. He graduated from Wuhan Technical University of Surveying and Mapping in 1980, received Master of Science degree from Graduate School of Chinese Academy of Sciences in 1983, and Ph.D. from University at Buffalo in 1992. Hui Lin joined CUHK in 1993 with major research interests in satellite remote sensing, virtual geographic environments (VGE), and urban and historic GIS. He was elected academician of International Eurasia Academy of Sciences in 1995 and member of the Expert Committee of China Lunar Exploring Program in 2008. Hui Lin is the founding president of International Association of Chinese Professionals in Geographic Information Science (CPGIS) and the chief editor of the journal *Annals of GIS*.

Michael Batty is Bartlett Professor of Planning at University College London where he directs the Centre for Advanced Spatial Analysis (CASA, www.casa.ucl.ac.uk). He was Director of the National Center for Geographic Information and Analysis (NCGIA) at SUNY-Buffalo from 1990 to 1995 and Professor of City Planning at the University of Cardiff from 1979 to 1989. His work is on simulation in urban and regional systems, focussing on both aggregate and disaggregate land use transport models, cellular automata and agent based models. He has also worked on visualisation in geographic information science, on 3D-GIS and virtual cities, and on the development of fractal geometry, social physics and scaling in city systems. His most recent book is *Cities and Complexity* (MIT Press, 2005) and he is editor of the journal *Environment and Planning B*.

Contributors

Katy Appleton
School of Environmental Sciences, University of East Anglia, Norwich, UK
Email: k.appleton@uea.ac.uk

Masatoshi Arikawa
Center for Spatial Information Science, University of Tokyo, Japan
Email: arikawa@csis.u-tokyo.ac.jp

Michael Batty
Centre for Advanced Spatial Analysis, University College London, UK
Email: m.batty@ucl.ac.uk

Iestyn D. Bleasdale-Shepherd
Valve Software, Bellevue, WA, USA
Email: iestynne@gmail.com

Chris Brunsdon
Department of Geography, University of Leicester, UK
Email: cb179@le.ac.uk

Guoray Cai
College of Information Sciences and Technology, Pennsylvania State University, University Park, PA, USA
Email: cai@ist.psu.edu

Andrew Crooks
Centre for Advanced Spatial Analysis, University College London, UK
Email: andrew.crooks@ucl.ac.uk

Jack Dangermond
President, ESRI Inc., Redlands, CA
Email: jdangermond@esri.com

Jürgen Döllner
Hasso Plattner Institute, University of Potsdam, Germany
Email: doellner@hpi.uni-potsdam.de

Hideyuki Fujita
Center for Spatial Information Science, University of Tokyo, Japan
Email: fujita@csis.u-tokyo.ac.jp

Dan Gonen
Department of Zoology, Tel Aviv University, Israel
Email: gonendan@post.tau.ac.il

Michael F. Goodchild
Department of Geography, University of California, Santa Barbara, CA, USA
Email: good@ncgia.ucsb.edu

Jianhua Gong
Institute of Remote Sensing Applications, CAS, Beijing, China
Email: jhgong@irsa.ac.cn

Yi Gong
Centre for Advanced Spatial Analysis, University College London, UK
Email: y.gong@ucl.ac.uk

Subhrajit Guhathakurta
School of Planning, Arizona State University, Phoenix, AZ, USA
Email: Subhro.guha@asu.edu

Hans Hagen
Department of Computer Science, University of Kaiserslautern, Germany
Email: hagen@informatik.uni-kl.de

Yoshitugu Hayashi
Faculty of Engineering, Nagoya University, Japan
Email: sugihara1128@yahoo.co.jp

Florian Höpel
Department of Computer Science, University of Kaiserslautern, Germany
Email: f_hoepel@informatik.uni-kl.de

Qingwu Hu
School of Remote Sensing and Information Engineering, Wuhan University, China
Email: qwhu@263.net

Mingxiang Huang
Institute of Remote Sensing Applications, CAS, Beijing, China
Email: hmx780202@163.com

Andrew Hudson-Smith
Centre for Advanced Spatial Analysis, University College London, UK
Email: andyhudsonsmith@googlemail.com

Yuzuru Isoda
College of Asia Pacific Studies, Ritsumeikan Asia Pacific University, Beppu City, Oita, Japan
Email: yuzuruisoda@gmail.com

Mathias Jahnke
Department of Cartography, Technical University of Munich, Germany
Email: Mathias.Jahnke@bv.tum.de

Claire H. Jarvis
Department of Geography, University of Leicester, UK
Email: chj2@le.ac.uk

Mitsuhiko Kawakami
Graduate School of Natural Science and Technology, Kanazawa University, Japan
Email: kawakami@t.kanazawa-u.ac.jp

Tatsunori Kawasumi
College of Letters, Ritsumeikan University, Kyoto, Japan
Email: kawasumi@lt.ritsumei.ac.jp

Jan Eric Kyprianidis
Hasso Plattner Institute, University of Potsdam, Germany
Email: kyprianidis@hpi.uni-potsdam.de

Deren Li
State Key Laboratory of Information Engineering in Surveying, Mapping and Remote Sensing, Wuhan University, China
Email: drli@whu.edu.cn

Jing Li
Department of Geography, University of Leicester, UK
Email: jl185@le.ac.uk

Wenhang Li
Institute of Remote Sensing Applications, CAS, Beijing, China
Email: mylihang@163.com

Yabin Li
Institute of Remote Sensing Applications, CAS, Beijing, China
Email: lybin_2001@163.com

Hui Lin
Institute of Space and Earth Information Science, Chinese University of Hong Kong, China
Email: huilin@cuhk.edu.hk

Andrew Lovett
School of Environmental Sciences, University of East Anglia, Norwich, UK
Email: a.lovett@uea.ac.uk

Roger Mackett
Centre for Transport Studies, University College London, UK
Email: rlm@transport.ucl.ac.uk

Liqiu Meng
Department of Cartography, Technical University of Munich, Germany
Email: meng@bv.tum.de

Ariane Middel
Department of Computer Science, University of Kaiserslautern, Germany
Email: middel@rhrk.uni-kl.de

Tomoki Nakaya
College of Letters, Ritsumeikan University, Japan
Email: nakaya@lt.ritsumei.ac.jp

Peter-Scott Olech
Department of Computer Science, Univer-

sity of Kaiserslautern, Germany
Email: olech@informatik.uni-kl.de

Philip Paar
Zuse Institute, Berlin, Germany
Email: paar@lenne3d.com

Donna J. Peuquet
Department of Geography, Pennsylvania State University, University Park, PA, USA
Email: peuquet@ems.psu.edu

Juval Portugali
ESLab (Environmental Simulation Laboratory), Tel Aviv University, Israel
Email: juval@post.tau.ac.il

Lutz Ross
Institute of Landscape Architecture and Environmental Planning, Technical University of Berlin, Germany
Email: lutz.ross@tu-berlin.de

Asaf Roz
ESLab (Environmental Simulation Laboratory), Tel Aviv University, Israel
Email: asaf@eslab.tau.il

Oz Shalom
ESLab (Environmental Simulation Laboratory), Tel Aviv University, Israel
Email: ozshalom@post.tau.ac.il

Zhenfeng Shao
State Key Laboratory of Information Engineering in Surveying, Mapping and Remote Sensing, Wuhan University, China
Email: shaozhenfeng@163.com

Zhenjiang Shen
Graduate School of Natural Science and Technology, Kanazawa University, Japan
Email: gisvrshen@hotmail.co.jp

Ifan D. H. Shepherd
Business School, Middlesex University, London, UK
Email: I.Shepherd@mdx.ac.uk

Jingjing Shi
Institute of Space and Earth Information Science, Chinese University of Hong Kong, China
Email: sjingjing@yahoo.com

Kenichi Sugihara
Faculty of Business Administration, Gifu Keizai University, Japan
Email: sugihara1128@yahoo.co.jp

C. Vincent Tao
Microsoft Virtual Earth Business Unit, Redmond, WA, USA
Email: tao@yorku.ca

Masayasu Tsunekawa
Graduate School of Natural Science and Technology, Kanazawa University, Japan
Email: gisvrshen@hotmail.co.jp

Mi Wang
State Key Laboratory of Information Engineering in Surveying, Mapping and Remote Sensing, Wuhan University, China
Email: wangmi@lmars.whu.edu.cn

Lei Wu
Institute of Space and Earth Information Science, Chinese University of Hong Kong, China
Email: wulei@cuhk.edu.hk

Kazuki Yanagisawa
Graduate School of Urban Environmental Sciences, Tokyo Metropolitan University, Japan
Email: kazuki_05190@yahoo.co.jp

Keiji Yano
College of Letters, Ritsumeikan University, Kyoto, Japan
Email: yano@lt.ritsumei.ac.jp

Tohru Yoshikawa
Graduate School of Urban Environmental Sciences, Tokyo Metropolitan University, Japan
Email: tyoshik@f5.dion.ne.jp

Foreword
The Future of GIS and the Web

Geographic information systems are now deeply embedded in contemporary information infrastructure in ways that would have seemed incredibly to those like myself who were involved in their initial development nearly half a century ago. No one ever expected these systems to become universally applicable with literally millions of professionals engaged in their use. Furthermore, the use of GIS in routine queries is hidden from many more users whose decisions are affected by them in countless ways as such systems become every more widely used in queries that require details of location. None of this was anticipated because the idea that computer memories could be associated with graphics was largely unknown to potential users before the PC was developed in the 1970s and only when graphical user interfaces became standard in the early 1990s did GIS propel itself into the public consciousness along with many other kinds of graphic systems involving design, drawing and painting.

Much of this revolution has been driven by miniaturisation and exponentially increasing speeds of computer hardware, specifically memories, and this has led to a massive proliferation of machines from handheld devices to various kinds of supercomputer. But dissemination is also by driven by the convergence of computers and communications where computer hardware, software and data are increasingly accessible from remote locations and where the network is all important in accessing these resources. In this, mapping and GIS play an increasingly important part. The development of systems at many different levels from research and professional use to routine queries by non expert users who require map data, represent a burgeoning of map-based technologies that are beginning to empower many groups that hitherto have not been exposed to geographic information.

This book collects together a set of key papers that define the current momentum that is driving GIS. In particular, the various contributions illustrate the cutting edge of GI Science as well as new applications of GIS with the processing and delivery of geographic information through the web and handheld devices forming two major directions to these developments. The idea that geographic information can be made available through web-based services, using Web 2.0 technologies that network many millions of people together, has formed a major research thrust in software development over the last decade. But the notion that these web-based systems can be used to collect information of a voluntary kind through methods of crowd sourcing is also an exciting and widely unanticipated development that is driving the field. The creation of map data through systems such as **OpenStreetMap** is providing an entirely new source of user-created information which promises to open up GIS again in ways that were largely unanticipated a generation or more ago. As these services gain ground, new business models are being invented that merge proprietary and non-proprietary systems and novel ways of integrating diverse software through many different process of software development from map hacks to open systems architectures are being invented.

At ESRI, we are ahead of the wave on many of these developments and see the opening up of geographic information in the ways reported in this book as nothing but healthy for the field. Moreover, the focus here on the use of GIS in developing design is something that has a special place in my own thinking about the future of the field. In this book, there are papers dealing with ways in which 3D environments can be constructed using 3D-GIS which was pioneered in the various **ArcView** products and is now an essential part of the **ArcGIS** toolkit. But the coupling of these systems to various forms of CAD and the development of neat and fast ways of rendering such environments is adding to our arsenal of technologies that are taking GIS into areas that hitherto were the domain of professionals that did not think in terms of geography, or even of design. We see GIS as an essential component in constructing the environments that represent the virtual geographies theme of this book that ties all the papers together. In some senses, such environments are the natural consequence of linking GIS to other technologies that deal with information, design, and service provision, and this will undoubtedly grow as it becomes ever easier to integrate diverse software and data across the web.

In providing a foreword to this book, I want to stress that virtual geographic environments or VGE as the editors call them, are essential to using GIS in design. Before a design for a city or landscape, for example, can be produced, an environment must be created through GIS which is then fashioned into a form where users have access to it, first to enhance their understanding through exploration, and then to enable them to change various components in the effort to solve problems that can realise better designs. We have talked about sketch planning in this fashion for a long time but only now through the development of virtual city models and through new ways of enabling users to interact with geographic information using new screen technologies are we approaching the point where design is possible. We still need to specify design processes in stitching these diverse software for design together with GIS but we now stand at a threshold where such design using GIS is at last possible. Many of the contributions in this book illustrate the components that will be essential to future spatial design systems and I welcome this edited collection as it serves as yet another pointer to the future of our field. There is an exciting decade of development before us and many of the new directions to the future of GIS are both implicit and explicit in the contributions that follow.

President, ESRI Inc, Redlands, CA.
March 2009

Preface

Virtual Geographic Environments (VGEs) build on the development of geographic information systems (GIS) and geographic information science (GI Science) in ways that focus on how users are able to embed themselves within such systems. VGEs thus constitute environments in which the user interacts with the geographical or spatial system of interest through immersion in its spatial representation, the methods used to model it, and the processes used to utilize such environments in problem-solving. Virtual reality (VR) technologies, computer-aided design (CAD), games, animations, movies, indeed the whole plethora of multimedia with geo-coded data and geo-process models, can be enlisted to enrich and extend such environments.

Geographic information systems were first developed over 40 years ago with Roger Tomlinson being the first to define the term GIS as part of his early efforts in developing such a system for resource management in Canada. These early systems essentially tagged information to place and space through some system of geocoding but they were more than this in that processes that characterized the working of such systems were implicit in their organization. As these systems developed initially as mainframe applications through workstation technologies where graphics became rather more central through to desktop applications, geoprocessing has become central while user interaction has become ever more significant in the way such systems are built and applied. With the movement of GIS technologies to the web, particularly to web-based services as well as grid and cloud computing applications organized around highly decentralized and rather seamless interactions between remote servers and various clients, the notion of the geographic environment has become important in terms of the way users might interact with the system and with each other becoming central to the use of VR technologies of various kinds.

This has not only generated new forms of GIS but has promoted more experimental approaches to the use of GIS for problem-solving. In fact the pipeline from data input to geoprocessing to the visualization of outputs is being continually extended as new forms of GIS not only reach forward to users but back to new ways of collecting data at different scales of media from GPS on hand-held devices through to satellite remote sensing. Integration along and across the input-output pipeline is a feature that define many of the VGEs that are presented in this book.

In the first chapter, we present a primer on VGEs which identifies all their salient features and prepares the reader for the contributions which follow. In Part 1, we focus on the cutting edge of VGEs which relate to online environments and Web 2.0 services. The move to linking users to VGEs through the web is the predominant way in which the pipeline is being extended back to data and forward to applications and in the rest of the book, various contributions unpack the technical implications of these developments. Part 2 deals with virtual cities and virtual landscapes, where the focus is on the nuts and bolts of how 3D-GIS and

CAD is being used for new forms of representation and analysis with a particular focus on new forms of rapid rendering of complex scenes. Hand-held devices enter the discussion here for many of these virtual realities are being accessed and viewed in situations of rapid mobility using such devices. This is continued in Part 3 where user interface design for public participation is the focus while in Part 4, the contributions focus on the construction of mobile and networked VGEs. In the final part, mobility and dynamics in visualization are explored. Various contributions stress ways in which new media can capture the dynamics of process as VGEs move from static representations with dynamic user interaction to representations that capture much more dynamic and rapid change while at the same time embedding users every more closely to such environments.

The chapters in this volume were all presented at an international conference on developments in visualization and virtual environments in geographic information science, under the general title of Virtual Geographic Environments, the title of the current volume. The conference was held at the Chinese University of Hong Kong (CUHK) on January 7 and 8, 2008 and the papers in this volume represent a fairly faithful representation of the various material that was presented at the conference. The conference was sponsored by the K. C. Wong Education Foundation, and organized through an international advisory board co-chaired by the eminent academician Professor Chen Shupeng who since the meeting, has sadly passed away. We dedicate this volume to the memory of his many lasting contributions to remote sensing and its passage into GI Science which he was so instrumental in forging in his long and distinguished career.

We want to thank many people for making this book possible. Besides the members of our international advisory committee, the editors of this volume were co-chairs of an international program committee which was key in selecting papers and abstracts. In terms of the refereeing of papers and their translation into the chapters that are included here, many of these authors as well as colleagues who did not attend the meeting were engaged in the review process; we wish to thank Christophe Claramunt, Diansheng Guo, Bo Huang, Bin Li, Xiang Li, Feng-Tyan Lin, Lin Liu, Daniel Sui, Yichun Xie, Yubin Yang, A-xing Zhu, and Qing Zhu for also supporting us in these reviews. The local organizing committee which managed the conference so successfully was chaired by Hui Lin and it comprised Jianhua Gong, Bo Huang, Bin Jiang, PC Lai, Guonian Lu, Hanqiu Sun, Jinyeu Tsou, Xiong You and Qing Zhu.

In the preparation of this book, we must thank Andy Hudson-Smith of CASA, UCL for the image used for the book's cover, and Dr. Chaoyang Fang, Miss. Peiyao Zhang, and Mr. Lei Wu of CUHK who have helped in its editing process. Last but not least, we thank Jack Dangermond, President of ESRI, for his support and for setting the context in his Foreword.

Hui Lin
The Chinese University of Hong Kong

Michael Batty
University College London

14 February 2009

Contents

Part 4 Constructing Mobile and Networked Virtual Geographic Environments

Part 5 Mobility and Dynamics in Visualization

Chapter 1 Virtual Geographic Environments: A Primer

Hui Lin and Michael Batty

Introduction

Virtual Geographic Environments (VGEs) are built on the foundations of geographic information systems (GIS) and geographic information science(GI Science) in which considerable attention is paid to the user in terms of the manner in which they interact with the software[1]. User interaction is thus a key component of such systems. They can range from systems designed for professional and scientific use where spatial analytic functions are dominant, all the way to systems which aim to popularize and disseminate information about various geographies to the public at large. Included in the spectrum of usage and users are systems that contain a blend of new functions for accessing and searching for information which have not been available hitherto through more traditional media. These are best seen in the proliferation of hand held devices, usually mobile in some sense, which are bringing entirely new modes of routine query and search to the casual user.

The degree to which users are immersed in such systems defines major differences between VGEs. Frequently these involve the media used to construct and access the virtual environment. Systems range from simple desktop interaction which are dominated by local or remote software whose access is through the screen display to entirely immersive systems in which the user and machine are interwoven with one another in intricate and often convoluted ways, engaging not only in visual but in audio and tactile interactions. The term virtual reality (VR) emerged in the 1980s as a shorthand for artificial environments in which the user was able to 'suspend belief' by accepting that such an environment was real. Full immersion of the user in such an environment was the starting point and this led to wearable computers in various guises but the term quickly became more generic with anything that involved users interacting with systems that mimicked the 'real' as being 'virtual' in some sense. In fact the media used to generate virtual environments has also broadened as the term has moved into more general use. Initially, literally embedding oneself in the hardware through wearable modes of interaction and thence to wrap-around displays such as CAVEs and VR theatres were regarded as being the modus operandi of VR. But thence came the Web. VR moved back to the desktop where virtual presence came to be established as much through online experiences as through any development of tele-presence involving stand-alone computers[2].

The key drivers of VGEs involve the extent to which such environments embody realism.

This of course turns on what we define as real or on what users consider to be an appropriate experience of a system that purports to embody the sort of realism necessary for their purpose. The particular virtual realities that preoccupy the contributions to this book involve many different kinds of media that in some way reflect the kind of geography that is built in GIS and GI Science[3]. This implies that the systems in question embody some sense of location and place which is analytically tractable. It may be that the users of such systems are not particularly interested or expert in the kinds of analysis that feature in such systems but such analysis is definitely necessary in the construction of such systems and is represented through its software. The VGEs we will present also involve a range of media although most involve the Web (or the Grid) as a means of delivering the virtual environment to the desktop, or more generally to the client from some remote source (or server/s). There are examples here of more traditional virtual environments such as those accessed through the VR theater or through various interactive immersive gaming but most do not involve anything other than users interacting with other users either in real environment (around the desktop) or across the web as, for example, in virtual worlds or virtual design studios. In fact, our rather general definitions of virtual reality and virtual environments suggest that these systems are more like 'augmented realities' in which the real and the virtual are mixed in different blends. Here users interact with each other in traditional real terms while engaging with others who are remotely connected to such environments, represented as avatars, for example, in virtual worlds or as simulated actors from models which engage with real users in actual settings or real users in imaginary worlds. A cornucopia of realities is suggested by this imagery.

To progress our primer, we will organize VGEs into four key areas, dealing in turn with virtual cities and landscapes which are amongst the earliest and most traditional of such environments, and then questions of the human-computer interaction (HCI) which involve interfaces and user immersion. We then illustrate developments in virtual realities, virtual worlds, and games, progressing through the currently most dramatic of ways of relating users and GIS through Web 2.0 technologies and services which are rapidly dominating the way all of us are engaging with contemporary computers, computation, and communications. We will interweave this discussion with comments about the media used to communicate and interact with such realities which define the hardware, software and data that compose our characterizations of VGEs.

Virtual Cities and Virtual Landscapes

The traditional expression of reality in GIS has always been in terms of the representation of landscapes. GIS really emerged from a concern for representing and simulating landscapes and the many factors that determine their form—topography, vegetation, climate, human settlement, underlying geology, agriculture and so on. In fact the idea of (data) layers which dominated GIS came from landscape. From the earliest times in the Harvard Computer

Graphics Laboratory which spawned much of the software and expertise in use today, there was a concern with moving digital representation beyond 2D into 3D, imparting even more realism into the way such geographic information might be viewed and displayed[4].

Much of this early work involved vector graphics displays which did not store the picture or imagery in computer memory but simply traced out the image on the oscilloscope, drawing the media in a painfully slow manner with refresh constituting a redrawing of the same image. It was not until the advent of the PC that computer memory became cheap enough to enable pictures to be stored in memory, and thence peeked and poked to enable changes to be made directly on the display device. In fact, computer graphics and the entire graphical user interface is a product of the PC age with the earliest Apple machines leading the way in visual computing, games and in the earliest applications of CAD. Minicomputers too were fashioned to store pictures in memory and some of the earliest fly throughs of cities were produced by powerful (but largely non-interactive from the user's point of view) computers that produced pictures and wire frames[5]. In fact, computers which are now essentially desktop devices for accessing remote resources over networks which are stored on servers are largely visual devices for accessing information which is generally displayed in pictorial terms. *Virtual landscapes* were the forerunner of such displays in GIS with *virtual cities* being the product very largely of the computer-aided design (CAD) industry. Needless to say these two kinds of 3D environment are beginning to merge with GIS extending into 3D and bringing with it its concern for spatial data structures, query and search, and CAD moving back to GIS bringing with it is its focus on rendering and realism. There is enormous convergence taking place in these fields, particularly with respect to software and these are beginning to embrace other technologies such as those that feature in photogrammetry, image processing and remote sensing.

Although virtual cities were the forerunners of VGEs and many VGEs continue to embrace their form, the predominant form of interaction is with the single user sitting at the desktop operating a software system that enables the users to interact with data in the environment, through actual manipulation of the data and/or through query and search. Extensions to multi-user versions are still quite rare for the scale and size of such systems still means that for many to interact, offline versions of the media need to be produced. Movies thus figure significantly in being able to distribute such media to audiences larger than those that can sit in the theater or around the desktop to view and interact with the environment. In this book, the contributions that deal with virtual cities and landscapes tend to assume that many users will use these systems but the focus is more generally on how such environments are to be made as realistic as possible using automated techniques as well as how users can best interact with such media in terms of the interactions that are possible, either through the Web or on the desktop. This brings us quite naturally to HCI and user interaction which increasingly marks out VGEs for other forms of virtual reality.

User Interfaces and Immersion

Our working definition of a VGE is of an environment which is explicitly geographical but into which users are more deeply embedded in terms of the experiences they require than those which have hitherto characterized the uses of GIS. In this sense, the purposes of these environments need to be closely defined for these are not quite the same as those which define virtual cities or landscapes although they are quite close. The classic functions of GIS involve data representation and manipulation of different representations for purposes of analysis—for determining patterns in data and for making sense of these patterns. The role of query and search within such spatial data also provides a key set of functions but these tend to be for a more exploratory role than the earliest forms of spatial analysis which are deeply embodied within GIS. In fact this distinction between search or exploration, and analysis is one that has evolved GIS into two significantly different roles—the traditional one of somewhat passive spatial analysis relevant to strategic issues, and a more recent, more active one of the use of GIS for routine analysis and exploration, for query and search by users who simply require structured information for routine tasks. The line between the two of course is increasingly blurred but it is significant.

In interacting in these ways with geographic information, the degree of realism of such data, particularly for users who require routine information, is important. Moreover the way this realism is imparted to users is crucial if the information is to be useful and relevant for the purposes in hand and this involves the design of good interfaces. The field of cartography has had a long tradition in developing 2D maps that are useful and relevant to users under different conditions and there is a wealth of detail about how to design good maps. Much of this is being brought to bear on the design of VGEs. Several contributions here deal with how such knowledge can be exploited and progressed to enable users to interact better with both 2D and 3D environments which are fixed or mobile. In fact some of the greatest challenges are in the design of interfaces which are limited by the size of the display and the frequency of updating which, in turn, are best reflected in the design requirements of VGEs on mobile devices. As readers will see, mobility and dynamics tend to be key issues in the development and use of VGEs not only for user interfaces but also for the way information is acquired, displayed, refreshed, and animated.

Search at the local level is becoming extremely significant at the present time as mobile devices with GPS merge with phones and the Web. For example, the digital earth paradigm that crops up at several points in various contributions published here is driving forward the search for local content which might be user generated as well as synthesized from a variety of public and private sector sources. This is taxing our best interface designers who are required push as much as possible about the real onto the virtual screen of small display devices and to organize this data and its graphics in such a way that confusion is minimized. The same kind of problem exists at the other end of the spectrum where users of VGEs are involved in much more generic issues of public participation. These tend to be systems where immediate feedback and response is not required and thus the dictates of spontaneous and instant search are not needed. But the same kinds of problem about vast quantities of seemingly relevant data that need to be synthesized easily and cogently apply. To an extent, the media is independent of this kind of user interaction for routine information of a specific or general nature might be accessed on any of the device that can connect us to such data sources from the small scale mobile to the VR theater. But the media does structure what is possible in terms of interfaces and the challenge is to design interaction systems that are flexible and adaptable for different problems, devices, and patterns of digital connectivity.

The last problem we want to signal here is perhaps the least tractable and that is "how can traditional systems of GIS which build on GI Science and spatial analysis be made more virtual so that users can explore and analysis patterns in geographic data more effectively?" GIS has moved slowly from systems that are all embracing to tool boxes that involve users in plugging and playing all the way to modules that can be fashioned into rather different systems that interact with quite diverse software. Extensions that plug into agent-based modeling packages and into CAD, for example, are now available[6]. There is a drift towards the Web with non-proprietary tool boxes of GIS-like functions being quite widely available. But so far, there has been little attention to good interface design, apart from moves within the information visualization community which link closely to exploratory spatial data analysis. In fact one of the conclusions from this survey of VGEs is that much more attention should be given to traditional GIS and its interfaces so that users can fashion data for themselves in innovative ways that give them control over the interface. Just as more general and widely available software can be customized to the requirements of different users, we believe that the same kind of developments should occur in GIS. In this, VGEs can point the way.

Games and Virtual Worlds

One of the more surprising developments in GIS (and CAD) is coming from digital entertainment, from games and their online equivalents ranging from systems such as World of Warcraft to virtual worlds such as Second Life. Gaming brings with it the need for realism which can be part fictional as well as real in the real world sense. Realism, however, at the most detailed level poses enormous challenges for standards to which graphics and all other user senses must accord. Game players demand realism of a kind that is as good as it can get whereas those dealing with virtual cities built in CAD and GIS tend to accept a level of abstraction that implies a good deal less realism than would be acceptable for gamers. In fact, we do not have much idea of what level of realism is required for routine search on mobile devices or in terms of the interaction of the public with digital planning schemes and models. To an extent, the standards that prevail on the web and in graphical user interfaces generally are those that tend to be accepted by default. We still do not know how much better our VGEs would be were we to develop the sort of standards that occur in gaming.

In virtual worlds too, although the realism is good, given the kinds of objects that inhabit such worlds—people as avatars, buildings, landscapes, and so on, the general impression is one of a cartoon-like quality that tends to divert the serious user. As several of these contributions here suggest, games and virtual worlds provide enormous potential for VGEs but their interfaces tend to look amateurish and rather clunky. Given the memory requirements that need to be met and the need to scale such systems to millions of users, the overall, effects of such worlds are good although it will take a while before the needed realism can be reached. However this is only matter of time as servers and clients improve in their memories and as access via broadband communication continues to improve. Nevertheless, these kinds of games and worlds do show what is possible as they are instructive in pressing home the message that in the future, all such virtual realities will be online.

The requirement for online access has remained implicit in our discussions so far although we might be reaching the point where VGEs are explicitly networked constructions and where users engage with each other and with the environment in a mixture of local and online actions and interactions. In fact, of the twenty one contributions to this book that follow, there are only four that do not deal with online and remote access as integral to the environments that are discussed and even these are developed against a background of networked data and information that is essential to their construction. One of the problems of course in gaming and virtual worlds is that there are so many human and physical processes required that the graphics is bound to look artificial as designers trade off realism for efficiency of execution. These environments require iconic and symbolic models—based on geometric representation and simulation as well as mathematical and logical representations of the processes involved. These are important issues as it is here that we see the true merger of the geographic, the geometric, and the processes and behaviors that characterize the real world. For example, the sort of models developed for pedestrian motion as well as those for geometric integrity of virtual cities are merged in virtual worlds. Combine these with the needs of real users entering these worlds and partaking in debate and discussion where virtual geometries and background human and physical processes form part of the constantly updatable backcloth on which this action takes place, then this gives some idea of the challenges involved in this quest. As various contributors argue here, games and virtual worlds tax our abilities to abstract and represent while at the same time challenging us to make our interfaces better.

Web 2.0 Technologies, Digital Earth, and Neogeography

Our last topic area that weaves itself throughout all the contributions here is the first we broach in Part One of this book. This is the online world, the world of networks, communications, and dissemination and of course, of users. It is the world where users meet software whether it be simply before the desktop or workstation or more likely through some mix of the real and the virtual, the immediate and the remote. The Web which began in the early

1990s nearly two decades ago, was quickly fashioned through graphical user interfaces which are still called browsers and through which increasingly remote and standalone software is now being delivered and accessed. Most of the access to data and software through the web is relatively passive in form but we have moved into a phase of interactivity that lets users change the content of the information they are using on the fly, so to speak. They are able to add new content from diverse sources, improve the content if they see fit, or even abuse it, subject to the "wisdom of crowds" which acts a restraining force on what is produced and made available publicly. These systems, accessible through the Web still in the form of Web pages, are often referred to as web-based services although a more accurate depiction would be as fully interactive systems, in fact virtual environments, in which users can interact as though they are in the real world by making changes, additions or deletions as well as simply passively observing the information in question. Everything from Facebook to Wikipedia are part of such services. The key is to let users manipulate the content and this is the criterion that distinguishes the original Web, Web 1.0 from its contemporary equivalent, Web 2.0.

Geographical services such as Google Maps and the range of digital earths from Google and Microsoft amongst others that have recently appeared are not strictly Web 2.0 services. However, once introduced, these companies have enabled these frameworks so that users can customize them. In terms of Google Earth, the Google 3D Warehouse lets users upload their buildings and make them available for others to download. Google MyMaps lets users customize their maps and embed them in other services while also linking these to local search which in fact is one of the most important commercial rationales for such developments. Copyright and IPR issues of course pervade these kinds of open access but the open source movement and the notion that in the digital world, the marginal cost of information is near zero, has begun to dominate. Companies and agencies can thus afford to give 99 percent of these services away for free and simply make their money on the remaining 1 percent which they can customize for niche markets. The business model for such developments has a logic of necessity where it is easy and almost costless to distribute content, one which is highly empathetic to users who create their own content as well.

The world of crowd sourced information which is upon us is leading to the creation of geographic data and content from the most unlikely sources. Combined with GPS, an army of volunteers are creating products such as OpenStreetMap, a user sourced digital map (of the world) that can be added to at will and improved in much the same way that Wikipedia has become the world's most advanced and informative encyclopedia. This, combined with the ability to take various map products such as Google Maps and customize them into open GISs, is leading to the creation of a new kind of bottom up geography—*neogeography* as it has been labeled[7]—which opens mapping to the masses. VGEs to date have been more professional in scope but scattered throughout this book are examples where such systems are opening up to non-expert and non-professional users who are likely to change and improve the contents of these virtual environments in ways that are beyond our wildest dreams.

Therein likes the power of the Web; the rise of neogeography threatens to change the nature of the way we handle and abstract geographic space for ever.

The Prospects for Virtual Geographic Environments: The Contributions That Follow

We do not intend to write a conclusion to all the contributions in this book that follow. We feel they speak for themselves as a snapshot of where we stand in GIS and GI Science with respect to our needs and expertise in embracing the virtual world and using it creatively to advance our use and applications of geography. One of the features that will strike the reader who ventures beyond our initial foray in what VGEs are all about is that the organization of the five parts that comprise this book are somewhat fluid. Virtuality runs throughout the parts as does GIS for most of the participants emanate from the community of GIS scholars. But ideas about mobility and dynamics merge with ideas about the power of the web and the rise of neogeography. In some senses, we speculate that VGEs will be very different constructions a decade or more in the future for the whole question of realism will be handled much more cleverly and professionally. User interfaces will be better but there may well be as many gimmicks. The graphic effects in movies like "The Matrix" and "Minority Report", the "1984s" of the Twenty First Century, could well come to pass as mobile devices get wired into ourselves and as a plethora of information continues to bombard our senses. With all this, we may well see a return to or rather a clearer distinction between professional uses of GIS in terms of the software for spatial analysis and its more routine use as web-based services. This then is for the future. In the meantime, the contributions which follow provide some sense of the state of the art.

First off, we review the overall power of the Web with respect to the fashioning of virtual geographic environments. Goodchild, Hudson-Smith and Crooks sketch the growing impact of Web 2.0 with respect to map hacks, volunteered geographic information, and customized non-proprietary applications. Many of these are likely to change the face of conventional GIS as Peuquet suggests in her review of integrated space-time systems which are geared to good user interaction. Tao turns the argument around in his review of the need for local search which he then suggests is driving the evolution of new forms of geographic information systems which build on the concepts of the digital earth.

The second part returns to more conventional applications in the form of virtual cities and virtual landscapes. Yano and his colleagues show how their Virtual Kyoto models link space to time as users navigate the environment to gain some sense of this Japanese heritage. At the same time they show how buildings need to be generated routinely and automatically to build such environments. This is echoed in Middel and her colleagues Digital Phoenix project where the output is more abstract but converted into realistic pictures for participation using similar rendering and generative schemes. This theme is extended by Lovett and his colleagues who, looking at virtual landscapes, explore the realism imparted by their realization on different systems while Sugihara and Hayashi illustrate how really quite sophisticated generative schemes are necessary to get the detail of the virtual city to the point where it is acceptable for viewing by a wider public.

Many of these issues involve appropriate user interfaces for experts and the public in using VGEs and in the third part of the book, this becomes the focus. Meng presents some important concepts which enable users to interact with geographic objects effectively illustrating various examples from digital cartography. Cai does the same but in the quest to generate the wider context of a VGE which is web enabled while at the same time being implementable on mobile devices. Jahnke and his group return to the idea of generating vast arrays of content for virtual cities using non-photorealistic data which must be tuned for various kinds of user. The tone then switches to public participation with an integrated system for urban design and townscape analysis being illustrated by Shen and his team. Portugali's group then sketch how VR, CAD and GIS can be nested in a decision support system which enables commentary on the barrier being created between two very different communities in Israel and Palestine.

The fourth and fifth parts of the book tend to be a little more technical dealing with mobility and dynamics in VGEs in networked and mobile contexts from large to small scale. Li and his group first show how uncertainty in positioning needs to be transmitted to users of VR system while Yanagisawa and Yoshikawa quantify information being delivered to mobile hand held devices so that the information can be optimized in terms of its delivery and form. The debate then moves to slightly larger scales. Deren Li shows how his group are constructing VGEs from digital measurable images that form the essence of a digital earth, and then Jianhua and his colleagues present the network dynamics involved in building a VGE. This focus on mobility is continued in the fifth part, where the emphasis turns to small scale movements. Arikawa and his group show how small scale environments involved with movement need to be developed with specific natural language processing to yield good data on movement while Wu and his colleagues illustrate how to build a pedestrian model for entertainment events in down-town Hong Kong. Gong and Mackett focus on measuring actual walking behavior and linking this to visualization techniques. Finally, the Shepherds, somewhat rhetorically, pose the question "Can videogame technology replace GIS?" by which they mean that the motion technologies and dynamic modeling in such games might provide a paradigm both for GIS as it extends into the temporal domain and quite clearly for VGEs.

References

[1] Lin H and Zhu Q. Virtual Geographic Environments. In: Zlatanova S and Prosperi D (Editors). Large-scale 3D Data Integration: Challenges and Opportunities. Florida: CRC Press, 2005: 211–231.

[2] Batty M. Virtual Reality in Geographic Information Systems. In: Wilson J P and Fotheringham A S (Editors). The Handbook of Geographic Information Science. Oxford, UK: Blackwell Publishing,

2008: 317–334.

[3] Gong J and Lin H. A Collaborative Virtual Geographic Environment: Design and Development. In: Balram S and Dragicevic S (Editors). Collaborative Geographic Information Systems. Hershey, PA: IGI Global, 2006: 186–206.

[4] Chrisman N. Charting the Unknown: How Computer Mapping at Harvard Became GIS. Redlands, CA: ESRI Press, 2006.

[5] Batty M and Hudson-Smith A. Digital Cornucopias: Changing Conceptions of the Virtual City. Environment and Planning B, 2006, 33, 799–802.

[6] Lin H, Zhu J, Xu B, Lin W, Hu Y. A Virtual Geographic Environment for a Simulation of Air Pollution Dispersion in the Pearl River Delta (PRD) Region. In: Lee J and Zlatanova S (Editors). LNG&C: 3D Geo-Information Sciences. Berlin, DE: Springer, 2009: 3–13.

[7] Hudson-Smith A. Digital Geography: Geographic Visualisation for Urban Environments. Centre for Advanced Spatial Analysis, University College London, London, http://digitalurban.blogspot.com/[2008].

Part 1

Online Virtual Environments and Web 2.0 Technologies

Our primer in Chapter 1 argued that Virtual Geographic Environments (VGEs) embed Geographic Information Systems (GIS) into computer environments which comprise hardware, software, data and users where the user is given pride of place. The focus in such environments is largely on the way the user interacts with GIS. This distinguishes VGEs from CAD, from virtual worlds and from the plethora of graphics software and environments that have proliferated in the last two decade since our interaction with computers became predominantly visual. As the internet has become the main means of linking users to computers where accessing information in terms of the written word and communicable graphics is now largely digital, then it is no surprise that the cutting edge of linking users to geographic systems is now through software and data that is disseminated using the Web.

We thus start with exploring ways in which this cutting edge is driving forward VGEs. We begin with the second generation of the Web, called Web 2.0, web-based communities and hosted services such as socialnetworking sites, wikis and dynamic and shareable content which facilitate collaboration and sharing between users. Maps are central to these developments as we will see in the contributions below but so too are 3D visualizations of such geographies as well as the networks that support them. In fact the data pipeline that is associated with the traditional model of a computer and computing based on the trio of input-processing-output is being revolutionized by these developments. Traditional GIS involved users collecting data offline with analysis often beginning once such data had been input to the software, the focus being largely on analytical functions and subsequent display and interpretation of such analyses. But the Web is not limited to simply analysis and display although its predominant use is still for these purposes. If we are all online, then the Web provides its own logic for collecting data, for input with the process of user interaction moving both forward to analysis and decision but also backwards to data collection. VGEs are not only environments for analysis but also for data collection set within the wider context of establishing the wider context to analysis; this activity is increasingly referred to as crowd sourcing.

The four chapters constituting this first part of the book all deal with these developments. Goodchild first argues that Web 2.0 technologies are enabling the collective construction of virtual environments. Unlike the imagined worlds of sites such as Second Life, these are attempts at geographic realism, built as patchworks of individually volunteered information. As such they echo the objectives of national spatial data infrastructures, but with amateur individuals rather than professional agencies and corporations as the contributors. He provides and assesses examples of such projects, reviews the domains in which they appear most successful, and examines the issues that they raise concerning privacy, standards, verification, sustainability, and benefits.

Hudson-Smith and Crooks also survey the development of Web 2.0, specifically focusing on the wider environment of online computing which is increasingly referred to as The

Cloud, GeoWeb and Wikitecture. These, they argue, are revolutionizing the way in which we present, share and analyze geographic data. They outline and provide working examples of a suite of tools which they have developed based on web-based technologies for the social and geographical sciences which allow quick and easy visualization of geographic data in systems such as Google Maps, Google Earth, Crysis and Second Life using various kinds of software 'hacks'. Their tools address two issues, firstly that spatial data is still inherently difficult to share and visualize for the non-GIS trained academic or professional, and secondly that a geographic data social network has the potential to dramatically open up data sources for both the public and professional geographer. Their software, GMapCreator and MapTube, enables the intelligent visualization and sharing of spatial data and they conclude by detailing usage and outreach as well as providing insights into how such tools are already providing a significant impact on GIS.

Peuquet takes a more user-centric focus beginning with traditional GIS and emphasizing the need for integrating diverse digital environments. She argues that the availability of increasingly large and multidimensional data resources in digital form as well as the availability of advanced visualization software is enabling increasingly large and complex collections of observational data to be explored. But dealing with important current issues such as managing urban growth, predicting the spread of disease or understanding an evolving world economy requires more than a data-driven approach to derive insights beyond patterns and relationships, toward deeper understanding of the processes involved. In order to achieve this, an integrated approach is required that links observational evidence with stored information about generalized entities as concepts and functional rules about how those entities behave and interact. This also must be linked with a sophisticated visualization capability that gives the human user the ability to interact with that stored information in a flexible and holistic way. She focuses on recent advances in space-time analysis and representation, and how these can be integrated toward the next generation of tools for geospatial analysis, modeling and problem-solving with the Web central to her concern for integration

Tao concludes this foray into new digital environments by standing back and taking a look at the whole question of location, geography, search and the Web. He presents a perspective on the impact of location on Web evolution in terms of the impact of on-line mapping, the vision of digital earth, and the impact of emerging local search and local services on mass user communities. With many of these new innovative developments and services, it is important to understand the overall landscape of the technological change in the context of the digital earth. Although he does not drill down to the technical details, key developments are discussed. Along with the description of the Web and location, the evolution of on-line mapping services is presented and focal concepts of local search and social media are described. Finally the application examples developed under the Microsoft Virtual Earth program are illustrated. In this sense, Tao takes the argument into the realm of 3D visualization which dominates later contributions to this book and he raises the concept of a digital earth that is providing one of the key paradigms for VGEs at the present time.

Chapter 2 Virtual Geographic Environments as Collective Constructions

Michael F. Goodchild

Introduction

The early Web achieved its success by providing users with easy access to pages located on servers. Pages could contain simple text with embedded tags (html), or images in a range of formats. The resulting formatted displays had immediate appeal, particularly to users with very little in the way of computer experience or skills. Hyperlinks allowed rapid access to related material, and simple and universal standards for content made it possible to support user access with very "thin" clients. Later, functions were implemented to support forms, enabling users to contribute records to databases, as for example with on-line reservation systems, opening the potential for Internet-based commerce. The format constraints on such content were gradually removed, allowing the development of such popular sites as EBay. Most recently, the concept of user-generated content (UGC) has expanded to include wikis, blogs, and a range of other services that make it easy for the general user to contribute unstructured information to servers that in turn assemble, possibly filter, and render content. The popularity of UGC has led to the coining of the term *Web 2.0*, implying a radically revised conceptualization of the nature of the Web as a multi-way enterprise in which clients and servers are no longer clearly distinguished.

This development fits well with a trend that has been apparent in the world of geographic information (GI)—or information about the surface of the Earth—for the past two decades. In most developed countries the process of map-making has long been entrusted to national mapping agencies such as the US Geological Survey and the National Geospatial -Intelligence Agency. Fifty years ago the costs of equipping a map-making enterprise, including field surveyors, photogrammetrists, cartographers, and printers, along with their expensive machines, justified the concentration of such efforts in a single agency funded by the national government. The high cost of map-making also contributed to its emphasis on the relatively static features of the Earth's surface rather than the temporal dynamic. But a series of developments has led to a substantial erosion of this role, the empowerment of lower levels of government and even individuals, and a new set of opportunities to capture dynamics and change. They include the Global Positioning System, which has partially replaced the need for costly surveying equipment; satellite remote sensing, with its abundant supplies of timely, high-resolution imagery; Web mapping and imagery services which make it possible for individuals to generate their own overlays; and a host of sensors of various kinds. By the early 1990s it was possible for many farmers to have better information about their soils and crop yields than the US Department of Agriculture; for companies selling digital street maps to rely on local employees to provide updates; and for military commanders to receive and compile intelligence from personnel in the field to augment and in some cases replace information from central sources.

Moreover, governments have been less and less willing to shoulder the ever-increasing costs of acquiring, compiling, and disseminating geographic data[1]. In some countries they have resorted to various user-pay arrangements, in which the ultimate consumers of geographic information pay at least some of the cost of acquiring it, and more than the cost of reproduction. In other countries governments have simply abandoned the map-making enterprise entirely, or maintained it solely as support for national security and defense. Estes and Mooneyhan[2] have tracked this decline, arguing that mapping as a national government function peaked some decades ago.

In the US, the National Research Council's Mapping Science Committee saw the writing on the wall in the early 1990s, and outlined the concept of spatial data infrastructure[3]—the notion that map-making would evolve into a collective but distributed enterprise, involving agencies from the national to the local level, along with the private sector and citizenry. Maps and geographic information in general would be produced to nationally agreed standards, but as a patchwork whose pieces would be driven primarily by local needs, and thus would be updated more frequently, and produced to finer levels of detail, in areas of greatest activity, such as metropolitan areas. Given a server with appropriate tools, the various pieces of the patchwork could be fitted together, removing any obvious inconsistencies, and distributed over the Web; and the quality and lineage of each piece of the patchwork could be documented in metadata. The creation of the *National Spatial Data Infrastructure* (NSDI) was authorized by President Clinton under Executive Order 12906 in 1994, and has provided the policy umbrella for geographic information in the US for the past 15 years. Since then extensive activities have been initiated under the Federal Geographic Data Committee, data clearinghouses and geo-portals have been established, and numerous standards have been published. The Open Geospatial Consortium, a group of industrial, academic, and government partners, has been particularly instrumental in promoting interoperability and real-time access to data.

These developments made it possible at least partially to reverse the traditional radial, top-down flow of geographic information. Rather than a single national agency producing much of the nation's supply, it is now local governments, commercial firms, and even individuals who are most engaged in the production, as well as the consumption, of the NSDI. Spatial data infrastructure efforts have been initiated in almost all developed countries, often with the assistance of the Global Spatial Data Infrastructure community. Software for exploiting these resources is now commonly available, in the form of Web services as well as the more traditional stand-alone GIS.

This picture has recently been reinforced by a spate of efforts that fall under the general rubric of UGC and Web 2.0, and are directly related to the supply and consumption of geographic information. Among the hundreds of such sites is Wikimapia, a service that builds on the success of Wikipedia and attempts to "describe the whole world" by allowing contributors to post descriptions of any user-selected area, from an entire continent to a single house. Figure 1 shows the Wikimapia coverage of Medina, Saudi Arabia, and is typical of the density of coverage in much of the world. Wikimapia attempts to mimic Wikipedia in its quality-control mechanisms: individuals can edit each other's contributions, and entries are reviewed by a hierarchy of editors. At time of writing there were some 9 million entries in Wikimapia. The Flickr site allows individuals to post and geo-reference photographs, and to add information to them that may include links to other sources, such as Wikipedia; At time of writing it held on the order of 10^8 georeferenced photographs, including over 2,000 of the Australian Uluru or Ayer's Rock alone. Somewhat more ambitious is OpenStreetMap, an effort to build a free, public digital map of the globe at sufficient spatial resolution to act as a way-finding aid—a remarkable effort particularly in areas of the globe where no such digital maps exist, or where they are available only at exorbitant cost.

Figure 1 Wikimapia coverage of Medina, Saudi Arabia

Each rectangle has been contributed by a volunteer to denote a significant feature on the Earth's surface. In many cases the description entered with the feature, which can be exposed by clicking on the rectangle, is in Arabic script. It may include photos and links to other material.

Together, these sites and others like them constitute a remarkable voluntary effort to construct a virtual geographic environment. They rely on the apparent willingness of large numbers of individuals, distributed over the surface of the Earth, to spend significant amounts of their own time creating pieces of the patchwork based on their personal knowledge of local geography. It is particularly remarkable that those same individuals are increasingly reluctant to contribute to other public data gathering exercises, including the census and questionnaire surveys. A number of terms have become associated with the phenomenon in addition to Web 2.0 and UGC. The terms *crowdsourcing* and *collective intelligence* draw attention to the notion that the collective contributions of a number of individuals may be more reliable than those of any one individual. The term *volunteered geographic information* (VGI) refers specifically to geographic information, and to the contrast between the actions of amateurs and those of authoritative agencies, and I will use the term here. The term *asserted* geographic information draws attention to the fact that such information is not subject to the normal checks and quality control mechanisms of those agencies, while *neo-geography* emphasizes the contrast between this grass-roots phenomenon and the current state of the academic discipline of geography.

The VGI phenomenon raises significant and far-reaching issues that call out for research. Moreover, the changes outlined in this section have occurred at a pace that far exceeds society's ability to rethink and reorganize its institutions, with the result that current legacy arrangements are already well out of touch with the emerging reality. The remainder of this chapter examines the phenomenon, its drivers, and its implications in some detail. Section 2 identifies several basic concepts that underlie and may help to understand the nature of VGI. Section 3 attempts to develop the foundations of a research agenda, spanning issues of individual behavior that are normally considered part of social psychology, and the technical and domain-specific issues that would normally be identified as geographic information science (GI Science).

Basic Concepts

Humans as sensors

Recently a great deal of attention has been devoted to the concept of sensor networks. The observational objectives of Earth science, as well as the objectives of security and surveillance, can be addressed at least in part by the installation of networks of sensors across the geographic landscape. Commonly cited examples include the network of video monitors in many major cities, proposals to instrument the ocean and seabed with sensors in the interests of science and early warning of tsunamis, and networks of traffic sensors that can provide useful information to planners, as well as real-time pictures of congestion.

It is useful to distinguish three types of sensor networks. Most of the commonly cited examples fit the first, a network of static, inert sensors designed to capture specific measurements of their local environments. Less commonly cited are sensors carried by humans, vehicles, or animals. For example, much useful research is emerging from projects that have equipped children with sensors of air pollution, in an effort to understand the factors affecting asthma. A third type of sensor network, and in many ways the most interesting, consists of humans themselves, each equipped with some working subset of the five senses and with the intelligence to compile and interpret what they sense, and each free to rove the surface of the planet, at least locally. The human senses may be augmented with imaging devices, and information may be uplinked using cellular, satellite, or other communication technology. Web 2.0 provides the necessary means to assemble, integrate, and redistribute this information. However, existing institutional arrangements make it unlikely that the information will ascend up the hierarchy to be incorporated into the authoritative products of traditional mapping agencies.

Citizen science

The term *citizen science* is often used to describe communities or networks of citizens who act as observers in some domain of science. A perfect US example is the Christmas Bird Count, an effort to enlist amateur ornithologists in conducting a mid-winter census of bird populations that began over 100 years ago. Participants require a fairly high level of skill, and over the years a number of protocols have been established to ensure that the resulting data have high quality. Details on the program and its scientific applications to date can be found at http://www.audubon.org/bird/cbc/biblio.html. An international example is Project GLOBE, an effort to enlist school-children and their teachers in providing a world-wide source of high-quality atmospheric observations. As with the Christmas Bird Count, a number of protocols and training programs have been established to ensure quality, and to collect, synthesize, and re-distribute the results. Recently there has been much interest in new efforts along these lines, enabled by GI acquisition tools such as GPS, and the services of Web 2.0.

Both of these projects require a fair degree of training and expertise. This need for expertise would be a limiting factor in any effort to extend VGI to such comparatively sophisticated mapping themes as land use, land cover, or soil class. Other forms of VGI are much less demanding, however, particularly those associated with local place-names, streets, and other well-defined geographic features that are commonly familiar to neighborhood residents.

Participant populations

Sites such as Wikimapia are open to all, as are many other VGI efforts. The Christmas Bird Count and Project GLOBE, on the other hand, place restrictions on participation in order to ensure adequate expertise. The question of *who* may volunteer to contribute geographic information has much to do with the quality of the resulting information, and a range of possibilities exist. For many years companies producing digital street maps have relied on networks of local observers to provide rapid notice of new streets, changes of street names, *etc.*, paying them as part-time workers. Inrix is collecting tracks from hundreds of thousands of trucks and other fleets, processing and compiling the results as a source of real-time information on the state of congestion and other short-term factors affecting travel on road networks. Military personnel are important potential sources of geographic information about local battlefield conditions that can be used to augment what is available from central mapping and imagery sources. Many farmers now have elaborate systems for mapping and monitoring their fields and crops (*precision agriculture*), and constitute a potential source of data that are in many cases much more detailed and current than those available from central agricultural agencies. In essence, such developments contribute to a growing reversal of the traditional top-down, radial approach to the creation and dissemination of geographic information.

Early warning

Recent events such as the Indian Ocean tsunami or Hurricane Katrina have drawn attention to the importance of geographic information in all aspects of emergency management, and to the problems that arise in the immediate aftermath of the event before adequate overhead imagery becomes available for damage assessment and response planning[4]. Earth-observing satellites may not pass over the area affected by the event for several days. Images from satellites and aircraft may be obscured by clouds and smoke. Conditions on the ground may prevent the rapid downloading of digital imagery because of a lack of power, Internet connections, or computer hardware and software.

On the other hand the human population in the affected area is intelligent, familiar with the area, and increasingly able to report conditions through mobile phones or Internet connections, using voice, text, or pictures. To date there has been very little use of such novel data sources in these situations, in part because of an almost complete lack of the tools needed to collect, synthesize, verify, and redistribute the information, and in part because of a mistrust of information asserted by volunteers. However the potential to obtain almost immediate reports from geographically distributed observers on the ground will surely drive increased efforts to overcome these problems in the next few years. In recent events in the US, however, such as the wildfires in Southern California in late 2007, volunteered information was routinely quoted and used by emergency managers.

Research Issues

Why do people do this?

In the mid 1990s the US Federal Geographic Data Committee published its *Content Standards for Digital Geospatial Metadata*, a format for the description of geographic data sets that became one of the foundations of the NSDI. The project was very timely, given the rapid increase in the availability of geographic information via the Internet that occurred at that time. Metadata were seen as the key to effective processes of search, evaluation, and use of geographic information. Nevertheless, and despite numerous efforts and inducements, it remains very difficult to persuade those responsible for creating geographic data sets to provide adequate documentation. Even such a popular service as Google Earth has no way of informing its users of the quality of its various data layers, and it is not always easy to determine the date when any part of its image base was obtained. A recent news report concerned the apparent replacement of its coverage of New Orleans with pre-Katrina imagery, though its coverage of the Darfur region was being updated almost daily at the time of writing.

Given this evident reluctance to provide documentation, it is perhaps surprising that the opportunity to create UGC has engaged the interests of so many individuals. Why is it that citizens who have no obvious incentive are nevertheless willing to spend large amounts of time creating the content of Web sites? What kinds of people are more likely to participate, and what drives them to be accurate (or inaccurate)?

Self-promotion is clearly an important motivator of Internet activity, and in its extreme form can lead to the exhibitionism of personal web-cams. Despite the vast resources of the Web, it is still possible to believe that *someone* will be interested in one's personal site. The popularity of some blogs can be misread as suggesting that an audience exists for *any* blog. At a different level many users volunteer information to Web 2.0 sites as a convenient way of making it available to friends and relations, irrespective of the fact that it becomes available to all. This may underlie the popularity of sites such as Picasa, which allow contributors of personal photographs to point others to them, but it scarcely explains the popularity of Flickr or Wikimapia, where content is comparatively anonymous. Contributors to OpenStreetMap may derive a certain personal satisfaction from seeing their own contributions appear in the patchwork, and from watching the patchwork grow in coverage and detail, but there can be no question of self-promotion in this essentially anonymous project.

Another possible motivation stems from the ability of contributors of UGC to access and retrieve their own content. This is in sharp contrast to most government data-collection efforts, which result in personal data disappearing into the maw of government agencies, to reappear only in aggregated form. Contributors of UGC can check, edit, and monitor the information they provided, and thus feel a sense of personal ownership over the collective enterprise.

A research effort to determine the motivations behind the volunteering could follow several possible strategies. Questionnaires distributed to contributors might yield useful information, as might scanning of Web pages to find personal statements about motives. The literature in social psychology and related fields contains results of studies of volunteering, and might shed useful light. To date, however, very little is known about the specific motivations behind VGI.

Authority and assertion

The traditional mapping agencies have elaborate standards and specifications to govern the production of geographic information, and employ cartographers with documented qualifications. Over the years their products have acquired an authority that derives from each agency's reputation for quality. Google, on the other hand, has no such reputation in the geographic domain. Nevertheless users appear willing to ascribe authority to its products, perhaps because computerization carries authority *per se*, and perhaps because of the company's success in other areas, particularly its search engine. Yet there is ample evidence that the quality of Google's geographic products falls far short of those of the traditional mapping agencies. Google's imagery is often significantly mis-registered, and no information is provided on the quality, provenance, or semantics of the imagery or ancillary data. Thus services such as Google Earth fall far short of scientific norms, including replicability.

UGC is sometimes termed *asserted* information, in that its content is asserted by its creator without citation, reference, or other authority. The early days of the Internet were characterized by a certain altruism, a belief in the essential goodness of users, and there was little anticipation of the subversive phenomena of spam, viruses, and denial-of-service attacks that now pervade the network. Similarly many UGC efforts are driven by the kinds of altruism inherent in any voluntary community effort. Can we expect, then, a similar pattern of disillusionment as antisocial elements recognize and exploit the inevitable vulnerabilities? Will there be efforts to create fictitious landscapes, or to attack and bring down Web 2.0 servers? UGC is currently a somewhat exotic domain, but if and when users begin to rely on its services a growing pattern of efforts to undermine it seems inevitable.

The digital divide

Despite the apparent openness of UGC, it remains largely the preserve of those fortunate to have access to the Internet—and broadband access in particular. While a growing fraction of citizens in developed countries have such access, it is largely unavailable to the majority of the world's population who live in developing countries, either as users or as contributors. Moreover issues of language and alphabet also affect access even for those with broadband connections, since many Web 2.0 servers support only the Roman alphabet and English. In principle, much could be achieved through mobile phones, which often have the ability to connect to the Internet and to capture images, but the tools needed to exploit this limited environment as a source for UGC do not yet exist except in limited and isolated form. So while it was argued above that such limited tools were potentially significant in early warning, military intelligence, and emergency management, significant work still needs to be done to realize the potential.

Applications

Geographic information compiled as UGC has the potential to be a significant source of understanding of the surface of the Earth. It can be timely, a property that was particularly stressed in the discussion of early warning. By motivating individuals to act voluntarily, it is far cheaper than any alternative, and its products are almost invariably available to all (but see the earlier discussion of the digital divide). With sites such as Wikimapia one can learn a great deal about remote places, acquiring the kinds of information needed for planned activities on the ground. Sites such as OpenStreetMap often provide the cheapest source of geographic information, and sometimes the *only* source, particularly in areas where access to geographic information is regarded as an issue of national security. It is already clear in many fields that such informal mechanisms as blogs and UGC can act as very useful sources of military and commercial intelligence. The tools already exist to scan Web text searching for references to geographic places, and to geocode the results. Thus the most important value of UGC may lie in what it can tell about local activities in various geographic locations that go unnoticed by the world's media, and about life at a local level. It is in that area that UGC may offer the most interesting, lasting, and compelling value.

Ethics

VGI raises substantial issues of an ethical nature. Privacy questions have been raised repeatedly over some of the geospatial content of such Web sites as Google Earth, where fine-resolution imagery is capable of revealing remarkable detail. More recently the addition of street-level imagery, obtained from cameras mounted on moving vehicles, has occasionally allowed faces and license plates to be identified. In Canada, recent legislation is forcing the sponsors of such sites to blur key areas of some images when identity might be revealed. Privacy issues are also raised over the volunteering of information about others—what are the limits to what one individual can say about another?

Conclusion

One of the greatest impediments to the construction of virtual environments has been lack of data. While parts of the world are now well endowed with digital geographic information, particularly of a topographic nature, and while satellites now provide abundant imagery that is often of high spatial resolution, the general state of world mapping is in decline. Moreover the richest legacy of geographic information describes phenomena that are relatively static, and describes only their two-dimensional form. At this point there seems little prospect of this situation improving, at least as far as traditional, authoritative production mechanisms are concerned. Only in those areas dominated by the private sector, such as the production of digital street centerlines, is there much evidence of improvement. A recent study estimated that a full three-dimensional representation of all built structures on Earth would amount to some 3 petabytes of digital information, not in itself an unmanageable amount even by today's standards—but that its acquisition, compilation, and quality assurance would consume approximately 10% of US gross domestic product for 10 years given existing, intensively manual techniques.

By contrast, I have argued that the concept of VGI, and the mechanisms that support its production, provide an interesting and compelling alternative. It is local communities, not national mapping agencies, that are the ultimate custodians of geographic knowledge, and the evidence suggests that they are willing, under the right circumstances, to provide it in forms that can be readily compiled and distributed. Modernist ideas of central authority and a single truth are increasingly outdated in a world that is growing used to the notion of multiple viewpoints, and to technologies that make it easy to maintain large quantities of potentially inconsistent data. We now have the tools to achieve interoperability between alternative spellings and renderings of placenames and even multiple placenames, and are increasingly coming to expect our information technologies to reflect and support our multicultural, post-modern world. Web 2.0 provides the opportunity to rethink many of the most cherished concepts of geographic information production, and to engage with the richness of individual geographic knowledge.

Acknowledgements I am grateful to the many students and colleagues who share my interest in VGI, and particularly to the participants in the NCGIA Specialist Meeting on VGI, which was held in Santa Barbara December 13–14, 2007. Further details on the meeting, which was funded by the US Army Research Office and Los Alamos National Laboratory, are available at www.ncgia.ucsb.edu/projects/vgi.

References

[1] Goodchild M F, Fu P, Rich P. Sharing Geographic Information: An Assessment of the Geospatial One-Stop. Annals of the Association of American Geographers, 2007, 97: 249–265.

[2] Estes J E and Mooneyhan W. Of Maps and Myths. Photogrammetric Engineering and Remote Sensing, 1994, 60: 517–524.

[3] National Research Council. Toward a Coordinated Spatial Data Infrastructure for the Nation. Washington DC: National Academy Press, 1993.

[4] National Research Council. Successful Response Starts with a Map: Improving Geospatial Support for Disaster Management. Washington DC: National Academy Press, 2007.

Chapter 3 The Renaissance of Geographic Information: Neogeography, Gaming and Virtual Environments

Andrew Hudson-Smith and Andrew Crooks

Web 2.0, Google and Neogeography

The world of Geographic Information (GI) Science has changed. It has experienced exponential growth over the last twenty leading to fundamental changes to the field such as complex spatial analysis within a desktop environment. However, this has been dramatically altered with the rise of internet mapping. Open up Google Trends and a simple search comparison between GIS, Google Maps and Google Earth highlights the point: on April 23rd 2005, the number of searches for Google Maps overtook those for GIS as we illustrate in Figure 1. Using Google Trends is by no means a scientific analysis, for it simply means that Google Maps and Google Earth are higher in their search volumes, but it does point to one thing—the increased awareness of GI by the public at large. With this increased awareness has come the rise of volunteered geographic information, crowdsourcing, Neogeography and citizen science, amongst many other newly emerging terms linked to the geographic profession. On that basis, April 23rd, 2005 can be pin-pointed as the day the world of GI Science changed, a change that we explore throughout this chapter in terms of outreach, communication, visualization and analysis.

Figure 1 Google Trends, Google Maps vs. Google Earth vs. GIS

Hand in hand with this change is the rise of Web 2.0 from an underground movement to the driving force behind many Internet communications and data collection of the present day. The term is adapted from O'Reilly Media in 2004 to summarise the rise of services from web-based communities focusing on technologies of social networking, social book-marking, blogging, Wiki's and RSS/XML feeds[1].

An increasing amount of the information we now consume digitally is user created as is evidenced by sites such as YouTube, Facebook, Blogger, Flickr *etc*. Central to all of these applications is ease of use and the ability to communicate via freely available tools as they can be learnt quickly and effectively without immersion in professional activities[2], this has also been viewed as the cult of the amateur[3]. In short professional tools are entering the realm of the masses via Web 2.0 technologies and from this the professions' themselves are changing. This is happening across the board from economics in the form of Wikinomics through to the 3D modelling world with the Google Warehouse and Wikitecture[4]. And perhaps this is more notable in the world of GI via a plethora of new tools, techniques and services.

One company stands out amongst all in the creation and deployment of such services and we would argue that it has done more for the field of Geographic Information Systems (GIS) and the outreach of geography in general than any other. That company is Google. We realise that in professional GIS circles this is controversial and indeed threatening to long-term practice but the simple fact is that companies that are not geographic specialists are simply doing digital geography more effectively than the big GI players. However, it needs to be stressed that the majority of the features developed are tools for the visualization of GI. They do not provide any complex spatial analysis *per se*, merely a spatial database which is somewhat confusingly under-pinned by generating income through advertising. This is symptomatic of Web 2.0 whereby tagging not only the type of information but where such information is produced, who uses it, and at what time, is fast becoming *the* killer application[5].

As such, geography is undergoing somewhat of a renaissance and one that is becoming known as 'Neogeography'. The term derives from Eisnor[6], one of the founders of www.platial.com where she defines it (Neogeography) as '… a diverse set of practices that operate outside, or alongside, or in a manner of, the practices of professional geographers. Rather than making claims on scientific standards, methodologies of Neogeography tend towards the intuitive, expressive, personal, absurd, and/or/ artistic, but may just be idiosyncratic applications of *real* geographic techniques. This is not to say that these practices are of no use to the cartographic/geographic sciences, but that they just usually do not conform to the protocols of professional practice'.

Map MashUps: GMapCreater and MapTube

Perhaps one of the most widely known terms in recent geographical discussions is the (map) 'MashUp'. Originally used to describe the mixing together of musical tracks on DJ Danger

Mouse's The Grey Album, the term now refers to websites that weave data from different sources into a new integrated user service[7]. In many ways Neogeography and MashUps go hand in hand which brings us to our work and software tools we have developed in our group CASA at University College London as part of the National Centre for e-Social Science's project on Geographic Virtual Urban Environments (GeoVUE)[2]. Both the software and techniques developed have been built and released on a short development cycle to meet the demand for new easy-to-use geographic tools in light of the rise of the 'Google'esk' geographic environment.

In a short period of time we have created a notable base of both professional and non-professional users allowing for the creation of new data sets and new ways of sharing geographic information. We detail these tools in the following sections in relation to Web 2.0 and their inevitable impacts on GI Science.

Central to our digital tool kit is 'Google Map Creator (GMapCreator)' which we illustrate in Figure 2. GMapCreator was developed partly as a result of frustration behind how difficult it was to rasterize, share and view maps on top of the standard Google Map interface. Google Maps allows two levels of basic integration with geographic data, firstly through the 'Application Programming Interface (API)' meaning getting ones hands dirty with code and secondly the more user friendly 'My Maps' system. Google released 'My Maps' in 2007, essentially building on the same product as Google Maps but allowing users to create their own maps via a simple point and click interface. One can draw lines and shapes, embed text photos and videos—all using a simple drag and drop interface. My Maps is viewed as possibly one of the most important innovations in mapping since the development of GIS[5]. My Maps is typical Neogeography, it has opened up mapping to the masses but it is of little use to the more professional geo-community. As such our GMapCreator takes the tile structure and concept of Google Maps and applies it directly to GIS datasets.

A B C

Figure 2 GMapCreator: Google Map Creator

A: initial screen when GmapCreator is first opened. B: uploading a shapefile and selection the required level of zoom. C: selecting the scales for the thematic data.

Using a tile-based approach gets around the simple 'KML' export path which is restricted to a1 Mb file size (see Gibin *et al.*[8], for more information). GMapCreator allows users to take a vector data map, in the form of either a .shp or formatted .csv and export it in the correct projection as a series of raster tiles for display over Google Maps. This is a notable move away from the points, lines and polygons approach of KML towards a 256×256 pixel tile approach with the number of tiles relative to the required level of zoom. The use of the tile-based approach has the added benefit of avoiding infringement of many intellectual property rights as the raw information is never shared or placed online, it is pure data visualization.

Data visualization is the key where the aim is not to create an online GIS, but simply to find a way to view and share geographic data for the masses. GMapCreator has to date been downloaded over 9000 times and is under continuous development, most notably in our MapTube application to which we turn next.

MapTubes' tag line is 'a place to put maps'. It combines the generic idea of YouTube providing a portal for geographic data produced using the GMapCreator software. Maps are pulled into MapTube via an XML file which is automatically created whenever GMapCreator is run. Pulling data in from outside servers rather than holding them internally allows data creators to maintain ownership of data tiles and ultimately their distribution, which is seen as a nod back to data providers who are perhaps wary of the Web 2.0 environment. We illustrate the front page of MapTube in Figure 3. Maps can be arranged by popularity, recently viewed and by the latest uploads. Once a user registers with the site they are given free access to the GMapCreator software and the ability to upload map data.

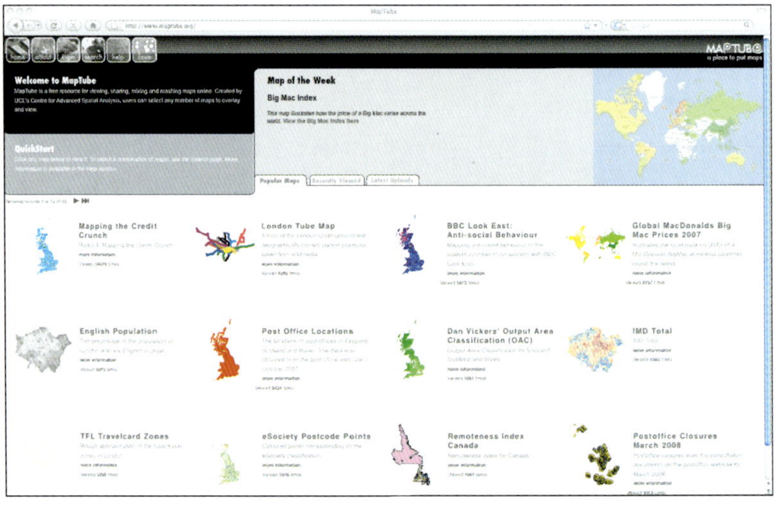

Figure 3 The home page of MapTube (http://www.maptube.org)

Online Mapping through Crowd Sourcing

Typical of Web 2.0 and the rise of the Neogeographer is user content, something we return to in 'Mapping the Credit Crunch' for BBC Radio 4 later. MapTube is typically reliant on this user generated content. Users can submit their own maps to MapTube via a simple user interface which we illustrate in Figure 4. Users are asked for several pieces of information to ensure as much information about the data and collection techniques used, which are subsequently documented and viewable on the site within the information tag attached to each map.

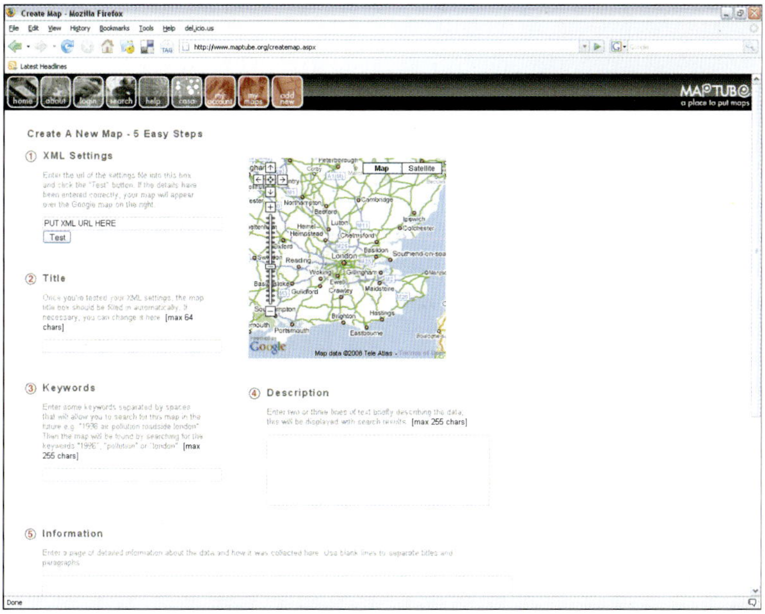

Figure 4 MapTube interface where members can add their own maps

Once the information is submitted, an 'icon' of the map is automatically created and the map is available to view, mix and 'mash' on the site. Any map on MapTube can be viewed on top of or in combination with any other map. This is of note as it is to date the only system online that allows such a simple integration of rasterized datasets. The ability to mix and match maps is illustrated in Figure 5. We have overlaid regeneration areas in London with the Index of Multiple Deprivation (IMD) and the percentage of Bangladeshi population. The ability to quickly and easily upload, view and compare datasets makes MapTube of particular interest to local councils, government-based organizations, academics and those requiring the ability for end users to view maps in a simple intuitive manner. Standard Keyhole

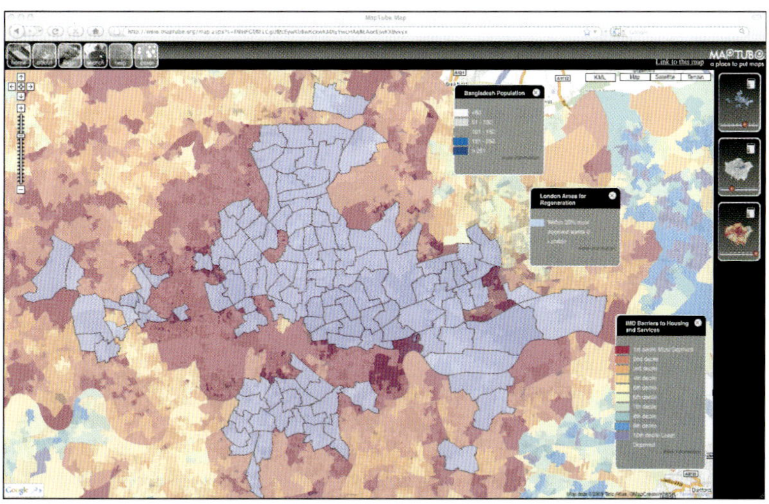

Figure 5 Viewing maps in MapTube: Bangladesh population, regeneration and IMD

Markup Language (KML) can also be overlaid on these maps, allowing any polygon output via either My Maps or a professional GIS package to be viewed via MapTube. In addition to this, KML files can be uploaded and saved in the same way as the standard datasets, providing a unique 'place to put maps'.

Not only does MapTube allow people to share and view other peoples' maps but it can also be used in more innovative ways. For example, as web surveys are often aspatial (*e.g.* surveymonkey.com), the ability to use GMapCreator and MapTube offers a simple solution to build spatial surveys for large areas. A pilot study was carried out as an experiment to create a mood map of the credit crunch within the United Kingdom in conjunction with BBC Radio 4 iPM show and NewsNight as illustrated in Figure 6A. Based on what is the singly most significant factor hurting the person the most about the credit crunch, participants were asked to enter the first part of their postcode (postcode district) so their responses could be geotagged choosing one of six options from: mortgage or rent, fuel, food prices, holidays, other, or the credit crunch is not affecting me, as shown in Figure 6B.

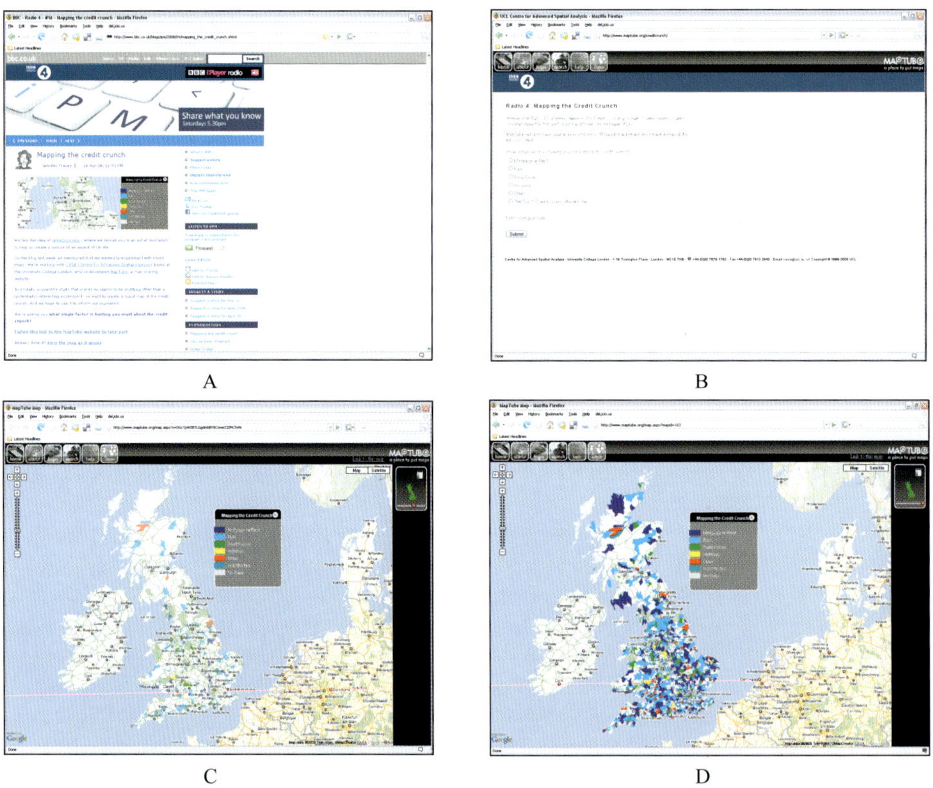

Figure 6 Mapping the credit crunch

A: link from the BBC Radio 4 iPM show. B: choosing the single factor hurting the person the most about the credit crunch and entering their postcode. C: map of the mood. D: map of the mood as more people enter their information.

No personal information was collected and participants were reassured that their actual locations could not be identified. This was ensured through the use of postcode district rather than the postcode unit or building address therefore preserving data confidentiality. Each response updated the database element of the underlying shapefile with GMapCreator running in the background every 30 minutes to create a new map which was subsequently updated on MapTube as shown in Figure 6C. Over time, as more participants entered information, the map went from blank to varying shades depending on what people were worried the most about in the postcode district as demonstrated in Figure 6D. Used in conjunction with MapTube, it allows participants and other users to take other information and lay the maps on top of one other. The potential of this approach for gathering spatial information is enormous. For example, it could easily be used to gather other information such as fear of household burglary, the quality of primary school education, access to local health facilities and so on.

Mapping the Credit Crunch represents one of the first near real-time geographic surveys of a nation's mood. As such the time element is also of importance as each response includes a time stamp allowing the nations mood to be visualized in both time and space. In excess of 21,000 people took part in the survey over a three-week period creating a unique and interesting dataset which is very much of its time. The Credit Crunch Map has since led to BBC Look East, the nightly news programme for East Anglia in association with BBC local radio, using the system to create a mood map of anti-social behavior. Using a similar data entry technique, viewers of BBC Look East were asked to answer a series of questions on their views on anti-social behavior at a postcode district level. The survey at the time of writing is on-going with 5000 plus respondents to date, Figure 7 illustrates its use as part of a news segment on BBC Look East.

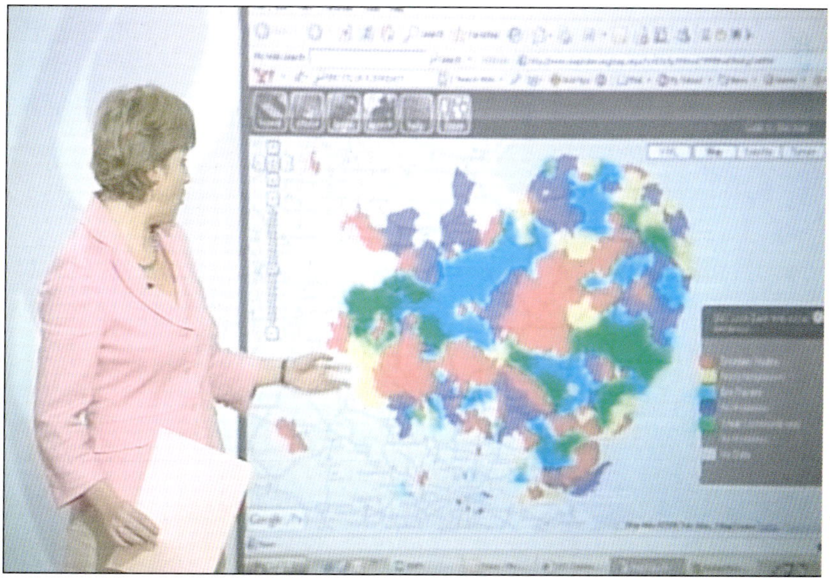

Figure 7 BBC Look East using MapTube

In a 'pre-Google' world, none of this would have been possible for the license to use the base map and aerial imagery would have been prohibitively costly and the behind-the-scenes

GIS would have been undoubtedly slow and cumbersome. We consider this in many senses to be Web 2.0 and Neogeography in action: free, easy to use and yet potentially very powerful in terms of GI Science.

Virtual Cities and Virtual Worlds

Returning to Figure 1 and the Google Trends search, of particular note is the rise of the digital earth with Google Earth perhaps being the most influential single release of geographic-based software to date in this current wave of new geographic software. Google Earth, Microsoft Virtual Earth and NASA's WorldWind to name but three are indicative of a new trend in geography, one where digital points, lines and polygons are moving into the third dimension. The ideology behind such developments can be linked back to David Gelernter[9] in his seminal book *Mirror Worlds: or the Day Software Puts the Universe in a Shoebox*. He defines "Mirror Worlds" as software models of some chunk of reality, some piece of the real world going on "outside your window" which can be represented digitally.

Gelernter predicted that a "software model" of your city, once setup, will be available (like a public park) ... it will sustain a million different views... each visitor will zoom in and pan around and roam through the model as he chooses[10]. Back in 1991 Mirror Worlds and the concept of the universe in a shoebox were fantasy leaning closer to the science fiction novel Snow Crash in which Neal Stephenson [11] defines life online as a "Metaverse":

> "As Hiro approaches the Street, he sees two young couples using their parents' computer for a double date in the Metaverse, climbing down out of Port Zero, which is the local port of entry and monorails top. He is not seeing real people of course. This is all part of the moving illustration drawn by his computer according to the specifications coming down the fiber-optic cable. These people are pieces of software called avatars".

Fast forward to the present day where swathes of the universe can be viewed via Google Sky and the Metaverse is close in definition to emerging virtual world systems such as ActiveWorlds and Second Life. Indeed the last year has in many ways been the year of digital geography, the rise of the digital earth, where advances in computer graphics cards and the free availability of modelling software has made it easier than ever to virtually build, tag and navigate three dimensional geographical space.

Through digital earth systems, the Geospatial Web can be seen as the foundation for something all together more powerful, a digital earth that mimics the real world, created by users at large via Web 2.0—Gelernter's vision of a Mirror World. Indeed although technology is moving at an ever increasing pace we are but at the beginning of a revolution in place and space. These new tools and techniques to communicate and visualize are providing a digital sandpit for geographers, GI specialists and the Neogeographer.

Side by side, with the development of geographic visualization using "digital earths" is a field often overlooked by the geographic profession—that of utilizing game engines for geo-visualization. Partly driving this emergence are games such as "Crysis" from Crytek. The game comes packaged with an editable "sandbox mode" allowing the main game to be stripped away and new models to be inserted and shared via the gaming 'modification' community. Game engines in many ways represent the cutting edge in desktop graphics technology and the ability to import models into game such as Crysis should not be under-estimated. The learning curve is slightly higher than importing a model into Google Earth but the results are much more realistic allowing flythroughs of large cities, landscapes and datascapes. Figure 8 illustrates our Virtual London model running in the Crysis engine allowing real-time lighting, shadow and object manipulation.

Figure 8 Virtual London model in the Crysis game engine

Our Virtual London model extends to every building inside the M25 and it is available free of charge to local authorities throughout the Greater London Authority boundaries. In essence, Virtual London is a professional 3D GIS model yet to run it in real-time with graphic details such as shadows and the ability to add-in/replace buildings within the city-scape. It is necessary to take it out of a 3D GIS and into a much lower priced game engine for only then can the model be freely navigated, manipulated, queried and indeed populated. Crysis at the time of writing retails for £24.99.

Sitting side by side yet somehow abstracted from mapping, gaming and digital earths is Second Life and other similar virtual environments. Second Life and their like are easy to dismiss as pure distraction and entertainment. Yet look under the lid of Second Life and it contains one of the most powerful geographical data visualization kits available. Second Life represents one of the most successful social/visual environments on the Internet. Launched in 2003 with little more than a few kilometres of simulated computer space, it now covers more than 750km^2. Created by Linden Labs, based on Linden Street in San Francisco, the world of Second Life has been created almost entirely by its users. The users have created a digital landscape primitive by primitive, with simple objects such as spheres, cubes and pyramids becoming the digital equivalent of bricks and mortar. In Figure 9 we

Figure 9 Editing the digital streetscape in Second Life

illustrate our avatar manipulating the digital streetscape. Each section of the building is based on a simple rectified image aligned as to create a high-resolution representation of the entire building. Adding objects is as simple as dragging shapes into the environment and aligning and texturing accordingly.

We have been developing a series of geographical visualization examples in Second Life on land kindly donated to us by the Nature Publishing Group. Our examples range from the 'global' view down to 'local' with an emphasis on real-time data feeds and data query techniques. Figure 10 illustrates a small section of our work.

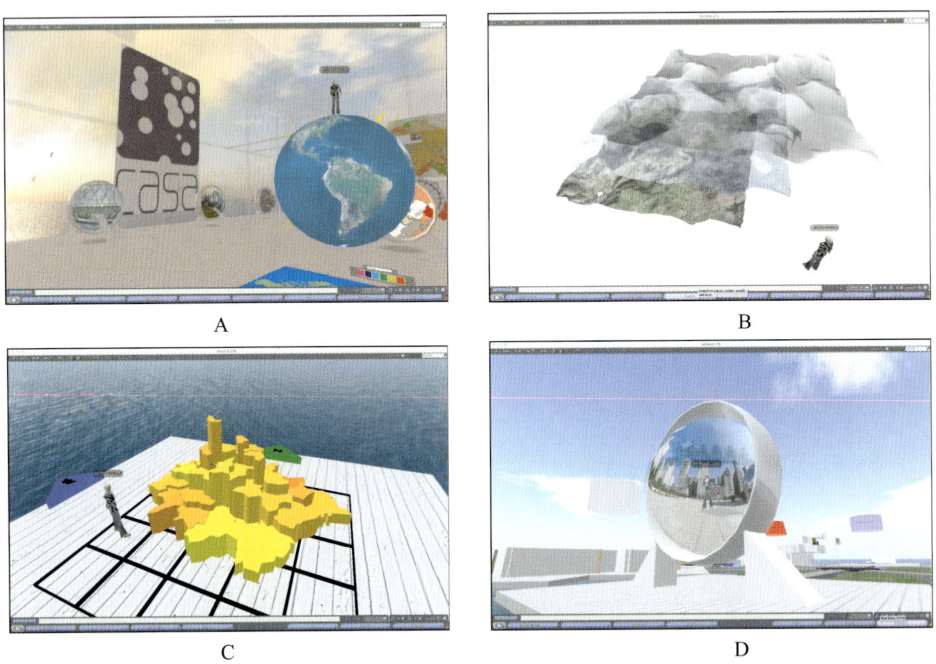

Figure 10 GeoVisualization in Second Life

A: digital earth with real-time weather data. B: multiple layer geographic surfaces. C: geographic data visualization, imported direct from ESRI ArcScene. D: urban sphere panoramic visualization.

Second Life is but the tip of an iceberg. It represents a glimpse into the future of digital geography, visualization and collaboration in virtual space. Overlapping with systems such as Google Earth and toying with the concept of populated space, within the next few years, the predictions are for a 'Second Earth': a merger of sorts of Google Earth and Second Life. These in turn move the digital toolkits we have today for constructing Mirror Worlds into a ParaVerse—a parallel virtual world geographically linked to the planet earth or other bodies in the physical universe. A working ParaVerse may seem some way off but a number of notable companies are working on the software as we type and a merging of the virtual earth combined with populated virtual space may be closer than we think. This will be a true geographic machine space where the real and virtual are populated and vice-versa.

Whither GI Science?

So where does this leave the world of GI Science? It leaves us with a wealth of new data, new ways to visualize it, tag it and query it in both two and three-dimensional space. April 23rd 2005 can be seen as the beginning of the first wave of a new age of geographic innovation. Coming waves include the 'GeoCloud' whereby data, software and tools are held on remote servers accessible regardless of machine or location. In many ways, this wave is already upon us with 'My Maps' existing in The Cloud. On site Geotagging is also of the moment and gaining in popularity via the iPhone 3G, Nokia N95 and alike with built in GPS and mapping tool kits. Combine The Cloud with location-based services and Web 2.0 Crowd Sourcing and you enter the phenomenon of OpenStreetMap, OpenAerialMap. Day by day new services and tools are emerging. The list is becoming endless.

Such tools open up a cornucopia of possibilities for the world of GI Science, especially for geovisualization and it is high time to embrace the Neogeographer, the data and perhaps more importantly the services they are creating. Welcome to the new world of geographic information Science.

Acknowledgements The authors would like to thank the team at CASA working on the GeoVUE project.

References

[1] Graham P. Web 2.0. http://www.paulgraham.com/web20.html [2008-8-5].

[2] Hudson-Smith A, Batty M, Milton R. The Neogeography of Virtual Cities: Digital Mirrors into a Recursive World. In Foth M (Editor). Handbook of Research on Urban Informatics: The Practice and Promise of the Real-Time City. Information Science Reference. Hershey, PA: IGI Global, 2008: 270–290.

[3] Keen A. The Cult of the Amateur: How Today's Internet is Killing Our Culture. New York: Currency, 2007.

[4] Tapscott D and Williams A D. Wikinomics: How Mass Collaboration Changes Everything. New York: Portfolio, 2006.

[5] Hudson-Smith A. Digital Geography: Geographic Visualisation for Urban Environments. London: Centre for Advanced Spatial Analysis, University College London, 2008.

[6] Eisnor D. Neogeography. http://www.platial.com [2008-4-23].

[7] Hof R D. Mix, Match, and Mutate "Mash-Ups"—Homespun Combinations of Mainstream Services—Are Altering the Net. Business Week, http://www.businessweek.com/magazine/ content/05_30/ b3944108_mz063.htm [2008-8-6].

[8] Gibin M, Singleton D, Mateos P, Longley P A. Exploratory Cartographic Visualisation of London using the Google Maps API. Applied Spatial Analysis and Policy, 2008, 1(2): 85–97.

[9] Gelernter D. Mirror Worlds: The Day Software Puts the Universe In a Shoebox... How it Will Happen and What It Will Mean? New York: Oxford University Press, 1991.

[10] Roush W. Second Earth. Technology Review, July/August 2007, 38-48, http://www.technologyreview.com/Infotech/18911/page1/ [2007].

[11] Stephenson N. Snow Crash, New York: Bantam Spectra, 1992.

Chapter 4 Toward Integrated Space-Time Analysis Environments

Donna J. Peuquet

Introduction

In our daily lives, we continuously wrestle with complex, chaotic and often contradictory or ambiguous information about the real world. This comes from direct sensory input (at least at a local scale) as well as from an array of indirect sources including books, newspapers and the World Wide Web. From this, we are able to continuously improve our understanding, to learn and generally to function better in our environment. Perceptions lead to cognitive information as interpreted observations that have been imbued with meaning. An accumulation of interpreted observations in turn leads to knowledge via linkages, as integrated understanding of how things work that may be applied to other contexts. In science, we similarly use data to produce new interpretations that advance knowledge and inform future investigation.

There has been much excitement over the recent introduction of free virtual globe software such as Google Earth, which allows the casual user as well as the earth scientist to fly through 3D virtual space. They can view any portion of the earth using live satellite data, add their own data if they wish, and interactively pan and zoom down to the street level using photographic scenes. This represents a particular form of cyberspace, blurring the distinction between sensory reality and a representation of it. There has also been much recent attention paid in the development of analysis methods in the area of Knowledge Discovery in Databases (KDD) and Data Mining (DM) for geospatial databases[1,2]. Such computational methods can search immense volumes of data for patterns very quickly and with mechanical accuracy.

These capabilities have also garnered much attention in GI Science. Many geospatial analysis techniques used in GIS are similarly observation-driven methods, following the belief that we should let the data speak for themselves in a theory-free process of pure induction. The learning process and development of real understanding, however, involves a significant amount of theory-driven investigation. This is true for both science and everyday cognition, as described above.

Theory-driven investigation in science is the domain of modeling and simulation. Existing examples for modeling geographic processes, such as UrbanSim for simulating urban systems[3] and DivGame for vegetation dynamics[4] tend to be stand-alone and focused on narrowly-defined contexts. Constraining the domain of a model—simulating specific components of a spatial process—is a standard approach for dealing with complex systems. The availability of massive data collections, particularly observational data through time, is also providing the opportunity for process models to be tested, verified, refined and even broadened on a scale not possible previously. This broadening includes building linkages between models of different types, such as urban simulation and global climate models[5, 6]. An example of such large-scale modeling enabled by the relatively recent wealth of data is the Long Term Ecological Research (LTER) project. But as emphasized in the LTER Project, the study of spatial process intrinsically involves time as well as space.

Within the GI Science community the integration of time into spatial databases needed for analysis of space-time processes has proven to be challenging. The added dimension of time significantly increases the complexity as well as the volume of a database. It is now generally acknowledged that storing large amounts of heterogeneous geospatial data requires a multi-representation approach based on the field-based/entity-based duality[7]. This has been extended for the inclusion of the temporal dimension into the what/where/when triad (representing entity-based, location-based and time-based representations respectively) that has been espoused by a number of authors[8-10]. In this scheme, each type of view is based upon a specific dimension within a balanced and interlinked structure.

But gaining new understanding about our environment involves *both* observational induction and modeling of spatio-temporal processes within an iterative and tightly-linked cycle. To accommodate both observation-driven and theory-driven components of the learning cycle within this multi-representation approach, observational data thus needs to be supplemented with higher-level information, extending the what/where/when scheme into a what/where/when + how and (ultimately) why, as shown in Figure 1. These higher-levels of information could potentially serve several purposes: first, to capture information or insights learned as the results of analyses; second, to inform subsequent analyses, including the interpretation of new data; and third, to provide a common and perhaps negotiated set of derived views on which collaborative analyses can be based.

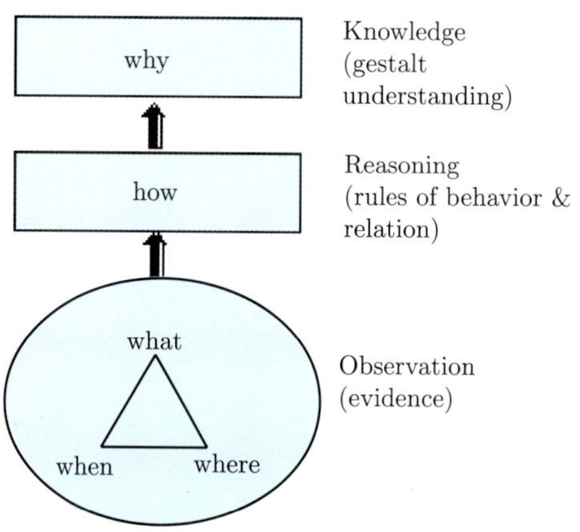

Figure 1 On-beyond data—higher levels of information need to be stored along with the data

This stored information, with different levels of understanding from observational data to knowledge, also needs to be integrated with an automated logic and reasoning system in order to utilize the information for deriving new inferences.

Storing Higher-level Information—From Data to Concepts

A key issue in storing and using higher levels of information in a computer environment is how to link these to the database. If the data represents observation, and higher-level information relates to how observed phenomena work as rules of behavior, then there needs to be an additional component that links those observations to stored behavioral rules. One approach is to use formalized ontologies as that link. Very generally speaking, formalized ontologies are a specification of entities, as concepts, types of entities, and the relationships among these entity types. This provides a structure in entity-based space and a means for interpreting data by mapping those data onto generalized concepts. For example, relating a specific observed instance of a lake in the database to the concept of "lake" and subsequent association of rivers, streams and drainage systems as concepts and their expected characteristics. This observation-concept association can then link to generalized rules regarding those types of entities as to their expected behaviors given their observed state and space-time relationships.

Ontologies are widely used in Database Management Systems as the means to provide common semantics for database integration and sharing. These formalized ontologies are often described conceptually and represented graphically as concept graphs. This approach provides a means of representing a level of abstraction above the observational data. Web Ontology Language (OWL) has quickly become the standard language for encoding ontologies, and is intended as a tool for providing a common understanding across the World Wide Web in the idea of the "semantic net" . OWL was designed to define classes and concepts as well as their properties and relationships, and to allow automated reasoning and learning about these classes and concepts[11].

Although formalized ontologies can play a key role in systems that combine machine learning with data exploration capabilities, there is currently a critical shortcoming that needs to be overcome with existing tools. In current implementations, ontologies may change through learning but there is no ontological history. Concepts and relationships are added, replaced or completely redefined, but the ontology is always represented in the most current state. It is critical, however, to be able to backtrack if mistakes were made by either the machine learning mechanism or the human user. It is also important to be able to backtrack if a major new insight requires rethinking current concepts. There is some recent research that shows how formalized ontologies might be extended to include ontological history, involving explicit storage of new and superseded concepts, these linkages, and description of how components originate[12].

From Form to Process

The fundamental mechanisms describing process—*i.e.*, *how* things change and evolve to affect the behaviors of the relevant entities and the overall form of the observed phenomenon—is not represented within any ontology. An ontology only represents abstracted entities (or concepts) relevant to that process and their structural interrelationships. A process, in contrast, can be defined as a function that acts on a domain and the individual entities within it. An example of a process would be urban growth, and relevant entity classes would be roads, housing, *etc*. Instances of specific relevant entities may or may not be present within the specific data set or space-time window being examined. A process could be simple and elemental (*e.g.*, move forward), or could be more complex (shoreline erosion), in which case, it contains a set of lower-level functions. Generally speaking, such a function set incorporates a hierarchically organized description of behaviors and their applicable context along with defined triggers of specific behaviors. At an elemental level, a function could be an algorithmic or mathematical transformation. At a higher level, these could be expressed as a set of interrelated rules.

There has been growing interest in intelligent software agents in GI Science as a specific tool in a variety of contexts, including simulation of land-use/cover change[13], wayfinding [14], and social simulations[15]. Software agents are not only a natural way to represent social aspects (institutions, societies, *etc.*) as well as natural aspects (climate, hydrology, *etc.*) of geospatial processes, but also a way to represent the coordinated dynamic behaviors of multiple entities constituting complex dynamic processes. Like components of any real-world process, agents also play roles of differing types, and these are often hierarchical in structure[16]. Communication among agents is an important element within agent-based systems for coordinating functions. The agent-based approach thus provides a machine-learning and simulation environment based on distributed knowledge representation.

Cognitive agents, as opposed to reactive agents, maintain an internal state (*e.g.*, goals, plans, and state information about the world), and behave by searching through a space of behaviors and cooperating with other agents (which in turn may affect their actions)[17]. Key properties of cognitive agents with these characteristics are that besides being *situated* (applicable, or active, within specific contexts) and *distributed* (maintain their own constrained view), they are *adaptive* (learn and improve through experience). Cognitive agents are also described as corresponding to the expertise of a human expert. Reactive agents tend to be more algorithmic in nature and are thus appropriate for dealing with low-level functions or elements of a process. A middle form (not fully cognitive) can also be useful as a flexible means of representing various, more complex process components. Such a multiple agent scheme provides a flexible means of representing knowledge about process in a modular form (captured through simulation and verified via linkage with observational data, or explicitly input by the user). Agents, in combination with ontologies, act as knowledge carriers to provide qualitative information, simulate/predict behaviors and behavioral patterns, and make large stores of observational data more easily retrievable, reusable, and shareable.

An integration of database, ontology and agent-based tools provides for a combined theory-driven/observation-driven environment for studying space-time dynamics. Observational data are stored together with abstracted information about generalized concepts, relationships and behaviors. The concepts, as a formalized ontology, provides the linkage to process representation in the form of an agent-based system. An agent-based approach, besides providing a flexible means of capturing knowledge about how things work, also provides a means of analyzing 'what-if' alternatives to aid the human analyst/decision maker via simulations and a fast means of conveying domain-specific knowledge to decision-makers in time-critical situations.

So, just as it is now acknowledged that storing large amounts of heterogeneous geospatial data requires a multi-representation framework, storing higher-level information concerning geospatial phenomena and their dynamics requires an extended multi-representation approach. Moreover, these multiple data and knowledge representations must be functionally interlinked. Initial research at Penn State[18] has shown that the use of multiple software agent types linked with an ontology and database within an overall what/when/where + how schema provides the needed combination of representational power.

Keeping the Human User in the Loop

Even with increasing sophistication and use of tools for representing derived concepts such as ontologies within the computer and the increasing sophistication and subsequent power of agent-based modeling, the human user and his/her analytical capabilities can never be left out of the loop. There are several reasons.

First, earth systems are 'open' and non-deterministic. The interactions among entities and their components within processes tend to be highly dynamic, interlinked, and with complex chains of cause-and-effect often involving both physical processes and human intervention. For example, environmental water quality is affected by urbanization (housing and road density, *etc*.) which is, in turn, affected by public policy and economics governing how and where things may be built at both local and regional scales.

Even with continuing advances in representation techniques toward multi-representation schemes and inclusion of higher-level knowledge, no model or representation, computer-based or otherwise, can possibly capture all of the factors that may be interlinked in what is really a system of systems—nor is it reasonable for practical reasons to want to do so. Interrelationships of entities and their attributes change in complex and inter-articulated ways over space and through time. While multi-scale time-based, location-based, and entity-based representations allow a means to reveal these patterns, multiple sets of space-time relationships may combine to obscure each other. This is where human judgment and intuition can be particularly powerful and effective in teasing these apart, in distinguishing real patterns from accidental artifacts, and identifying additional factors or influences that need to be considered.

Second, the data-gathering process inherently involves selection (via temporal and spatial resolution, *etc.*) and interpretation (via measurement, classification and encoding). As a result, observational data always provides a filtered and abstracted view of reality. This view often reflects a given disciplinary perspective or anticipated set of problems to be addressed. The data thus cannot really speak for themselves. Only the human user can interpret the specific view of reality presented by the data and understand the uses and limitations for a particular application context.

In the current era of exhaustive data sets collected automatically via sensors (*e.g.* RFID chips in merchandise, passports, *etc.*) or collected as artifacts of computer-based transactions, data go through an additional filtering and abstraction process before being passed on to programs or the user. The massive scale of these data, now measured in terms of petabytes, necessitates the use of reduction/simplification methods such as categorization and statistical generalization to make the data accessible interactively. The Mapreduce approach developed by Google to make individual queries efficient is an example[19]. This introduces an additional problem in that how the filtering and abstraction process is performed may be unavailable to the user because it is a proprietary process.

The third and most important reason to keep the user in the loop is that while computers are very good at formalized deduction and some forms of inductive inference, they are not good at creative reasoning. For humans it is the opposite. Computer-based analysis tools have emphasized induction (finding patterns and associations from observational data), as in the case of Knowledge Discovery and Data Mining techniques, and deduction (drawing conclusions from a set of previously known rules and relationships for verification and simulation, as in the case of agent-based techniques). The combination of these two forms of reasoning is the classical cycle of the scientific method as it is normally described. Such 'text book' descriptions, however, ignore the key step of discovery through creative reasoning.

Creative reasoning, what Pierce termed abduction, is a third form of reasoning and a distinctive kind of inference[20, 21]. Inference is reasoning based on observation. Inductive inference goes from the specific to the general without explanation. For example, Every day I have been in London it has rained. Therefore it rains every day in London. In contrast, abductive inference is conjecture or formulation of a hypothesis that can potentially explain a surprising observation. Unlike deduction and induction, this form of reasoning is non-deterministic and nonlinear. Abduction involves analogies and thought experiments in order to generate a completely new hypothesis. This is how entirely new concepts are formed and 'breakthroughs' in science are made. A classic example is the observation that the shapes of the east coast of South America and the west coast of Africa seem to fit together. A potential explanation is that the two continents were at one time connected and drifted apart at some time in the distant past. Continental Drift remained a hypothesis until observational evidence was found decades later.

Agent-based models have been built that can simulate very complex behaviors using known rules, and that can adapt to new situations by modifying those rules by extrapolating directly from observations of the specific phenomenon. They cannot develop new hypotheses for explaining those behaviors. Similarly in Knowledge Discovery and Data Mining tools, sophisticated clustering, pattern analysis, and other statistical techniques can uncover new and complex associations and relationships, even global level structures, but these come with no explanation. Such explanation, or hypothesis generation through generic abstraction, is the domain of creative reasoning.

The important issues relating to applications such as understanding urban growth, the spread of disease and other currently pressing social and environmental concerns thus center on the understanding of complexities that are far beyond the representational and learning capabilities of any stand-alone computer-based tool: How do the components of a process and the subcomponents of each relate to each other, and at varying spatio-temporal scales? While recent tools can provide realistic interactive visualizations and detect patterns in vast amounts of varied geospatial data, what can be actually learned from these tools in terms of deeper understanding of the phenomena under study requires the creative thinking of the human user.

Visualization offers a key capability to link the creative thinking capabilities of the user with the capabilities of the computing environment. While current visualization capabilities are most often used for data exploration, recent work in what is now known as geovisual analytics is attracting significant attention. Geovisual analytics is the development of sophisticated and coordinated methods for aiding the user to derive insights from graphical display of geographic data, as well as from abstracted information directly and in combination with statistical or modeling techniques[22, 23]. It is a less structured process that enables the use of analogues and visual thinking. Visual displays are important in the creative thinking process in that they enable the human analyst to bypass constraints inherent in linguistic, mathematical and similar representations. They can also allow attention to focus on salient aspects that can then be directly manipulated. The displays can be pictorial or highly diagrammatic. This allows analogical thinking as structural components are separated (or generalized) and mapped onto a different form.

Use of sophisticated visualization capabilities linked to stored knowledge and theory-driven process modeling capabilities as well as to multidimensional observational data can also be seen as providing the 'why' element as shown in Figure 1. But then, how can these be stored in a coordinated way so that views can be linked and coordinated?

These components combined into a single, integrated environment would provide a powerful exploration, analysis and simulation environment far beyond the capabilities of current GIS software. The issue is really what should GIS become to continue living up to its name—Geographic Information Systems. The exciting part of this is that we already have the individual components, more or less. While technologies relative to the individual components are progressing rapidly, a key factor will be to develop the necessary linkages among them that will make the necessary level of integration among them possible.

On a broad conceptual level, this requires effectively combining the capacity, speed, and mechanical accuracy of the modern computing environment with the insight and ad hoc judgment capabilities of the human user. Integrating the broad range of capabilities needed for a tightly integrated environment while still providing flexibility for change as tools continue to advance requires a modular approach. A modular approach also allows easier customization for differing application and problem solving contexts. The following section briefly describes STNexus; a prototype example of an integrated, space-time analysis environment built at Penn State using this approach.

A Prototype Example

STNexus combines pre-existing components to provide a new database facility for encoding time varying geospatial data along with high-level, domain specific knowledge and a multiview and highly interactive visualization environment. The conceptual unifying factor is the use of the triad multi-representation scheme of what/where/when views.

The triad database model was implemented using Berkeley DB, a database development toolkit. A major difficulty in implementing such a multi-representation scheme suitable for a large, real-world database is how to deal with the potential storage redundancy related to multiple views. This problem was overcome in STNexus using a multiple indexing strategy as described in Ref. [24]. The visualization environment utilizes the same what/where/when scheme for showing these multiple views of the data graphically.

STNexus Uses a visualization builder called Improvise. This visualization toolkit supports multiple, linked views. STNexus allows ad hoc database queries to be generated via direct interaction with the visualization environment. This eliminates the need for formulating database requests via a database query language, which can be difficult to express when the user is not sure what they are looking for. In addition, multi-factor queries can be particularly hard to express and often do not produce correct results.

The application example shown in Figure 2 is displaying visitation patterns of guests at a hotel in central Pennsylvania at the turn of the last century. The objective was to uncover interrelationships and patterns in complex data involving the movements of individuals and groups of people over varying spans of time. Such interrelationships and patterns would reveal insights about how commerce was conducted and how social groups interacted in the rural United States before the introduction of the automobile and modern highways. As seen in Figure 2, there are multiple views that graphically show the data in various types of dimension-specific (who/where/when) displays. The most important component of this view is the array of (mostly colored) circles and squares with numbers. This is essentially a calendar view of the number of visitors to a specific hotel on each day within a selected time interval. A circle indicates a weekday and a square indicates either a Saturday or Sunday.

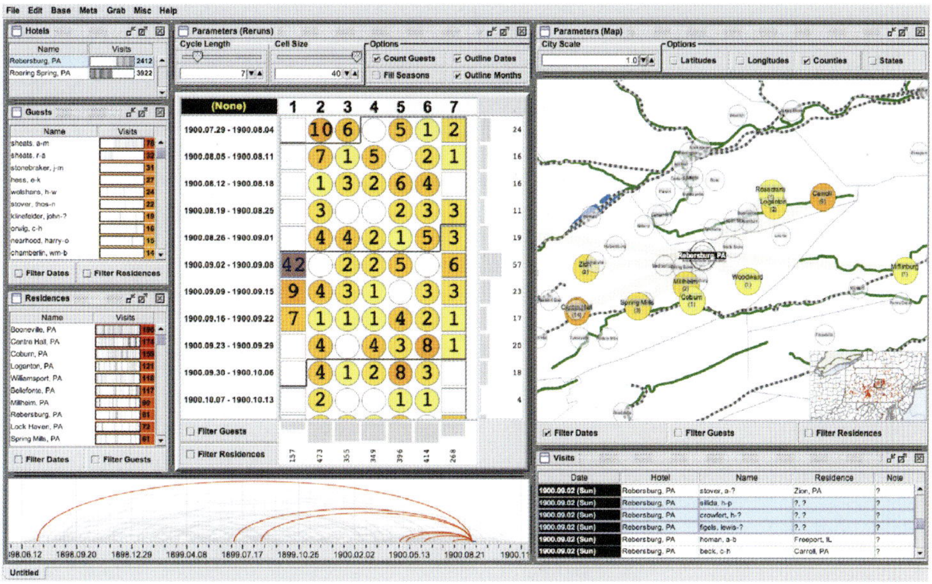

Figure 2 Sample screen shot of STNexus showing hotel registry data, views selected for data exploration

The number inside of the circle or square is the number of guests registered at the hotel on that day. The darker the color, the greater the number of guests. This coloring allows a quick visual means for seeing temporal patterns. Circles and squares that are white with no number indicate days on which the hotel had no registered guests. As shown by the plain numbers at the top, each row shows a seven-day sequence. Since temporal patterns often don't occur in seven-day repeat cycles, the number of days in each row can be lengthened or shortened interactively. Temporal patterns can be easily seen when colors and numbers line-up.

The two lists on the left of the display, as entity-based views, provide the names of the hotel guests in the upper window, and the names of places listed as residences for any guest in the lower window. The small box to the left of each name gives a linear frequency graph of the visits made by each guest, or the place from which a guest came. On the right side of the display, the map window, or location-based view, shows the locations of the towns surrounding the selected hotel. The location of the hotel is shown as the grey circle toward the middle of circles indicate towns listed as places of residence for visitors of the hotel. The greater the number of visitors from a town, the darker the coloring.

All views can be 'filtered' to show only selected times, people, or places. In Figure 2, for example, a time-line view in the bottom left shows the dates of the previous visit for each of the people and dates selected from the list of visits shown on the lower right. Figure 3 shows a different set of views for the same application, anticipating use of higher-level knowledge. The graph in the upper right shows a simple ontology using a concept map tool called ConceptVISTA. This particular example was developed anticipating the integration of additional data based on store receipts for a store in the same town as the hotel selected in the example shown in Figure 2.

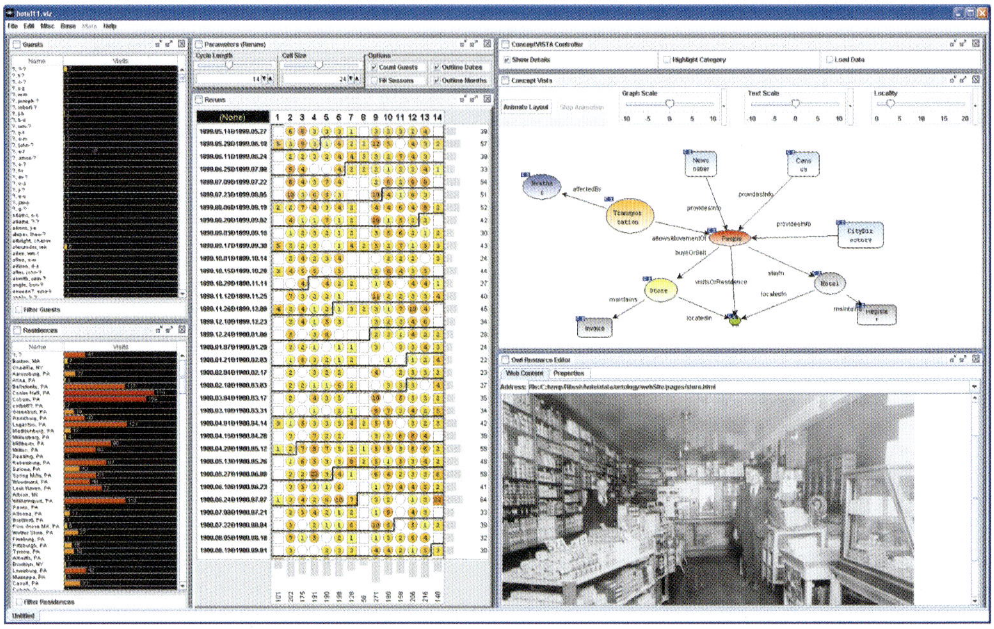

Figure 3 STNexus showing hotel registry data with a view showing ontology and another showing photo of an example instance (store in this location and time period)

Beyond GIS?

Is all of this possible beyond GIS? It is certainly beyond the current capabilities of existing commercial systems, although GIS have database, visualization and analysis as essential components. The new visualization capabilities becoming available go far beyond the mapping paradigm included in current GIS and beyond the imagistic approach of recent virtual globe software. But although these tools are more specifically attuned to a data-rich environment, each has been developed independently. Sophisticated tools for higher-level knowledge representation and reasoning have also been developed independent of GIS.

A flexible visual interface that integrates these more recently developed data-driven and theory-driven tools would provide the means for guiding steps in analyses in a highly interactive manner, assisting the user in the cycle of exploration, discovery, theorizing, confirmation and simulation, and in turn guiding subsequent data exploration. From a GIS standpoint, we need to think in terms of interoperability of software capabilities and tools instead of monolithic GI Systems.

There has been evolutionary progress toward modular systems. ESRI's current approach to offer varying capabilities as add-ons is a step in this direction. Traditional GIS representation and analysis capabilities have been extended with the ability to represent higher-level information and the provision of reasoning capabilities for modeling and simulation through the use of agents in ESRI's Agent Analyst. But we need to go far beyond this to truly modular systems where the user themselves can easily combine various tools. This not only allows easy customization, but also allows easy addition or substitution of tools as needs change or as more advanced capabilities become available in specific areas.

Achieving a truly modular environment where independently-developed tools can be easily integrated without the need for significant reprogramming requires an overall conceptual framework. This framework is needed to provide options and guidelines for internal (storage) and external (visual) forms for information as well as progressive levels of abstraction. In addition to an overall conceptual framework for guiding tool construction, agreed-upon software mechanisms are also needed for linking and coordinating qualitatively different kinds of information. What this really describes is the open system concept, developed in the late 1970s and early 1980s for sharing advancements within the Unix operating system environment and was adopted by the Open Geospatial Consortium (OGC).

Acknowledgements The STNexus prototype is the result of a collaborative team effort. The visualization component described here is primarily thanks to the combined efforts of STNexus project team members including Chris Weaver, the author of Improvise as well as team member, Anthony Robinson and Alan MacEachren in close collaboration with David Fyfe and Deryck Holdsworth.

References

[1] Miller H J and Han J (Editors). Geographic Data Mining and Knowledge Discovery, London: Taylor & Francis, 2001.

[2] Yuan M, Buttenfield B P, Gahegan M N, *et al.* Geospatial Data Mining and Knowledge Discovery. In: McMaster R and Usery L (Editors). Research Challenges in Geographic Information Science. Boca Raton: CRC Press, 2005: 365–388.

[3] Waddell P. Urbansim: Modeling Urban Development for Land Use, Transportation, and Environmental Planning. Journal of the American Planning Association, 2002, 68: 297–314.

[4] Alonso D and Sole R V. The DivGame simulator: A Stochastic Cellular Automata Model of Rainforest Dynamics. Ecological Modelling, 2000, 133: 131–141.

[5] Grimm N B, Faeth S H, Golubiewski N E, *et al.* Global Change and the Ecology of Cities. Science, 2008, 319: 756–760.

[6] Grimm N B, Grove J M, Pickett S T A, *et al.* Integrated Approaches to Long-Term Studies of Urban Ecological Systems. Bioscience, 2000, 50: 571–584.

[7] Couclelis H. People Manipulate Objects (but Cultivate Fields): Beyond The Raster-Vector Debate in GIS. In: Frank A U, Campari I, and Formentini U (Editors). International Conference GIS-From Space to Territory: Theories and Methods of Spatio-Temporal Reasoning. Pisa, Italy: Springer-Verlag, 1992: 65–77.

[8] Andrienko N, Andrienko G, Gatalsky P. Exploratory Spatio-Temporal Visualization: An Analytical Review. Journal of Visual Languages and Computing, 2003, 14: 503–541.

[9] Peuquet D J. It ' s About Time: A Conceptual Framework for the Representation of Temporal Dynamics in Geographic Information Systems. Annals of the Association of American Geographers, 1994,

84: 441–461.

[10] Yuan M. Wildfire Conceptual Modeling for Building GIS Space-Time Models. GIS/LIS'94, ACSM, Phoenix, 1994, 860–869.

[11] Aref M M and Zhou Z. The Ontology Web Language (OWL) for a Multi-Agent Understating System. International Conference on Integration of Knowledge Intensive Multi-Agent Systems (KIMAS 2005), IEEE, 2005.

[12] Brodaric B. Geo-Pragmatics For The Geospatial Semantic Web. Transactions in GIS, 2007,11: 453–477.

[13] Parker D C, Berger T, Manson S M. Agent-Based Models of Land-Use/Land-Cover Change. LUCC Report Series No 6, LUCC Focus 1 Office Anthropological Center for Training and Research on Global Environmental Change, Indiana University, 2001.

[14] Raubal M. Ontology and Epistemology for Agent-Based Wayfinding Simulation. International Journal of Geographical Information Science, 2001, 15: 653–665.

[15] Conte R, Edmonds B, Moss S, *et al.* Sociology and Social Theory in Agent Based Social Simulation: A Symposium. Computational and Mathematical Organization Theory, 2001,7: 183–205.

[16] Jennings N R. On Agent-Based Software Engineering. Artificial Intelligence, 2000, 117: 277–296.

[17] Ferber J. Multi-Agent Systems: An Introduction to Distributed Artificial Intelligence, Addison-Wesley, New York.for Knowledge-Oriented Representation: Embracing Social Rules in GIS. International Journal of Geographical Information Science, 1999.

[18] Yu C Q and Peuquet D J. A Geoagent-based Framework for Knowledge-oriented Representation: Embracing Social Rules in GIS. International Journal of Geographical Information Science, 2008, 9999(1):1–38.

[19] Anderson C. The Petabyte Age: Because More Isn' t Just More—More Is Different. Wired Magazine, 2008,16(07).

[20] Hartshorne C, Weiss P, Burks A (Editors). Collected Papers of Charles Sanders Peirce. Cambridge, MA: Harvard University Press.

[21] Nersessian N J. Model-Based Reasoning in Conceptual Change. In: Magnani L, Nersessian N J, Thagard P (Editors). Model-Based Reasoning in Scientific Discovery. New York: Kluwer Academic/Plenum Publishers, 1999: 5–22.

[22] Frihida A, Marceau D J, Thériault M. Extracting and Visualizing Individual Space-Time Paths: An Integration of GIS and KDD in Transport Demand Modeling. Cartography and Geographic Information Science, 2004,31: 19–28.

[23] Guo D. Visual Analytics Of Spatial Interaction Patterns For Pandemic Decision Support. International Journal of Geographic Information Science, 2007, 21: 859–877.

[24] Dou J and Peuquet D J. Flexible Indexing of Spatiotemporal Information with Multiple Views. Geocomputation'05, Ann Arbor, MI, 2005.

Chapter 5 Empowering the Web with Location

C. Vincent Tao

Introduction

Driven by some fascinating innovations in online services, we are experiencing fundamental changes in the way we think, communicate, live and do business. Among many drivers that have made profound impacts on the emergence of the Web as a computing and social platform, geospatial technology is probably the least expected. We are stepping into an era where the Digital Earth[1)] vision is no longer a science fiction but something we are approaching and even have partially achieved. The rapid evolution in this path is beyond the projection of the science and business community. Driven by technological innovations such as Google Earth and Microsoft Virtual Earth, the market is experiencing an increasing adoption by the mass users. This is the first time in history that the geospatial industry is recognized as part of the mainstream of information and communication technology (ICT).

From online mapping to virtual worlds, from location-based services (LBS) to local commerce (L-commerce), from local search to location-based advertising (LBA), location became an indispensable part of the Web evolution. On the other hand, the Web is in transition from its 1.0 to 2.0. Long-tail phenomena[1] have made an impact on the almost every aspect of business, government and academia work.

Given disruptive technology advancements, product innovations and services proliferation, it is important to understand the overall landscape change in the location sector. The chapter is organized as following. We firstly discuss an interesting relationship between the Web and location, and the essence of location in the context of the information organization. We then analyze the three phases of the evolution of online mapping services. Two key concepts of modern online mapping, namely, local search and social media, are discussed. Finally, we use the Microsoft Virtual Earth program to illustrate the latest developments in location powered Web services.

The Web and Location

The Web has progressed along its own natural path driven by hundreds of millions of Web users, authors, and publishers. There is no grand plan for the Web, neither in the past nor in its future. As an open world-wide platform, the Web allows us to explore the most fascinat-

---

1) Al Gore (1998), The Digital Earth: Understanding our planet in the 21st century, given at the California Science Center, Los Angeles, California, on January 31, 1998.

ing user experiences in knowledge sharing and discovery, communication and social networking, commerce and entertainment. The Web has evolved to the era, popularly known as 'Web 2.0'. The core of Web 2.0 is the notion of 'the Internet as platform'. This is a revolutionary paradigm shift in the Internet computing and knowledge sharing[1)].

Web 1.0 and location—"the death of distance"

Location is a fundamental element in the evolution of human society. We all understand that communication and transportation are critical to social and economic development. At the Web 1.0 stage, the most influential factor of the Internet was in communication for the Internet, as an open super highway, enables the connection of any point to point regardless of the location. It supports almost unlimited data flow and transport from one location to the other. On the Internet, location becomes transparent and the 'distance' becomes less meaningful. This can be observed from the following facts:

- Networking and interacting with people no longer entirely dependent on the location, through services such as e-mail, instant messaging;
- Internet search allows people to find information and documents from Websites anywhere, information sharing and discovery becomes universal; and
- Electronic commerce (e-commerce) via the Internet has proven to be one of the most powerful business models, removing the barrier of physical distances in business transactions. Users can purchase items, like books, cameras, or even cars, online without any geographic restrictions.

Since users mainly consumed the information from the Web, ease of access is far important than any other usage. Location is no longer as important as it was and the concept of 'the death of distance'[2] is widely adopted.

Web 2.0 and location—"location matters more"

The Web itself is spatially organized. IP addresses, hubs and servers etc are all location relevant. Though the Web makes location transparent in many ways, location that is indispensable in people's lives and this has emerged as one of driving factors to the Web 2.0 evolution. Our user survey[2)] shows that 72% of participating users prefer that their home, work office and their shopping locations are within the distance of a 20 min driving. From this we can infer that location still matters most to people's daily lives. Search is so far the most popular activity on the Internet. It is reported by the comScore Networks[3)] that 63% of US Internet users performed a local search online in July'06, a 43% increase over July'05. Local search is growing at a faster rate than general Web search. Local search grew 24% and

1) http://www.oreillynet.com/pub/a/oreilly/tim/news/2005/09/30/what-is-web-20.html.

2) Microsoft Internal Market Research Study Report on LBS, 2007.

3) http://www.comscore.com/press/.

general Web search grew 14% in 2007 compared to the previous year. Finding and sharing of location related information and data on the Internet or from mobile devices has already become one of top Internet activities.

The market penetration of location enabled devices, such as GPS-Phones and personal navigation devices (PND) for in-car and outdoor uses has tripled. Sales of PNDs rose to more than 30 million units in 2007 from 12 million in 2006. Both the user base and the revenue in personal navigation, friend finder, enterprise LBS like fleet management, asset tracking etc are increasing at ever faster rates of growth. The world-wide revenue of LBS, by Garner research[3], will reach to 8 billion dollars in 2011.

One of the most interesting developments of Web 2.0 is the emergence of the user generated content(UGC), where users are voluntarily contributing their contents for public sharing. Web 2.0 is all about harnessing collective intelligence and personalization. Many web services are purely operated based on the UGC, like Yelp.com, CitySearch.com *etc.* Even road vector maps can be generated by thousands of volunteers who carry the GPS devices to collect, annotate and process the road data. The OpenStreetMap project (http://openstreetmap.org) has demonstrated the possibility of this approach. UGC and its network efforts are some of core concepts in Web 2.0.

Local commerce is proliferating. It is observed that the long-tail advertisers, *i.e.*, small and medium sized enterprises (SMEs), are the beneficiaries to such advertising approaches. SMEs require more cost-effective, time-sensitive and audience-targeted advertising vehicle to promote their business and sales. Traditional media like TV or newspapers do not fit well given uncertainty in the advertising medium. Connecting locals to local businesses is all that L-commerce is about. Though the academic research in this area is lacking, location-based advertising (LBA) technology is gaining great attention by both large and small industry players.

Essence of location

The essence of location is finally being understood. From an information perspective, location is nothing but one dimension of our data index like a keyword that can be used to organize the massive data and information on the Web. As "*what, who, when and where*" are associated with any Internet queries, location that addresses 'where' questions is fundamental to any data and information organization. The Web has already become our social platform on which Web users not only act as information consumers but also information producers. Advanced data indexing, aggregation and organization techniques are required to deal with the massive and often fragmented data produced and contributed by the Web user community. Location is now considered as a more precise and more effective means to organize Web information in the Web 2.0 era.

Evolution of Online Mapping

Figure 1 shows a user survey report based on a percentage of respondents[1)] from the USA: 75% of respondents are using Internet for "*getting maps and directions*", ranked second

1) From iCross, US consumer study report, 2006.

75% of respondents are using Internet for "*getting maps and directions*", ranked second among all other Internet user activities, ahead of "*looking for news*" and "*shopping*" *etc.* categories. It is worth noting that this figure does not show the frequency of the usage in which "*looking for news*" is more frequent behavior than 'getting maps and directions'. The figure does indicate that this feature is a "must-have". Figure 2 is a user survey result for the most wanted mobile services. "*Location-based specific information services*" is ranked the second most wanted features on mobile devices. These surveys show that maps and location-based information are very valuable to both Internet and mobile users.

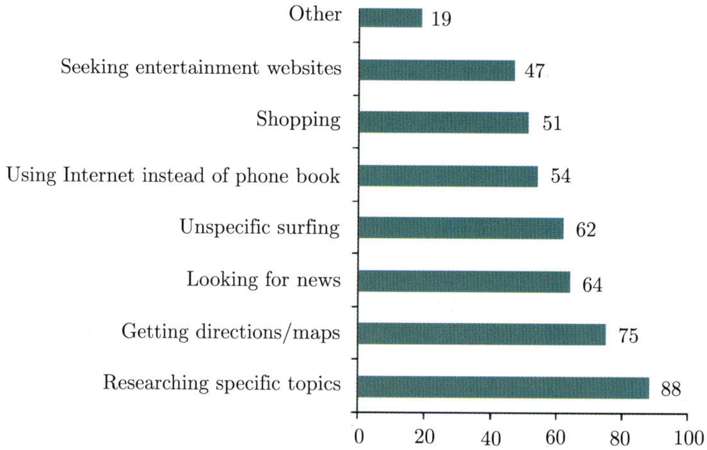

Figure 1 Popular Internet user activities based on % of respondents

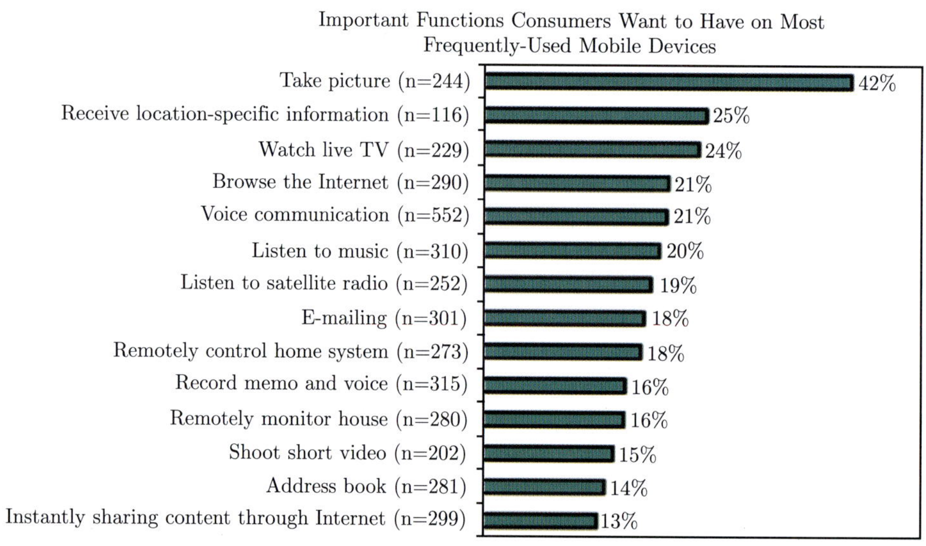

Figure 2 Most wanted mobile services based on a user survey

Source: *Mobile Entertainment Platforms and Services*, a survey of 2, 112 Internet users ©2005 Parks Associates and Google Earth, Microsoft Virtual Earth, and others *etc.* are used by millions of people around the world. More than just eyeballs, online mapping technology has evolved from map search to local search and to local services. I will group these technology developments into these three phases.

Map search

In this phase, mapping services are mainly designed to help users find the location on a map and provide the direction and navigation information from point A to point B. A typical user scenario is, for example, "*locate a restaurant on map*". To facilitate locating a local business, yellow page (YP) data, *i.e.*, business listings, are ingested in the map search engine in order to support a query for a specific location. As a result, the user can find a location of the restaurant via its phone number or its name without a need to remember the detailed restaurant address. Combining maps and the associated YP data has enriched users' mapping experiences to search locations on maps. This phase of the development is called 'map search', a typical online mapping service application. It is worth noting that map search is generally designed against a structured location database.

Local search

Local search is the use of specialized Internet search engines that allow users to submit location constrained searches or queries. Typical local search queries include not only information about "what" (such as keywords, a business category, *etc.*) but also "where" information, such as a street address, city name, postal code, *etc.* Local search needs to be intelligent and be able to detect a user's location intent. For example, one may submit a search query, "*where is the best restaurant within this map window*". This is not a typical query that can be handled by a conventional map search engine. In order to address this query, a typical fuzz query, the local search engine needs to crawl, and aggregate users' ratings and reviews data for restaurants, identify user's past search behaviors to understand his/her intention about "*best*", *i.e.*, inexpensive or luxury or romantic, and finally provide a list of ranked 'best' restaurants on the map via a sophisticated ranking algorithm. Generally, local search is developed to deal with unstructured location constrained queries. It requires a specialized and intelligent search engine algorithm along with rich and comprehensive local contents including maps, YP data, Web documents and user generated content. In most applications, like Google Maps or Microsoft Live Search Maps, local search is architectured on top of a search engine, a mapping engine, and a large local content repository. Since local search is more dependent on the quality of local content, user generated content becomes critical to any local search service. For example, data such as "*where are the best fishing spots*", "*where are the most scenic biking routes*", *etc.*, can only be collected effectively via user contributions.

Local services

The result of local search queries may possibly lead to an action. For example, searching for a dentist's office may lead to a dental appointment; finding the best hotel may lead to a hotel a dentist's office may lead to a dental appointment; finding the best hotel may lead to a hotel room reservation. User survey data has shown that local search activities are the ones that would most likely result in off-line actions. 70% of US users are performing online research in shopping products but purchase them off-line[1]. Bridging the gap between the online and off-line activities is considered one of the hottest research and commercial opportunities. In this context, local services refer to location-based Web services that address a user's search goals by bridging the online search and offline actions. For example, the dentist office calendar can be made online when the dentist's location on map is found. One can book a dental appointment directly online. A hotel's booking system is connected to a mapping site so one can compare the pricing of hotels and reserve the hotel online.

Local services are tightly integrated with the mapping, local search as well as e-commerce capabilities. It meets the needs of online users by offering end-to-end services. E-commerce or transaction-based payments can be an attractive business model for this kind of services in addition to advertisement-based monetization. We are still at an early stage of the development of local services both in terms of technologies and business models.

Local Search

As discussed in the previous section, local search essentially performs an intelligent query understanding and content mining analysis in order to answer location constrained queries. The principle of local search is similar to the general keywords based search, sometimes called general Web search. Most local search engines are built upon or integrated with general search engines.

Local search engine

These are the core components for a local search engine as we show in Figure 3.

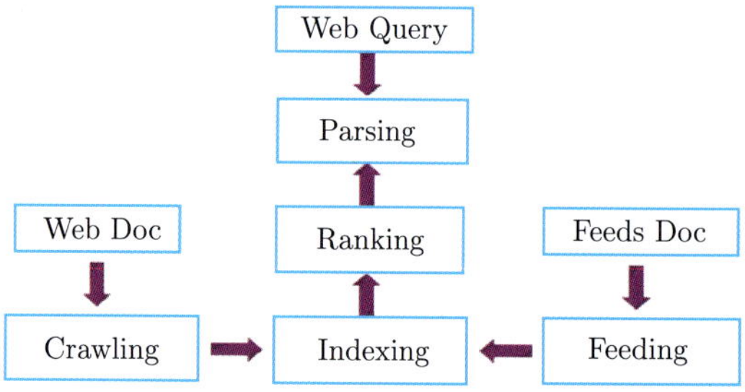

Figure 3 Local search process

- Feeding: retrieve documents from a feeds database. Feeds are structured local contents such as business listings, POIs, user rating and reviewing data, local events *etc.* These data sets are generally provided from the content vendors.

1) ComScore press release, Sept 28, 2006: http://www.comscore.com/press/release.asp?press=1017.

semi-structured documents extracted from public Web sites.

- Storage: this is a centralized repository that stores the local related content and the indexing data. It is worth noting that an ultra large and efficient storage system is needed.
- Indexing: the content is indexed in advance to allow a fast ranking of the search results.
- Ranking: the ranking algorithm is also a core to any search engine technology as it determines the relevance of search results, *i.e.*, ranking of the search result listings.
- Parsing: a Web interface is required for entering queries and presenting answers. Intelligent parsing is key to detect the intent of any user queries.

Relevance and performance are the two primary quality measures to any search engine technology. For local search, the relevance of results can be affected by the content quality (breadth and depth) in feeds and crawled data, smart parsing algorithm in detecting a user's local intent and the ranking machine. The following session briefly describes how location-based ranking is performed.

Ranking and relevance

Technically, ranking consists of two components: *static ranking* (query-independent ranking) and *dynamic ranking* (query-dependent ranking). For example, "*find a restaurant within a specified polygon*" is a query-dependent question, while "*find the best rated Chinese restaurant in New York City*" is a query-independent question, *i.e.*, the answer to this query is independent of who is asking this question. For local search, dynamic ranking is applied to queries that are tied to a specific geographic area (or an area of interest), zoom level, query classification, *etc*. The ranks are generated at runtime. Static ranking is used for pre-ranking the content including Web documents, location popularity, user ratings, category popularity *etc*. The ranking results are as one of the inputs into the dynamic ranking phase.

A good static ranking algorithm is crucial to search results[4] as it affects:

- Relevance: the static rank of a page provides a general indicator to the overall quality of the page. This is a useful input to the dynamic ranking algorithm.
- Efficiency: the search engine's index is ordered by static rank. By traversing the index from high quality to low-quality pages, the dynamic ranker may abort the search when it determines that no later page will have as high a dynamic rank as those already found.
- Crawl Priority: Search engines need a way to prioritize their crawled pages and to determine which pages to recrawl, and how frequent. The static rank of a page can be used to determine this prioritization.

There are several known algorithms in static ranking with the PageRank developed by Google one of the best known in this field. PageRank ranks all pages in a collection and is then used as a static part of a query evaluation. PageRank is intended to measure the authority of a Webpage on the basis that high authority pages have other high authority pages linked to them. The Hypertext-Induced Topic Selection (HITS) model developed by IBM is also a known method; HITS is intended to be applied to just the subgraph of pages retrieved with a query, and perhaps some of their neighbors. HITS is also referred to as the hubs and authority model: a hub is a Web page that is viewed as a reliable source for links to other Web pages, whereas an authority is viewed as a reliable content page itself. Generally speaking, good hubs should point to good authorities and visa versa[5].

Microsoft has developed a machine learning based static ranking approach, called Rank-Net. As a generic problem of classification and regression, letting $\boldsymbol{X}=\{\boldsymbol{x}_i\}$ be a collection of feature vectors (typically, a feature is any real valued number), and $\boldsymbol{Y}=\{y_i\}$ be a collection of associated classes, where y_i is the class of the object described by a feature vector $\boldsymbol{x}_i$. The classification problem is to learn a function f that maps $y_i=f(\boldsymbol{x}_i)$, for all i. When y_i is real-valued as well, this is called regression. Static ranking can be considered as a regression problem. If let $\boldsymbol{x}_i$ represent features of page i, and y_i be a value (say, the rank) for each page, we could learn a regression function that mapped each page's features to their rank. Recent work on this *ranking* problem directly attempts to optimize the ordering of the objects, rather than the value assigned to them. For these, let $\boldsymbol{Z}=\{<i,j>\}$ be a collection of pairs of items, where item i should be assigned a higher value than item j. The goal of the ranking problem, then, is to learn a function f such that, $f(\boldsymbol{x}_i) > f(\boldsymbol{x}_j)$. The details of the method can be found in Richardson *et al.* [4]

Social Media

Without local knowledge contributed from local people, local search is nothing more than a lookup from a phone book and a map. We understand that the vast amounts of local knowledge are kept by local people and are mostly not in any published media (*e.g.*, Web documents). Traditional media like TVs, Radios and newspapers are a limited way to share, and publish such long-tail local contents, for example, "*LuLu Hair Salon is less crowded on Thursday*", "*Jerry is an experienced landscaper*", *etc*.

There is no single definition on what social media is[6]. It actually takes many forms, including text, images, audio, video, map and blogs. The key concept of social media is its ability to allow people to share contents, opinions, insights, experiences and anything. Compared to the traditional media where contents are created by editors, user generated contents are the most valuable assets in social media. A few prominent examples of social media applications are, Wiki sites, like Wikipedia; Social networking sites, *e.g.*, MySpace and Facebook; Video sharing sites, YouTube *etc*. Many believe that social media in its many and varied forms is clearly one of the most important trends in Web 2.0 where information is democratized by users generated content and shared knowledge.

We have realized that the word of mouth is perhaps the most effective marketing approach as people do trust their friends and colleagues to make good recommendations. There is considerable empirical data to support this. With network effects, a virtuous circle can be built to get the information distributed much faster than by any other conventional non network means. It literally took just one year to get MySpace and YouTube known to millions of people. The dynamics of social networks cannot be underestimated. Most local search sites are now going social. The local-social paradigm is a powerful way to enrich user experiences in online mapping, local search and local services.

The map is being evolved as social media where users can associate the location on a map to their contents such as photos, ratings and reviews, events, travel experiences, *etc.* There is an organic relationship between social media and local contents. A new form called "social directories" has grown exponentially in the last couple of years when compared to the yellow page databases. With many easy-to-use Web tools, users are contributing enormously about their local knowledge, resulting in some highly compressed local guide information. Like TV and newspaper, map can be essentially a media that carries rich, comprehensive and timely local information to serve local users.

An Application and Demonstration—Microsoft Virtual Earth

In the following section, application examples are presented to illustrate the concepts discussed in the previous sessions and we will also illustrate these with some enabling technologies for Microsoft Virtual Earth.

Virtual Earth introduction

Microsoft Virtual Earth program started in 2005. As part of Microsoft's strategic investment program, it has gained a great attraction internally and externally. Virtual Earth's mission is to extend the existing online mapping experience to immersive local search, infotainment and social networking. It is an indispensable set of services and user experiences to connect people to local content, connect people to local businesses and to connect people to people. For consumer users, Virtual Earth is an online mapping and local search service portal, called Live Search Maps at *maps.live.com.* It is a rich 2D/3D Web experience that is designed to engage the mass user community. Virtual Earth pioneered the first Web Browser based on 2D and 3D mapping and exploration experiences so users do not have to download a separate desktop application to run and explore the 3D globe. Virtual Earth is also available through any popular mobile phone devices via *wls.live.com.* For enterprise users and developers, Virtual Earth is a software and service platform with a set of services and API's allowing developers to integrate Virtual Earth into their own Web or desktop applications.

The Virtual Earth program is aimed to engage users to explore, discover, and share location-based information and services. There are three distinctive features of this program:

- Convergence of search, sharing, creation and exploration of local information (places, businesses, directions, events and people) in an integrated online environment;

- Immersive experience in the exploration of high resolution and cartographically appealing maps, images and 3D models, and their associated location information (traffic, weather, rating and reviewing *etc.*); and
- Social media platform to enable community engagement via a suite of community services and across device support (communications, mails, blogs and mobile *etc.*).

Virtual Earth core technologies

Virtual Earth technology has undergone remarkable development in the last decade with some of the core components of the Virtual Earth system and services as follows.

Content acquisition and aggregation. Content is 'king' to any online services. Content can be generally grouped into: first spatial content: providing basic geographic reference layers such as maps, aerial or satellite images, oblique images, streetside images and 3D city models; and second non-spatial content: referring to broad location-related contents including business listings, points of interests or landmarks, weather, traffic, as well as other rich media like community photos, audios *etc.*

These contents can be acquired in many different ways: licensing or syndication partnerships from the content providers and aggregators, machine-based crawling from the Web and public information repositories, in-house edited or programmed contents. There are several technical issues worth noting for the content development strategy: automated geospatial mapping, enablement of user generated contents, and location tagging/association.

In Virtual Earth, multiple-level realistic representations of the globe are offered through aerial and satellite images, oblique images and streetside images and even community photos. Collecting and integrating of many levels of geospatial content at a global level is non-trivial. Automated processing, georeferencing and updating of these contents is key to the development of a reliable and scalable geospatial reference framework for a digital earth. Figure 4 illustrates an automated process pipeline starting from raw aerial images, to segmented feature images, to 3D surface recognition and finally to textured 3D buildings. There are about 300 urban cities with 3D texture building models released publicly on the Live Search Maps site.

Democratization of information has clearly affected the content acquisition strategy. We need to engage users and enable them to participate, while with a care to control the spam, and filter and aggregate the already fragmented and proliferated data. Associating contents to a specific location is a fundamental process for any local search and local services. The increasing use of location aware devices accelerates the growth of location contents. Nokia expects to sell 35 million GPS-capable phones. There is no doubt that mobile devices will become the primary means for collecting, and tagging location contents. It is worth mentioning that the use of location tags in user generated contents is growing amazingly fast. 72 million photos in Flickr are already geotagged[7]. The Semantic Web aims to address the

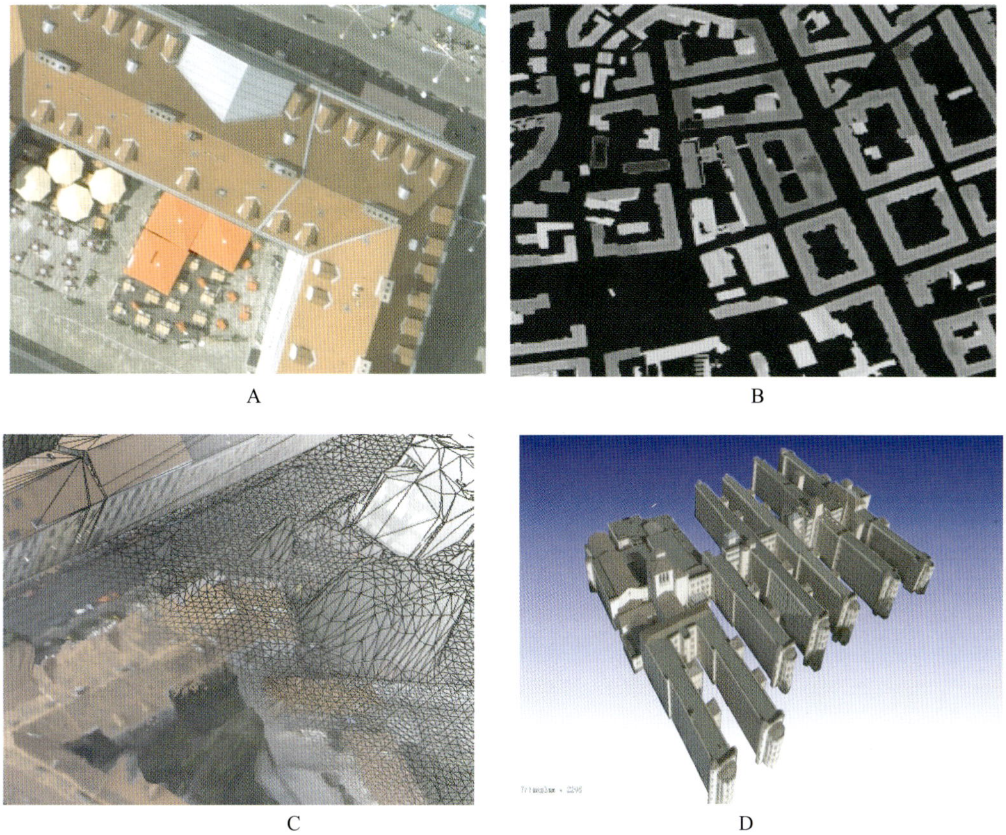

Figure 4 An automated pipeline

A: raw high resolution aerial images. B: automatic segmented image features. C: automatic reconstructed 3D mesh model. D: automatic generated textured 3D building models.

generic tagging problem from a top-down level and there is already evidence that such a 'semantic' aware Web may come from the grass-roots[8]. Location tagging is one of these examples.

Cloud computing architecture. The wide usage of online services is largely contributed by its 'free to users' business model and the high performance cloud computing infrastructure. The cloud computing infrastructure consists of reliable services delivered through next-generation data centers that are built on compute and storage virtualization technologies. The services are accessible anywhere in the world, with the cloud appearing as a single point of access for all the computing needs of consumers. In Virtual Earth, investments in data centers, high performance and distributed computing, data streaming and compression, reliability and redundancy *etc.* are enormous. Processing of petabyte level of data alone is already challenging. Making them available at run-time and processing millions of queries in seconds or minutes requires a scalable and reliable services oriented cloud computing architecture.

Online Virtual Earth Services. Virtual Earth provides a rich set of online services including maps and directions, local search, community collection, local business services and visual exploration. Visual exploration is a basic element for map-based data discovery and sharing. Several technologies have been developed in Virtual Earth. Figure 5 shows a Web interface of Virtual Earth where a digitally represented 3D city including textured 3D buildings and animated models is rendered in a Web browser environment. Images, textures, geometries and location contents are streamed to the browser on the fly. With this immersive 3D experience, visual exploration becomes a true infotainment, *i.e.*, combination of information exploration and online entertainment. One of the innovative features in Virtual Earth is its large scale deployment of the birds' eye images, *i.e.*, oblique images. From Figure 6, one can see that the oblique images are much more appealing to general users than the conventional ortho-images.

Figure 5 Virtual Earth 3D within the Web browser

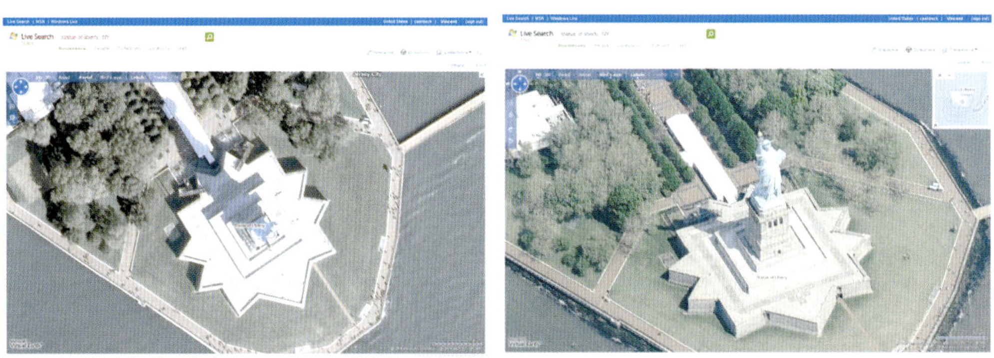

Figure 6 Conventional ortho-image vs oblique image

In order to offer a more realistic local experience, streetside images were used in a Virtual Earth street view prototype. Figure 7 shows an interface for exploring streetside images. Streetside images carry much more local relevant content such as bus stops, store front images, parking lots *etc*. The potential usage is still under investigation.

Figures 8 to 10 show some interesting services provided in Virtual Earth. Figure 8 is a real-time traffic service where the traffic data such as road speed condition and congestion, road maintenance information *etc*. are updated every 5 minutes (this service is available only

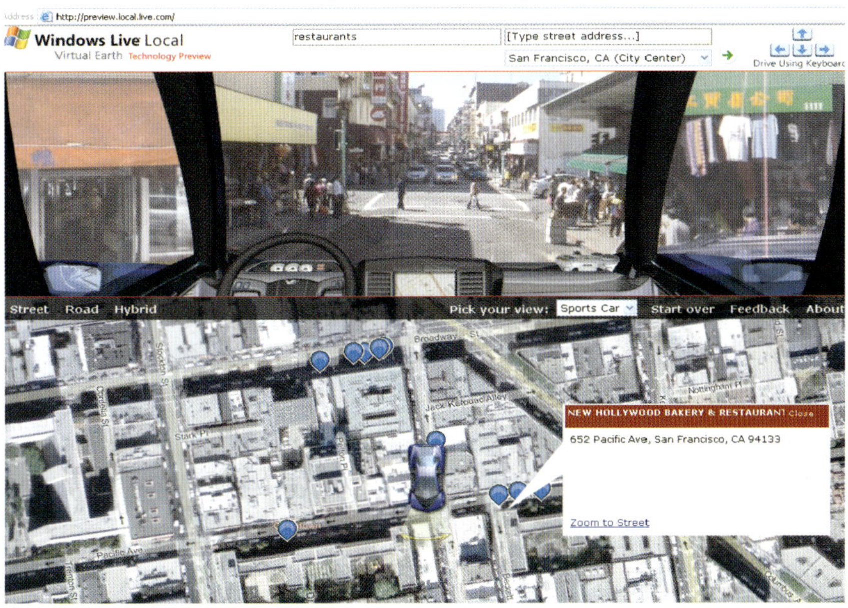

Figure 7 Infotainment experience with streetside images

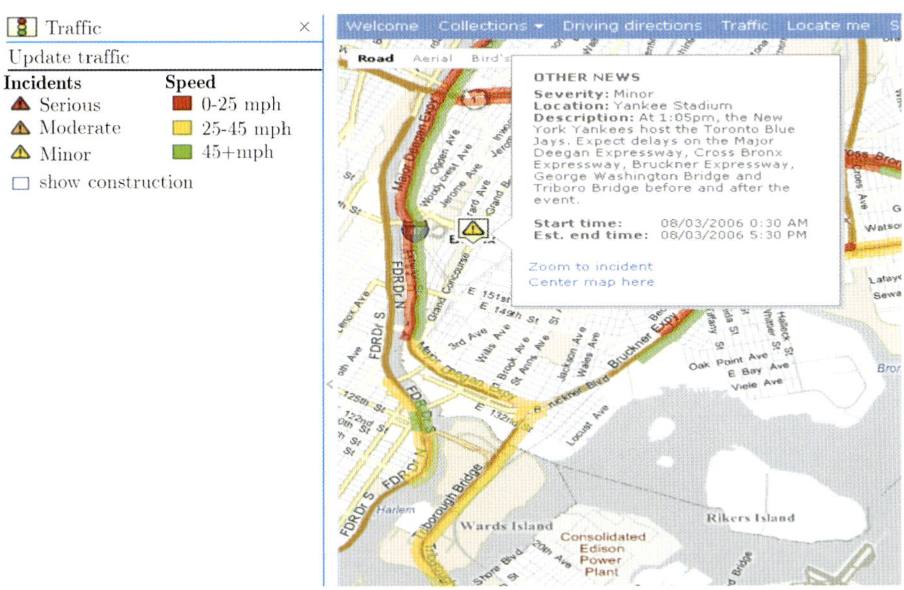

Figure 8 Real-time traffic data overlaid on the road map

in the United States). Red, yellow and green lines illustrate the current traffic condition. Figure 9 is a local business service, called "Click-to-Call". When a business, *e.g.*, a hotel, is found on the map, the "Click-to-Call" feature allows you to link your phone to the hotel re-

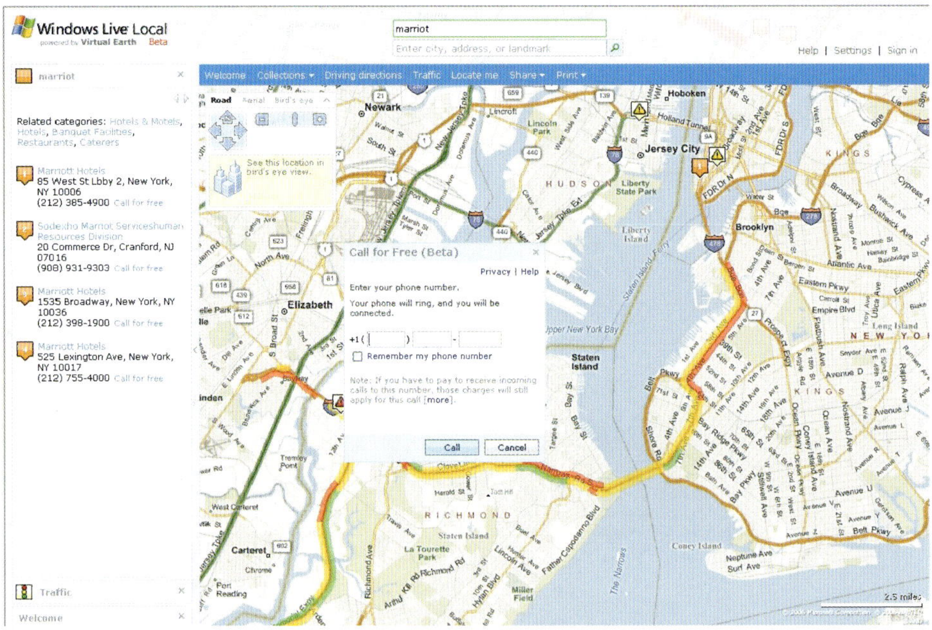

Figure 9 A local service example, free "Click-to-call"

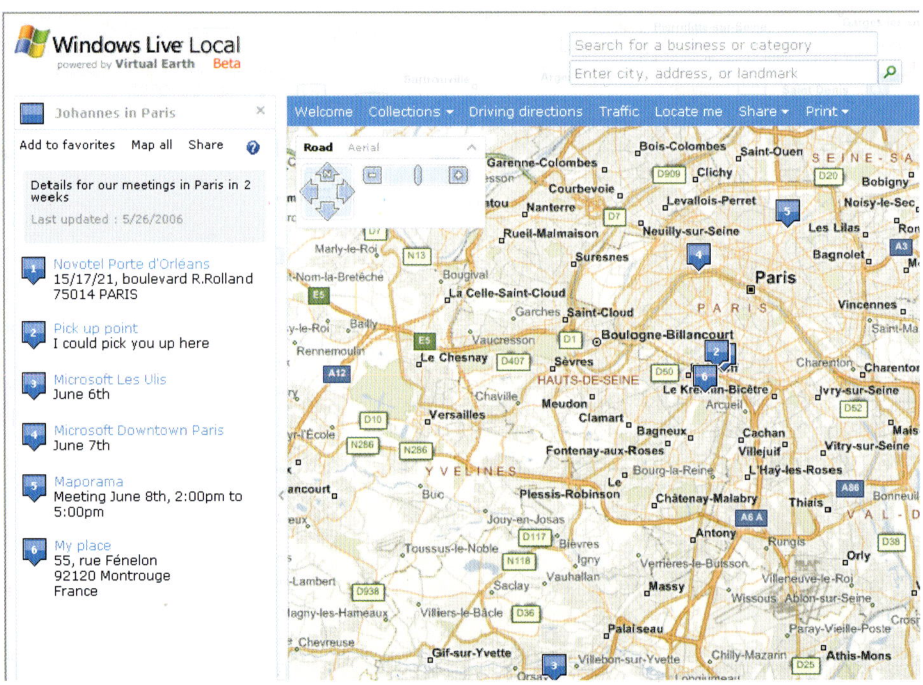

Figure 10 User generated content via a collection service

servation desk. As discussed, this is a typical local service scenario where the user's local search is connected to the ultimate action. Such a service benefits both end users and the business. Figure 10 is a community collection service. Collection is a central repository for user generated contents. In this example, Johannes (user) planned his Paris trip by annotating his stops on the map such as his hotels, business to visit, airport *etc*. Johannes associates his working documents, photos, comments etc to this collection. The collections can be accessible anywhere and can be shared with his friends, colleagues or the public. He can also invite others to comment, edit, and review his collections.

Conclusions

Location powered Web services such as map search, local search and local services are recognized as part of mainstream Internet activities. The essence of location is nothing but a means to organize our information and data. 'What, where, when and who' are associated with any information queries. From an information perspective, location is one dimensional element in indexing the information and data. To the geospatial community, there is a fundamental conceptual change in the notion that we are moving from organizing spatial information to organizing information spatially. In other words, we are not just dealing with maps or images, rather we are dealing with any data or information that are associated with location.

There are many new developments in this area. Among them, democratization of information, in particular, location content and vector maps, has proven to be effective to the wide adoption of local search and services. The map is being evolved as social media that connects local users to local businesses, local events and local contents. As mobile devices become ubiquitous, one would expect some exciting services coming from the mobile world or the PC Web and mobile combined. There is no doubt that location plays an even more important role in the mobile world. This is an exciting time for geospatial professionals, researchers and practitioners as the opportunities are enormous for exploring and realizing the power of location.

References

[1] Anderson C. The Long Tail, Why the Future of Business is Selling Less of More. New York: Hyperion, 2006.

[2] Cairncross F. The Death of Distance: How the Communications Revolution is Changing Our Lives. Boston, MA: Harvard Business School Press, 2001.

[3] Zimmermann A. Location-based Services, Subscriber and Revenue Forecast, 2006–2011, Garner Research Report. 2008.

[4] Richardson M, Prakash A, Brill E. Beyond PageRank: Machine Learning for Static Ranking. In: Proceedings of the 15th international conference on world wide web. http://research.microsoft.com/

en-us/um/people/mattri/papers/www2006/staticrank.pdf [2006].

[5] Borodin A, Roberts G, Rosenthal J, Tsaparas P. Link Analysis Ranking: Algorithms, Theory, and Experiments. ACM Transactions Inter Tech, 2005, 5: 231–297.

[6] Cook T and Hopkins L. Social Media: An Introduction to the Power of "Web2.0". http://trevorcook.typepad.com/Weblog/files/CookHopkins-SocialMediaWhitePaper-2007.pdf [2007].

[7] Turner A and Forrest B. Where 2.0: The State of the Geospatial Web. An O'Reilly Radar Report, http://radar.oreilly.com/research/ [2008].

[8] Cardoso J. Semantic Web Services: Theory, Tools and Application. The Idea Group, Information Science Reference, New York , 2007.

Part 2

Virtual Cities and Virtual Landscapes

Involving users in virtual environments raises the question of the extent to which the user's experience is appropriate and this involves the question of realism. The first virtual environments with a distinct geographic flavor emerged from computer-aided design (CAD) with its focus on the third dimension (3D). During the last decade, those working with 2D maps have sought to embrace the third dimension, initially many years ago in GIS through the simulation of quasi-3D in terms of landscapes. The origins of GIS in landscape architecture are well known and much functionality such as view shed analysis and buffering can be traced to a concern for representing and simulating landscapes. But recently urban GIS has moved into 3D. There is in fact a remarkable convergence occurring between GIS and CAD with GIS taking on greater and greater issues of realism with CAD taking on questions of data structure and spatial query. In some sense, the merger of GIS and CAD is a match made in heaven for there is enormous synergy that is waiting to be exploited. But at the same time, this convergence is being driven by the movement of the media from the workstation and desktop onto the Web. In fact for a decade or more, web-based forms of 3D environment based on software such as VRML (the Virtual Reality Markup Language) have been available. The software used in the chapters in this part of the book attest to the wide range of possible environments that can now be fashioned and although our concern is to give the reader some sense of these, our focus here is much more on the extend to which such environments can be made realistic.

Realism is thus the watchword and in terms of spatial data, this means new and different data inputs from those that have traditionally characterized GIS. In fact the rather painstaking addition to detail using hand-crafted but nevertheless digital means, has largely given way to either detailed rendering using photo-imagery or to procedures for generating 'typical' 3D content and we will focus very much on the latter here. Generative systems for developing the 3D content of cities for example has been used by several of the authors writing here. Once an urban model moves beyond a handful of buildings, then it is necessary to use automatic and routine methods for generating content. Geometric issues such as modeling roof shapes and similar intricate detail is still quite problematic but generating realism using fractal and shape grammar techniques is well advanced and some of the examples illustrate here make great use of these techniques.

The other feature of concern is the linking of process models of cities and landscapes—symbolic models based on articulating urban and other processes in terms of algebraic or decision rules forms—with their representation not simply as maps which has been the stuff of spatial analysis for many decades but with their realization in 3D. Moreover showing how models might be linked to different temporal slices and processes is also of concern picking up on the space time ideas that Peuquet developed in the previous section. Last but not least although our concern for users is played down a little in this part of the book, it is evidently clear that such ideas are central to the whole notion of virtual cities and virtual landscapes for their rationale to disseminate to those other than their designers is the key motivation for such work.

Kanji Yano and his colleagues spell out all these themes in their model of Kyoto which is not only a convergence of CAD and GIS but also one which embraces the temporal dimension. Virtual Kyoto is a virtual time-space environment created on a desktop and the internet for the purpose of investigating the past, present and future of the historical city of Kyoto. Using cutting-edge technologies in GIS and VR, the model is a 4D-GIS that comprises a series of 3D-GIS at different points in time. The 3D modeling begins with present Kyoto and then goes back to the past, including the period just after and just before World War II, the Taisho and Meiji eras, the pre-modern Edo era and finally back to Hein era when the city of Kyoto was founded in the late 8th century. Creating the data involved archiving geo-referenced materials such as current digital maps, old topographic maps, cadastral maps, aerial photos, picture maps, street photos, landscape paintings, archaeological sites data, and historical documents. Its database comprises all existing buildings including Macias (traditional town houses), early modern buildings, shrines and temples including historical and cultural heritages. 3D VR models of the buildings have been created using generative techniques and land use and landscape changes over the study periods have been simulated. Virtual Kyoto is an infrastructure which lets the authors define various digitally archived materials associated with the city, thus enabling wide dissemination of Kyoto's subtle and sophisticated forms of cultures and arts to the world across the internet. Their web-based system provides a user-friendly interface to explore historical materials of cultures and arts in the geographical context of Kyoto with its historical landscapes.

Middel and her colleagues then show us how to build a 3D model of Phoenix but one which is built on very different principles from Virtual Kyoto. Urban simulation models are now extensively used in urban planning for estimating future population and job distributions to measure the impact of planning decisions on future urban environments. Simulated population projections usually result in large, macro-scale, and multivariate geospatial data sets. Millions of records have to be processed, stored, and visualized to help planners explore and analyze complex population patterns. The authors introduce a database driven framework for visualizing geospatial multivariate simulation data from UrbanSim, a software-based generative simulation model for the analysis of future household and building types. The designed framework is extendable and aims at integrating empirical-stochastic methods with multi-dimensional visualization techniques developed for information visualization and cartography, in this sense showing how important it is to provide such routine generation if very large scale model outputs are to be visualized appropriately.

In the third contribution to this part, we shift to landscape visualization where the focus is on evaluating the realism of real-time visualization techniques for the communication of large scale vegetation planting scenarios. In these problems, the landscape might change dramatically and thus it is essential that good techniques for modeling the realism of such scenes are available. Lovett and his team compare three contrasting systems in the context of visualizing the potential changes in an English rural landscape under increased planting of 'miscanthus grass' which is a type of energy crop. The results highlight the importance of issues such as 'level of detail' management and demonstrate that some real-time models can now rival the level of feature detail possible with rendered still images. Nevertheless, there is a continued need for research to evaluate the practical benefits of using such virtual environments in decision-making contexts and to extend interactive capabilities in ways that would provide users with improved 'what if' capabilities.

In the last chapter, an integrated GIS and CAD system is developed for the automatic generation of 3D building models. Sugihara and Hayashi, using building polygons, propose a GIS and computer graphics integrated system to generate 3D building models where the polygons are building contours or footprints on the digital map stored in a GIS. A 3D urban model including these 3D building provides an information infrastructure that can be utilized in several fields such as landscape evaluation, urban planning, building design and so on. In order to automate the laborious process of constructing the building models, the system aims at creating 3D building models automatically from building polygons. Most building polygons' edges meet at right angles, that is, they are orthogonal polygons, and thus a complicated orthogonal polygon can be partitioned into a set of rectangles. In order to partition an orthogonal polygon, they propose a polygon expression which leads to a partitioning scheme based on identifying vertices which form a dividing line (DL) with respect to roof shapes and related detail. After partitioning, the integrated system places rectangular roofs and box-shaped building bodies on these rectangles. In this way, they develop rather good models for the basic gable roof and apply these to models of an ancient Japanese temple and a pagoda.

Chapter 6 Virtual Kyoto as 4D-GIS

Keiji Yano, Tomoki Nakaya, Yuzuru Isoda and Tatsunori Kawasumi

Introduction

Large amounts of geo-spatial data have been accumulated since the GIS revolution of the late 1980s, supported by advances in IT technology, new survey methods, and high-resolution satellite imagery[1]. Such evolutions in GIS environments have enhanced 3D-GIS and Virtual Reality (VR) models in terms of precision and quality. Our Virtual Kyoto project uses the cutting-edge 3D-GIS/VR technologies to visualize, analyze and restore the historical urban landscapes of Kyoto [2, 3].

This Virtual Kyoto project has been a part of "Kyoto Art Entertainment Innovation Research (2002–2006)" at Ritsumeikan University driven by the 21st Century COE (Center of Excellence) program funded by the Ministry of Education, Culture, Sports, Science and Technology (MEXT), and now handed over to the Global COE program (2007–2011) entitled "Digital Humanities Center for Japanese Cultures and Arts". It aims at reconstruction and visualization of historical landscapes in Kyoto using 4D-GIS of Kyoto, which is the 3D-GIS of Kyoto with a time dimension. Virtual Kyoto is built on a 4D-GIS database that becomes a container of digitally archived materials, such as Japanese drawings and performing arts that are being digitally archived recently. The contents of the database can be explored by a 'real time' interactive system, so that users have the capability of navigating freely in the 4D space. We have realized an on-line access to the virtual city of Kyoto over the internet. This chapter introduces our Virtual Kyoto and suggests the utilization of the 3D model for digital humanities research.

What is Virtual Kyoto?

In this chapter, we define virtual cities as 3D urban models using VR technologies and having various services, functions, and information contents with realistic buildings and allowing social interactions in the same way as in real cities[4, 5]. Crucial aspects in constructing 3D virtual cities are threefold[6]: 1) Level of reality (amount of contents); 2) Data input (retrieval of height and appearance) and 3) Functionality (as practical and/or analytical tools). As the intensity of their geometrical content increases, automatic generation of these 3D models will become more difficult and cost of creation will increase. In terms of the level of reality aspect, Shiode[6] defines six levels of virtual city: 1) 2D maps and orthogonalized aerial photos; 2) Image base rendering (panoramic 2D image pasted on virtual 3D space);

3) Prismatic building block models (3D models based on building footprint and height); 4) Block modeling with image-based texture mapping (the above model with UV mapped textures); 5) Models with architectural details and roof morphology; and 6) Full volumetric CAD models (3D model using CAD data for individual buildings). In practice, it is possible to mix different levels of 3D models to create a virtual city. In this project, we have created 'prismatic building block models' for the whole city of Kyoto, by gathering as much 2D geographical information as possible, incorporating more elaborate models with architectural details and roof morphology and some full volumetric CAD models for important architectures.

Kyoto, the historical capital of Japan founded in 794 A.D. has a large number of historical architecture such as temples, shrines, machiyas (townhouses) and early modern buildings of the pre-war period. This is a rare case in Japan in that the city avoided much of the World War II damage. Unlike other restoration projects, our strategy in restoring historical Kyoto is to first model the present day Kyoto as a virtual city, and then replace the newer buildings with the previous buildings going back to the past. Street networks and topography are also altered for periods where we have evidence that they were significantly different from today.

As a consequence, we have created a 3D model of the whole city for several time slices. We integrate the models into 4D-GIS and have named it as Virtual Kyoto. 4D-GIS is a conventional GIS with vertical and temporal dimensions added. It has the capabilities of navigating across space as well as through time; a facility for location-based enquiry is available that searches not only the contemporary features but also objects that existed and events that occurred somewhere in its history at a particular location. The system supports a number of new areas of geographic inquiry, especially in its ability to represent the implications of historic conditions for current and future policy in three dimensions over time.

In order to efficiently create 3D models in the whole city of Kyoto, we adopted MAP CUBE® data for basic components of Virtual Kyoto reflecting the present day Kyoto. MAP

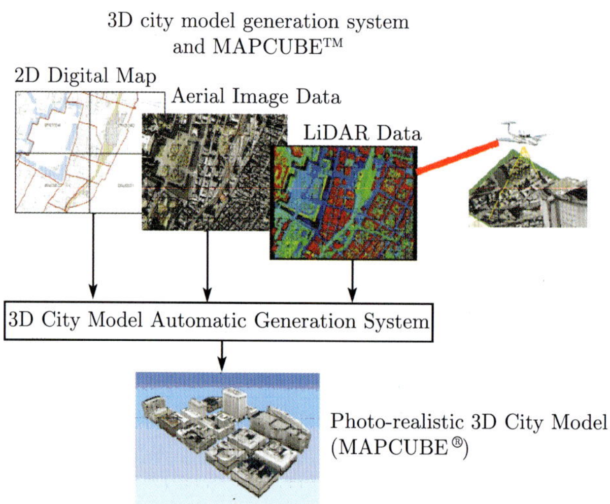

Figure 1 MAPCUBE® data generation

CUBE® data is composed of prismatic building blocks that are created based on high-precision laser-profiler data (from PASCO, Co.) and building footprint data (from Increment P, Co., or iPC) using the most up-to-date technology of automated generation of prismatic models developed by CAD Center Corp. The 3D models are created by extruding building footprint data with air-borne laser-profiler data which take height values at an interval of 2.5m within an error of 15cm as in Figures 1 and 2. MAP CUBE® data also has some full volumetric 3D models for buildings which act as landmarks in Kyoto.

Figure 2 3D city model of Kyoto

The Virtual City of Present Kyoto

We first modeled the present-day Kyoto before restoring past Kyoto, because unlike other cities in Japan, Kyoto was exempt from wartime air raids, and thus retained numerous building from the pre-war period such as machiyas, temples, shrines and western style buildings. There are four major elements of historical architectures in Virtual Kyoto for the present day as discussed below.

Machiyas

Machiyas, or historical townhouses as shown in Figure 3 are of special importance to our restoration project because most of urbanized Kyoto consisted of them in the pre-modern

Figure 3 Machiyas in Kyoto

period. Machiyas, most of which had been built in between the 18th century and World War II, have been decreasing rapidly in recent decades. However, they are still dominant elements in the urban landscape of Kyoto. The research first aimed at the construction of 2D-GIS of those elements through large scale field surveys and then the 3D-GIS was built, based on these constructions.

We use Machiya databases from 'Machiya Community-building Survey (*Kyo-machiya Machizukuri Chosa*)'[7] and 'Toyota Fund Citizen Survey of Machiyas (*Toyota Zaidan Niyoru Shimin Chosa)*' in 1995 and 1996, for the base of 2D-GIS of Machiyas. The two databases combined ('Machiya Exterior Surveys' hereafter) cover the area in central Kyoto that had been urbanized by the beginning of the 20th century. In the Machiya Exterior Surveys, investigators visited all buildings in the study area and examined whether a building is a machiya, and recorded the type, condition and use of the premise. The surveys identified about 28,000 dwelling units of machiya (Figure 4). Since the initial aim of Machiya Exterior Surveys had been to count the existing machiyas, the geo-referencing of identified machiyas

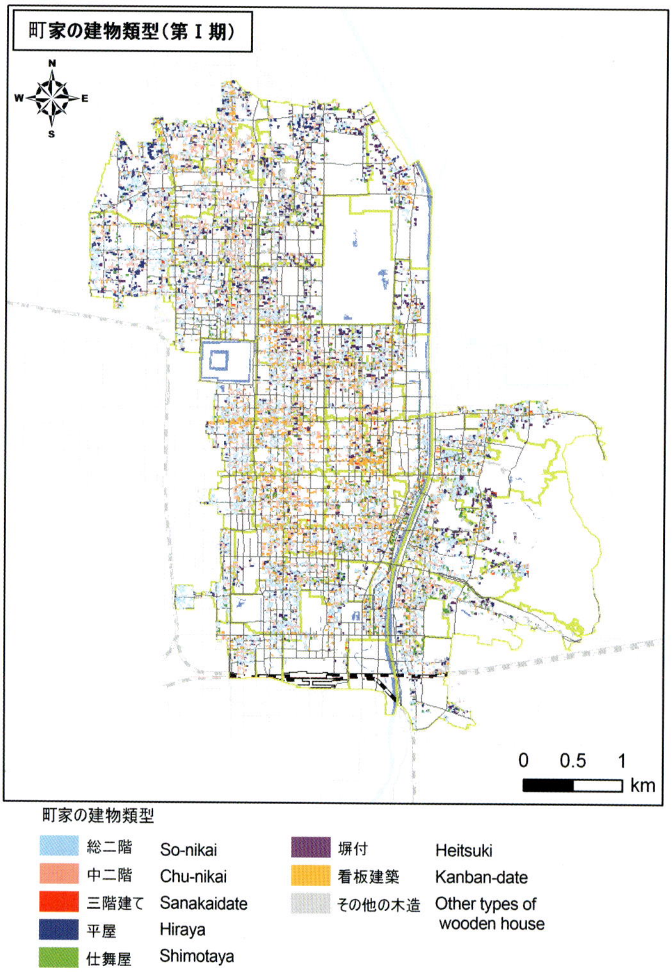

Figure 4　Distribution of machiyas by type

is incomplete. Therefore we matched the data to building footprint data from large scale residential maps by iPC, and we have done a supplementary survey for unmatched records with the help of Kyo-machiya Saisei Kenkyukai, an NPO who initiated the original survey. The changes in machiyas since 1995/8 until today are also conducted in the supplementary survey, and we will continue monitoring the changes in machiyas [8].

Considering that there are still so many machiyas in Kyoto, a computer program for automatic generation of machiya 3D models has been developed in Microsoft Excel VBA macro. "Machiya VR Generation Macro" retrieves the coordinates and attributes of a machiya from GIS database, applies one of seven machiya type, resizes the model matching to the width and depth of the building lot, rotates and places the model in the VR space, and this is repeated for all machiyas in the database as in Figure 5.

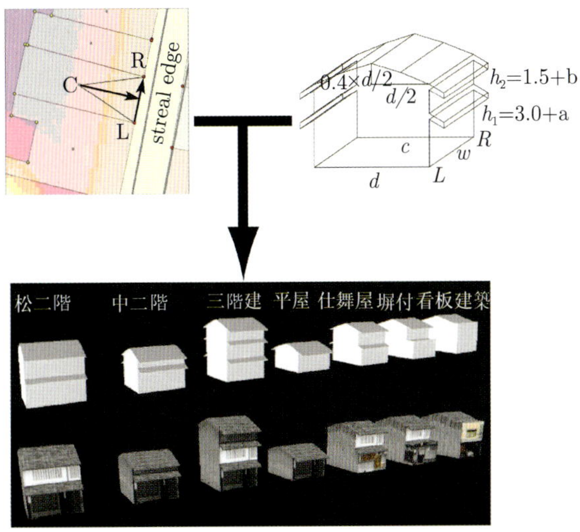

Machiya models with textures

Figure 5 Machiya VR generation macro

The advantage of this macro is that it is possible to visualize urban landscape with information only on building footprints and machiya types. In another words, past urban landscapes can be restored according to different scenarios of past urban land use and machiya type composition as shown in Figure 6. Although this is not a precise restoration of individual building, this would be a reasonable approximation of the past urban landscape as a whole. It is also possible to model machiyas from blueprint, or laser-scanning existing machiyas, and this is done on an ad hoc basis for the more important machiyas.

Modern architectures and architectural heritage

Kyoto has been the cultural capital of Japan and consequently houses the largest collection of architectural heritages in the country. There are 375 designated or registered historical buildings, and the city itself constitutes the cultural legacy. Kyoto has approximately

Figure 6 Machiya models in Virtual Kyoto

1,300 temples and 350 shrines as illustrated in Figure 7. Among them, 16 temples and shrines, and the Nijo Castle are designated as World Heritage sites, and the areas around them are subject to development restriction to preserve the landscape.

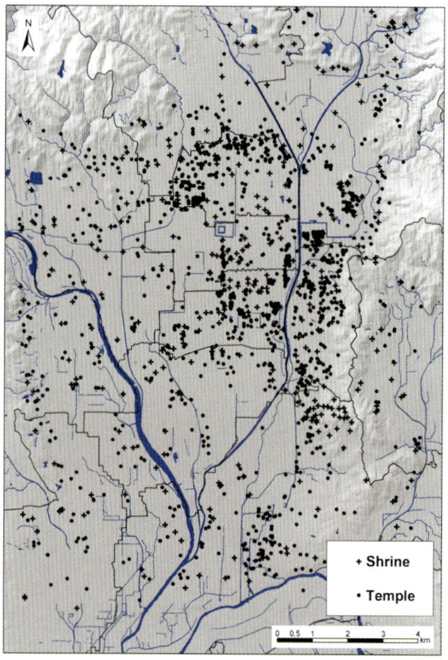

Figure 7 Distribution of shrines and temples in Kyoto

There are also about 2,000 early modern buildings, based on the recent survey by the Kyoto City. These buildings were designed and built just when Japanese architects began to incorporate Western techniques into their thinking. Representative buildings of this category are made of brick, but many different types of modern buildings including reinforced concrete buildings, wooden structures, and buildings that combined traditional wood and modern materials have also been developed in Kyoto as shown in Figure 8.

Each individual building in these categories has its own unique character, and thus cannot be well represented by the prismatic model or automatically generated models. Several detailed VR models of those existing are being modeled using CG/VR software as in Figure 9.

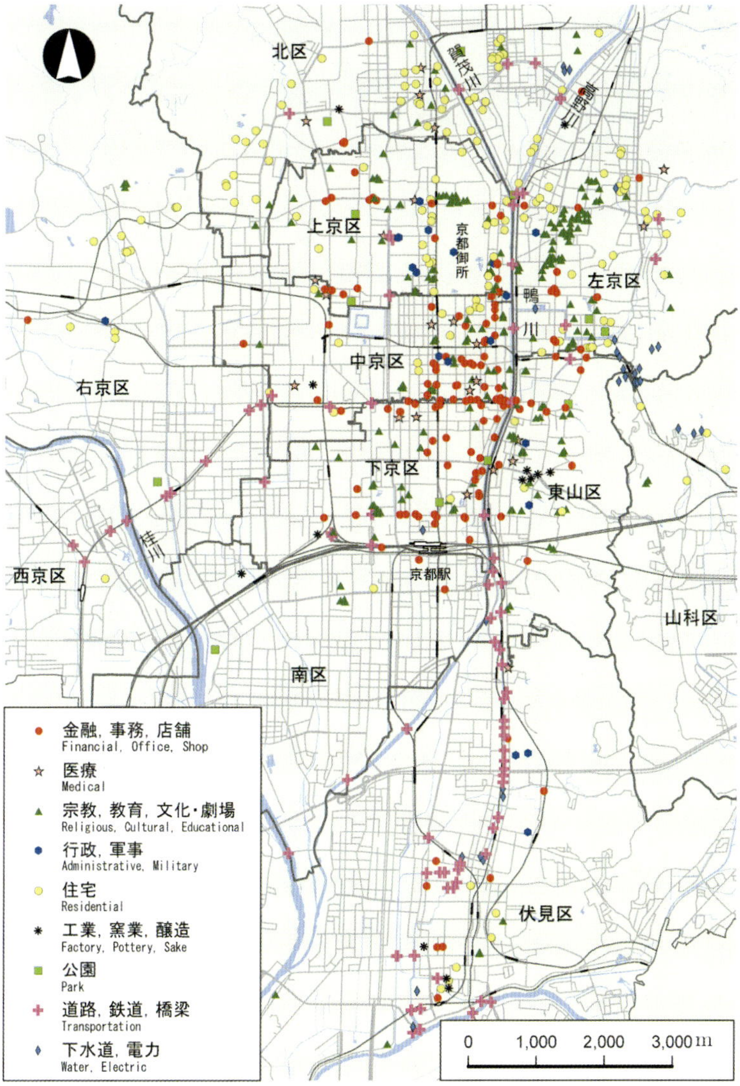

Figure 8 Distribution of early modern buildings in Kyoto by type

A. Kiyomizu-temple

B. Nijo castle

Figure 9 Detailed VR model examples

Minamiza theater

Minamiza theater was originally constructed in 17th century on Shijo Street, close to the place where the first Kabuki was performed in the early 17th century. Although there were seven major theatres along Shijo Street in 17th–18th century, Minamiza theatre solely remains on the street today. In order to situate digitally archived traditional dance in the digitally constructed landscape, modeling the theater has been an important issue in our project, A detailed 3D model that includes the interior of the building was created based on blueprints. Along with the theatre, we also created a detailed model of the street to the theatre (Shijo Street), allowing walk-through experience of the street toward the entrance of the theater, the reception and into the hall of Minamiza as illustrated in Figure 10. In this VR environment, it is planned to stage virtual Kabuki performance or traditional dance, created by 'motion captures' of real performers.

Figure 10 Minamiza theater

Gion Festival and Yamahoko float

The visual elements that comprise the historical urban landscape is not limited to buildings. We attempt to model landscapes of Gion Festival that originated in the mid 9th century, which has evolved to take its current form by the mid 14th century and continued until today with minor interruptions. During the festival, 32 Yamahokos (floats) representing each local neighborhood parade in the city centre, including Shijo Street. Yamahoko parade of Gion Festival becomes a symbolic landscape of Kyoto during the festival period. Yamahokos of similar scale have been enshrined over the centuries[9], and thus are one of the important visual elements that existed throughout our study period.

At present, four VR models of Yamahokos (Kankoboko, Funeboko, Naginataboko and Kitakannonyama) have been created by laser scanning detailed miniatures and digital images of the real Yamahokos taken by digital cameras during the festival, as well as by manual modeling based on blue prints as in Figure 11. These 3D Yamahoko models placed in the VR space have already become a reference object to the changes in urban landscape over the years.

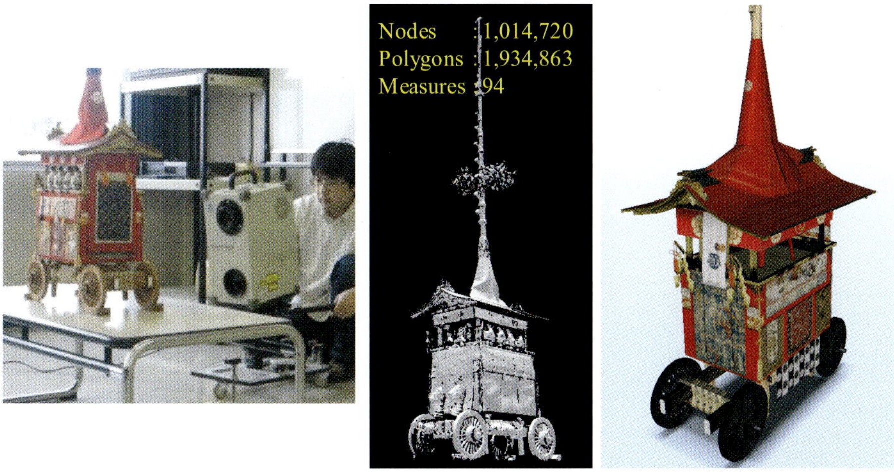

Figure 11 Laser scanning of the miniature Yamahoko and its VR model

Three Dimensional Reconstruction of Kyoto of the Past

In this research, 3D reconstructions of Kyoto at various time slices have been and are being done, starting from the present to the past; including times soon after and before World War II, the Taisho and Meiji eras (early 20th to late 19th century), the Edo era (17th to late 19th century), and finally up to Heian era (late 8th to 12th century). The 5 time slices are discussed in turn below.

Landscape changes before and after World War II

We examined aerial photographs to trace the changes of the distribution of machiyas in the center of the City since the late 1920s. Machiyas employ gable roofs covered with roof tiles. The observation of aerial photographs makes possible identifying machiyas as different from other types of house. Six sets of aerial photographs taken at approximately 13 years of intervals since the 1928 have been used as in Figure 12. The photos were scanned and rectified to fit to surveyed maps. Then gable roofs were traced to create the building footprint data of machiyas. Using GIS, we first traced machiyas row by row against 1928 and 1946 photographs. We then examined photographs of later years and eliminated the machiyas which no longer existed. In this way, we identified changes in the machiyas distributions for each year for which we had photographs. The result indicates the rapid changes of machiyas distribution in six stages since 1928. Approximately 7,400 machiyas existed in this central area in 1928, 6,200 in 1946, 5,900 in 1961, 4,500 in 1974, 2,900 in 1987 and 1,800 in 2000 as shown in Figure 13.

Kyoto during the Meiji and Taisho eras

Information on the allocation of urban land in the past is obtained from past cadastral maps. The oldest cadastral map available is that of 1884 owned by the Kyoto Prefectural Library and Archive, but it lacks a register. Therefore, we have used cadastral map 1912 (scale 1/1,500 – 1/1,300) owned by the Ritsumeikan University Library. We have matched the past cadastral maps to current map-based on the shape of neighborhood communities rather than based on streets and intersections. This is because streets in Kyoto are substantially broadened and thus unsuitable for referencing. After this process we traced each parcel to examine the land division in the beginning of the last century as in Figure 14.

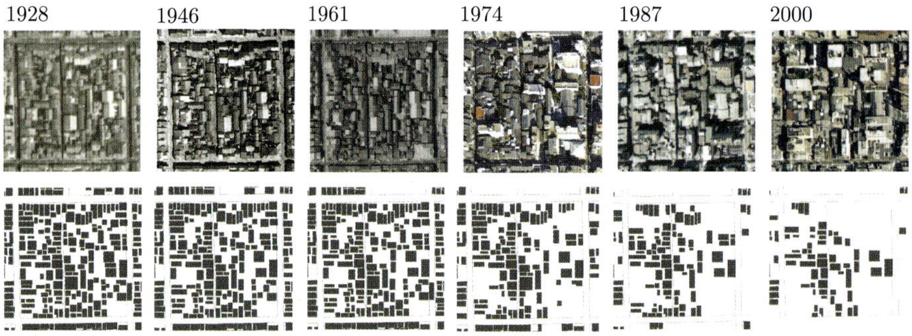

Figure 12 Identification of machiyas in the aerial photographs

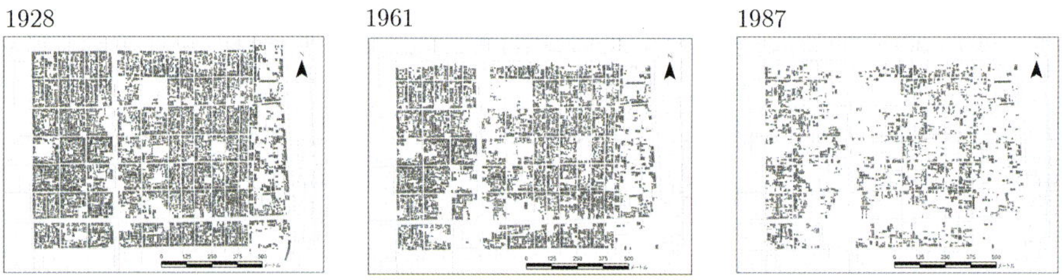

Figure 13 Changes in number of machiyas in central Kyoto, 1928–1987

Figure 14 GIS-based Kyoto cadastral map in 1912

The landscape in 1910s has been visualized based on digitized cadastral maps of 1912 and "Machiya VR Generation Macro." Figure 15 presumes that one house is built on one plot and constructs a virtual image of streets built up with nothing but machiyas and we get a good sense of how different the Kyoto of Taisho was compared to the present day.

Figure 15 Urban landscapes in the early 20th century based on the Kyoto cadastral map

Kyoto in early Edo era

No photographs or charts exist to provide us with objective measures of Kyoto in the Edo era. However, the Edo era was a time when culture flourished, and many paintings, drawings and documents from that era survive. Some Ukiyoe woodblock prints depict landscapes of Edo era Kyoto. The series of paintings called Rakuchu-Rakugai-Zu (Views Inside and Outside Kyoto) produced in the late 14th century and after give us a detailed account of the bustling city.

Kan'ei-go-Manji-mae-Rakuchu-Ezu, drawn in 1640s and presently owned by Kyoto University, is a considerably accurate early map of Kyoto. It has the size of 636cm by 282cm and shows names of towns and streets, land use, widths of streets, widths and lengths of blocks and major building lots and names of landowners. So we reconstructed early 17th century Kyoto based on this old map base. The map was scanned and rectified to fit to the modern map[10]. After these geometrical adjustments, streets, blocks and major building lots were traced, and their attribute data were input on a 2D-GIS system as in Figure 16. We reconstruct the street scenes of Edo period Kyoto by placing machiya 3D models on places where Rakuchu-Ezu indicated townsmen's district and lord's mansions on the lord's parcels, and on temples in the temple lands. Moreover we were able to reconstruct a model of the Nijo castle keep, which no longer exists, based on remaining blueprints as in Figure 17.

Virtual Heiankyo

When Kyoto was founded in 794 A.D. as the capital of Japan, the city was then called Heiankyo. It was built based on a rectangular street grid extending 4km from East to West and 5km from North to South. Extended excavations of Heiankyo archaeological sites have led to the accumulation of much information on the city's early environment and scenery. Many historical documents of Heian era Kyoto remain, providing valuable resources on which to base a reconstruction of Heiankyo. Based on all of the archaeological evidence and historical manuscripts collected, scholars have compiled a number of analog databases of the

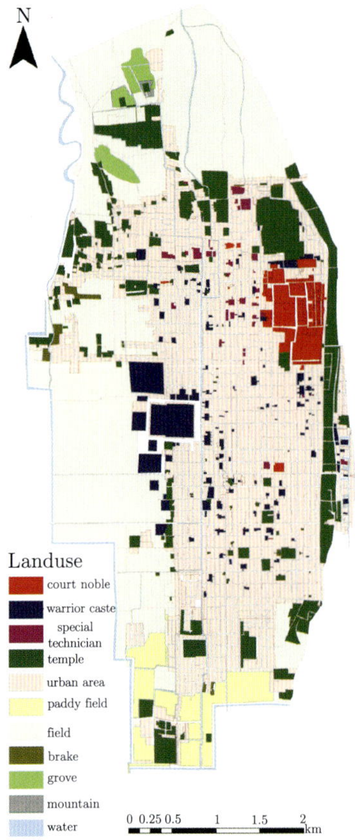

Figure 16　Kyoto land use map in 17th century based on Rakuchu-Ezu

Figure 17　Kyoto street scenes in the early Edo era

Heiankyo's environment and land use. The authors have started the reconstruction of Heiankyo, based on available historical documents and information.

The topographical data has been reconstructed using excavation and geological and archaeological results. Buildings of this era does not exist, but street blocks and buildings have been modeled using 3D CAD based on blueprints for miniature model of Heiankyo which was made by the Kyoto City celebrating its 1200th anniversary. The 3D CAD models have been automatically located according to the land use of the time as in Figure 18.

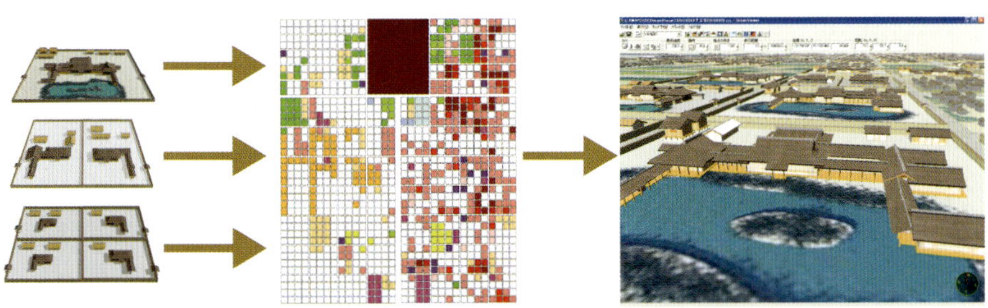

Combination of 3D CG and 2D-GIS　　The display by VR software

Figure 18　Virtual Heiankyo

Time Traveling in Virtual Kyoto

So far we have created 3D VR models of urban landscapes of Kyoto for each of several time periods. In this section we would time travel in the virtual time-space of Kyoto as 4D-GIS.

Urban landscape changes through Virtual Kyoto

We can see the changs in urban landscape through Virtual Kyoto. Based on the data of machiyas identified by aerial photographs in section 4.1, VR data of machiyas of different years were automatically generated by "Machiya VR Generation Macro" and then placed in to the Virtual Kyoto as we illustrate in Figure 19. It shows the ways in which machiyas were replaced with modern buildings during the post-war period. It clearly shows that machiyas facing major streets disappeared first, and this disappearance gradually expanded inwards to the street blocks. Modern high-rise buildings have become more and more dominant in the urban landscape.

Figure 19 Changes of the urban landscape in the central Kyoto

Virtual Gion Festival

The models of Yamahoko floats were placed in the virtual time and space of Kyoto streets. We then place floats in a virtual early 20th century scene, parading them from the West, with Higashiyama Mountains as a background, and allowing us to see their huge size against the backdrop of low buildings as in Figure 20. It clearly demonstrates that Yamahoko, buried among the high-rise buildings today, had been an extremely salient object in the pre-modern days. The visibility of the mountains surrounding Kyoto was also very different. We also lined up and paraded some floats through the streets of Kyoto as illustrated in Figure 21.

Figure 20　Landscape changes of Gion Festival between the present and early 20th century

Figure 21　Virtual Gion Festival of the present and the late 1920s

Fly through over Virtual Kyoto at the different time slices

Drawing on leading edge GIS and VR technologies, we took up the challenge of creating 3D maps of Kyoto for several time slices. As a result the Virtual Kyoto we produced is a 4D-GIS that includes two and three dimensional GISs of various historical ages. By adding a time dimension to our GIS, we have created an unprecedented GIS time machine allowing travel within time and space in a manner not hitherto possible. We can freely fly through the Virtual Kyoto. Figure 22 shows the urban landscapes of Kyoto from the same view point at present, Edo era and Heian era.

Figure 22 Fly-through in Virtual Kyoto

Virtual Kyoto: A 3D Map on the Web

With the rapid global expansion of broadband communications, the potential for a web-based Virtual Kyoto has been greatly increased. At its foundation, Virtual Kyoto is constructed from a 3D-GIS that links the present and the past and displays the city in a high-resolution virtual reality, but on the web this kind of material needs to be displayed using 3D web-based GIS technology.

To develop a web-based 3D-GIS we use UrbanViewer for Web^{TM} from CAD Center Corporation which is specifically designed to distribute a 3D urban model on the web. This system allows users of DSL-level broadband to view Virtual Kyoto and to interact actively using widely available browsers like Microsoft Internet Explorer. Users can enjoy very realistic walk-throughs of Kyoto neighborhoods as well as experience fly throughs of the city from a bird's-eye perspective. Further, it includes functionality to search for geographic sites by place name or building type.

The model of Kyoto consists of building, landform and other shapes, and texture files; together this results in an extremely large amount of data. In order to distribute this huge data we need to use the following techniques: Data compression, Scalable topographic data, Texture files efficiently archived for application at different resolutions, Level of Detail (LOD) displayed relative to scale and Data Streaming. Incorporation of these techniques has allowed to distribute Virtual Kyoto over the Web.

We have constructed a web-based 3D model that demonstrates the unique characteristics of Kyoto scenery, including the landform with texture that can make use of LOD technology. We have also created a detailed model of the buildings and streets along Shijo high street that visitors can walk through and even enter the interior of the Minamiza Theater. In addition, we created a reconstruction of the same area of the 1930s. This version of Virtual Kyoto works together with the 3D city model of Heian era Kyoto, "Virtual Heiankyo on the

Web". Keeping the same geographical position, users can switch the time period for any walk through scene just by clicking a button.

The web-based version of Virtual Kyoto presents detailed models of temples, shrines, castles and other monuments that are UNESCO world heritage sites (Figure 23). A list of famous places of Kyoto is prepared on the left side of the screen so that users can instantly move their view points to the specified places. Users can also view the city from the selected site. Choosing a building on the map, viewer can call up web-search results that provide more information about it.

Figure 23 Web-based Virtual Kyoto

Conclusion: Towards the Use of GIS in Digital Humanities

Our research will continue to reconstruct the historical city of Kyoto starting from the present going back to the past making available large varieties of digital content which constitute urban landscapes at different time periods. At the final phase of this research, we intend to employ Virtual Kyoto as a platform to integrate a large collection of digital archives of arts and cultures in Kyoto. We are expecting that Virtual Kyoto will be a base for a newly launched project, "Digital Humanities Center for Japanese Arts and Cultures" in Ritsumeikan University.

According to Gregory[11], Digital Humanities (also known as Humanities Computing or Computing in the Humanities) is a broad and rapidly growing inter-disciplinary field. It is concerned with using computational techniques to do the following: 1) Create databases concerned with documents or artifacts relevant to the humanities. This involves capturing, structuring, documenting, preserving and disseminating such data; 2) Develop generic methodologies to provide new insights into these datasets; and 3) Conduct new scholarship on these databases to increase our understanding of disciplines across the humanities. The last is the most important of the three but also perhaps the most neglected as it requires inter-disciplinary collaboration between experts in technologies and methodologies on the one hand, and academics with specific research questions who want to use information technologies on the other. GIS with a time-dimension would provide a key tool in digital humanities because it can deal with any theme with a spatial and temporal dimension.

The capability of our web-based 4D-GIS would form a new base for museology (museum studies). We can put various digital contents in our 4D-GIS. The system would provide user-friendly interfaces to explore historical materials involving the arts and entertainment in the geographical context of Kyoto. Our experience to date with Virtual Kyoto has revealed considerable demand for assisting urban planning in Kyoto for the city has serious issues concerning historical landscape conservation. Our Virtual Kyoto which delivers location-based historical knowledge would effectively support a decision support system for alternative urban futures for Kyoto by contextualizing planning in its historical setting.

Acknowledgements The authors wish to thank the City of Kyoto, and NPO Machiya-saisei-kenkyukai, who allowed us to use Machiya Exterior Surveys results, as well as to Shochiku Co., Ltd., who allowed us to use design drawings of Minamiza theater. All brand names and product names are trademarks or registered trademarks of their respective companies.

We also thank our colleagues Yutaka Takase and Dai Kawahara for technical assistance in 3D rendering, Keigo Matsuoka and Manabu Inoue for administering field surveys, Toshikazu Seto and Takashi Kirimura for online dissemination, and Akihiro Tsukamoto for compiling historical materials of the Edo period

The project has also received financial support from the 21st century COE program (Kyoto art entertainment innovation research), the Global COE program (Digital Humanities Center for Japanese Cultures and Arts) and the Academic Frontier Project (Disaster Mitigation of Urban Cultural Heritage) by MEXT (Ministry of Education, Culture, Sports, Science and Technology).

References

[1] Longley P A, Goodchild M F, Maguire D J, Rhind D W. Geographic Information Systems and Science.New York: John Wiley, 2001.

[2] Yano K, Isoda Y, Nakaya T, Kawasumi T, Matsuoka K, Y. Takase, Kawahara D, Kawahara N, Inoue M, Tsukamoto A, Kirimura T. Creating Virtual Time-Space of Kyoto. E-journal GEO, 2006,1: 12–21.

[3] Yano K, Nakaya T, Isoda Y (Editors) . Virtual Kyoto. Tokyo: Nakanishiya, 2007.

[4] Dodge M, Smith A, Doyle S. Urban Science. GIS EUROPE, 1997, 6: 26–29.

[5] Fisher P and Unwin D (Editors). Virtual Reality in Geography. London: Taylor & Francis, 2002.

[6] Shiode N. 3D Urban Models: Recent Developments in the Digital Modeling of Urban Environments in Three Dimensions. GeoJournal, 2001, 52: 263–269.

[7] Kyoto City. Kyo-machiya Machizukuri Chosa Shukeikekka (Machiya Community-building Survey Report). Kyoto, Japan: City Planning Bureau, City of Kyoto, 1999.

[8] Kawahara N, Yano K, Koga S, Takase Y, Kawasumi T, Inoue M, Kawahara D, Iwakiri S. 4jigenGIS wo Mochiita Kyo-machiya Monitaringu Sisutemu no Kochiku (Creating Kyo-Machiya Monitoring System Using 4D GIS). Minzoku Kenchiku, 2003, 124: 13–22.

[9] Ashikaga K (Editor). Kyoto Rekishi Atrasu (Historical Atlas of Kyoto). Chuo Kouron Sha, Kyoto, Japan.

[10] Tsukamoto A, Isoda Y. Analysis of Local Distortions In A Historical Map "Rakuchu-Ezu".Theories and Applications of GIS, 2007, 15: 111–121.

[11] Gregory I. What is Digital Humanities? http://www.lancs.ac.uk/staff/gregoryi/ [2007-6-20].

Chapter 7 Visualizing Future 3-Dimensional Neighborhoods in Phoenix: An Application Incorporating Empirical Methods with Computational Graphics*

Ariane Middel, Subhrajit Guhathakurta, Hans Hagen, Peter-Scott Olech and Florian Höpel

Introduction

Modeling urban futures has been of considerable interest of late in computer graphics and urban planning. This interest is driven by a number of developments, in particular by increased computing power and significant advances in the field of integrated urban and environmental models[1]. Growing awareness for sustainability amplifies the demand for planning support systems that analyze and visualize future carbon footprints, air quality, traffic volumes, energy intensity, and other indicators. Visualizing critical information about how cities and urban environments might develop in the next one or two decades is an effective tool for analyzing impacts of today's planning decisions on future environments.

Recent efforts in computer graphics aim to automate the complex and expensive task of generating and visualizing realistic cityscapes. Various tools and techniques are constantly being developed for automatically detecting and reconstructing buildings from remotely sensed imagery to generate 3D city models[2-4] and to undertake building energy calculations[5]. While photogrammetric reconstruction approaches allow for generating existing and past stages of urban development, they provide little information on future cityscapes. In this context, procedural modeling techniques have been subject to active research lately. Procedural methods algorithmically generate arbitrary geometries from a predefined rule-set. Key techniques for procedural visualization include agent-based modeling approaches[6] and grammar-based approaches[7-11].

In general, procedural models are suitable for applications where the visualization of look-alike cities is adequate and the empirical validity of the environmental model is less relevant. To date, procedural approaches are not suitable for the visualization of urban simulation data since they lack the capability to model the impacts of human behavior on urban developments. Furthermore, procedural models digitally create a highly complex visual reality that might impede visual data mining. A realistic representation helps analysts to

* Supported by the German Research Foundation (DFG, Grant No.1131).

perceive urban environments in a more tangible way, but at the same time it distracts from the actual task of analyzing and comprehending the underlying complex data structures. The real world is too complex to assimilate at once; therefore, we need abstraction to help us interpret it[12]. The specific nature of visualization and cognition is often enhanced through the use of abstract symbols, and not through realistic renderings. Therefore, we propose an integration of abstract and realistic 3D visualization for the visualization of multidimensional urban simulation data.

In this chapter, we close the gap between the simulation and visualization of multidimensional geospatial data on future urban environments. First, we present a statistical method for estimating future residential building types on a neighborhood scale from demographic data. Future building types are crucial for many sustainability metrics, but are not supported by current simulation models. We use multinomial logistic regression (MNL) to map dwelling types derived from household characteristics. We predict building types for the Phoenix metropolitan area based on demographic data from *UrbanSim* output. Then, we visualize the resulting multidimensional simulation data as abstract symbols on top of Google Earth™. While several techniques have been explored to generate three-dimensional building objects including procedural or template-based database methods, our research provides an empirical model that allows a visual synthesis of future neighborhood characteristics in realistic urban contexts.

Current Visualization Approaches

Procedural models based on grammars have been developed to generate architecture[7], building facades[8], ancient Roman sites[9], and large-scale 3D cityscapes[10]. Procedural city modeling allows for the generation of realistic-looking urban environments, but to date it lacks underlying empirically-based models to generate meaningful projections of urban growth patterns that can be calibrated to simulate real conditions. However, future population growth and household location is of great interest to urban planners. Planning agencies in most industrialized countries are now mandated to provide official projections on future urban growth to monitor land development. Therefore, geographers and other urban scholars have pursued research in modeling complex urban systems to guide land-use decisions and growth management strategies. Several commercially available GIS-based planning support systems, such as *What If?*[13] and *CommunityVIZ*[14] are being widely used in academic and policy contexts. *UrbanSim*[15] is another open-source planning tool used for analyzing long-term effects of land use and transportation policies on household and job locations. It provides a platform for generating different urban scenarios based on current trends and specified policy choices, for example to examine impacts of low-density urban sprawl or to evaluate the land use implications of transportation plans[16].

Typically, projections of urban simulation models are visualized with the help of GIS as color-coded polygons or icons on 2D maps (Figure 1). Since prediction models usually result in large-scale, highly attributed spatial data sets, the challenge is to find a multidimensional representation for the simulation results. The multi-attributed visualization should enable planners to easily compare different planning scenarios and to evaluate simulated impacts of different land use policies. In the early 1970s, Chernoff presented a technique to visualize trends in highly dimensional data by relating data to facial features[17]. Gradually over the years, new information visualization techniques were introduced, ranging from 2D scatterplots to 3D treemaps (Figure 2).

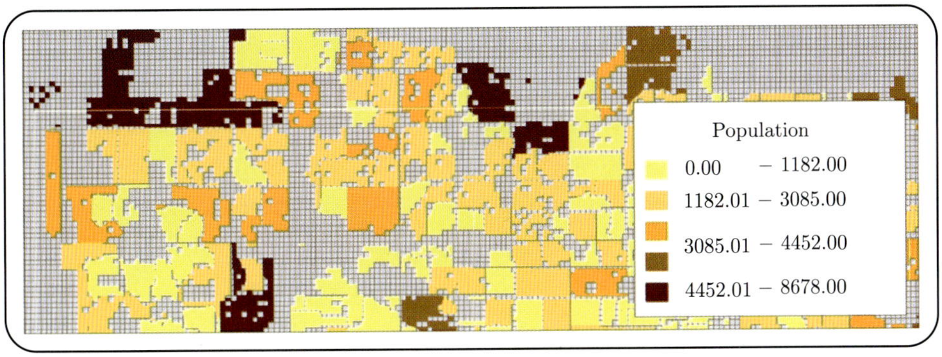

Figure 1 Example of GIS Visualization: population density in Phoenix

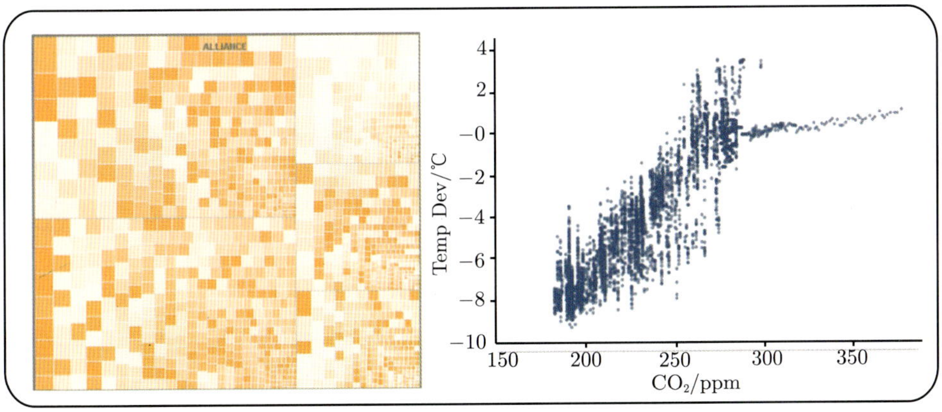

Figure 2 Examples for InfoVIS: treemap on character data in a World of Warcraft realm, scatterplot comparing CO_2 levels against temperature[18]

Whereas *InfoVIS* primarily deals with the display of large multivariate datasets, cartography is mainly concerned with representations constrained to a spatial domain. The cartographer Bertin established a basis for designing maps in his classical work "Semiology of Graphics"[19] where he identified a set of fundamental visual variables and defined graphical rules for their appropriate use. Since then, Bertin's concepts have been constantly modified and extended. Modern cartography transfers design knowledge from 2D paper maps to new media. On-screen interactive maps are designed to assist in visual data exploration and analyses[20]. Cartographic visualization is also extended to abstract and non-geographic data by spatialization[21].

Recently, efforts have emerged to combine techniques from both cartography and information visualization[22, 23]. Geographic visualization, *GeoVIS,* is a new, rapidly evolving domain, especially since the availability of geodata has increased considerably. In 1998, MacEachren compiled the first research agenda entitled "Visualization-Cartography for the 21st century"[24] and addressed *GeoVIS* research challenges. Since then, cartographic and *InfoVIS* techniques have frequently been applied to design integrated geovisualization tools. Latest advances include multivariate analyses with self-organizing maps[25, 26], studies on human activity patterns using 3D space-time paths, and bivariate maps for public health studies[27]. Most recent activities in geovisualization research are discussed by Kraak[28].

Pinnel *et al.*[29] conducted a study of visualization designs for urban modeling. They found that map-centered visualizations are the most useful portrayals for urban planning and analysis, since map layout encodes location information, which is crucial for decision-making. A map-based visualization approach, "The Indicator Browser", was designed by Schwartzman *et al.*[30] to display *UrbanSim* simulation results. The browser uses comparative visualizations of 2D maps to visualize complex multiple variables. This technique is however limited in communicating patterns across many dimensions at a higher geographic resolution.

Instead of encoding *n*-dimensional data in *n* 2D maps, Tominski *et al.*[31] follow a different approach to display monthly health data (Figure 3). They visualize time dependent multivariate disease information as 3D pencil and helix icons geocoded on a base map. This research helps to analyze complex patterns across multivariate, spatial, and temporal dimensions, but it lacks a powerful database and GIS functionality to manage, process, and distribute data.

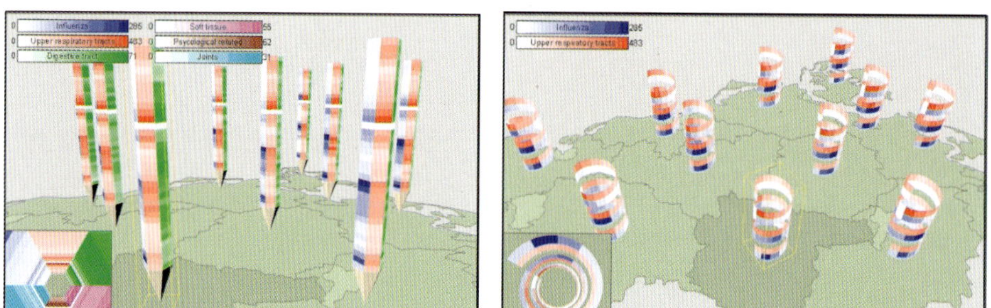

Figure 3　Visualization of monthly health data by means of 3D icons on a map[31]

In the following sections, we will present a database-driven simulation and visualization framework that uses abstract visualizations of geospatial data in a 3D context. Within our framework, multidimensional urban simulation data is encoded as specified graphic variables of different geometries and mapped to a Google Earth™ environment to effectively enhance visual thinking. The focus of research here is on designing visualizations applied specifically to the multidimensional simulation data at hand in order to incorporate knowledge into planning processes. We aim at facilitating exploratory visual analyses of simulation data by creating a web-based interactive geovisualization tool.

Methodological Framework

The framework we use for visualizing future urban neighborhoods is based on four components: First, the predicted households by type at a fine spatial resolution of 150 meters for the Phoenix metropolitan region is obtained with the help of the *UrbanSim* modeling and simulation environment. Second, we estimate the types of buildings that the households will occupy by using logistic regression methods applied to the 2000 Census of Population and Housing. This is achieved by intersecting census block level information with the grid cells layer generated for *UrbanSim* data. Third, we import the household data and the parameters of the empirical analysis into a PostgreSQL database management system, with the PostGIS module enabled. Finally, three-dimensional geometries and icons for visualizing the data are developed and implemented in Keyhole Markup Language (KML) for integration with Google Earth™.

Data modeling with *UrbanSim*

UrbanSim[15, 32] is a modeling environment that uses detailed data on households and employment to predict changes in their locations over time. It consists of several model components simulating different actors in the urban development process, including discrete choice models for household location and relocation, for business choice of employment location, and for developer choices of locations to develop built spaces for various distinct activities. An overview of the *UrbanSim* model development as well as a detailed description of implemented model components is given in Borning[33] and Waddell[15]. The open source simulation model was developed by a research group in the University of Washington and is currently implemented in the Digital Phoenix project[34] at Arizona State University to predict jobs and population distribution in the Phoenix Metropolitan area.

Estimation of residential building types

In order to investigate the effect of household characteristics on discrete types of housing, we use multinomial logistic regression. Building type distributions are based on demographic household information at a census block group scale. To derive homogeneous building categories by location we form clusters with typical building type distributions using k-means techniques in order to establish nominal categories for the regression model. We then model the log odds of the clustered neighborhood categories as a linear function of the household and demographic characteristics of these neighborhoods. This approach gives good results, but implies a possible correlation between the clustered building type categories. Therefore, we refine our model by estimating building types at grid cell scale. Estimating coefficients for a set of 150m×150m grid cells with demographic data synthesized from

Census[35] gives slightly better results than using synthesized household data. Finally, the estimated coefficients are applied to *UrbanSim* simulation output to derive predicted building types from household location.

Data management with PostgreSQL and PostGIS

PostgreSQL[36] is an open source object-relational database management system (ORDBMS). In contrast to relational database management systems, an ORDBMS is not limited to a pre-defined set of data types, which raises the level of abstraction. PostgreSQL natively supports inter alia arbitrary precision numerics, unlimited length text, arrays, and geometric primitives. In addition, it provides support for the integration of custom data types and methods into the database.

As geospatial extension to the PostgreSQL backend server, we use the PostGIS module[37]. PostGIS is an open source add-on developed by Refractions Research under the GNU General Public License and enables PostgreSQL to integrate spatial data structures into the database. PostGIS follows the Simple Features for SQL specification (SFS) from the Open Geospatial Consortium (OGC). The OGC is an international non-profit organization that is leading the development of standards for geospatial and location-based services[38]. As a Simple Features for SQL compliant spatial database, PostGIS includes the geometry types diagrammed in Figure 5 and provides functionality for spatially enabled SQL queries like distance between geographic objects, unions, calculation of perimeter, and buffering.

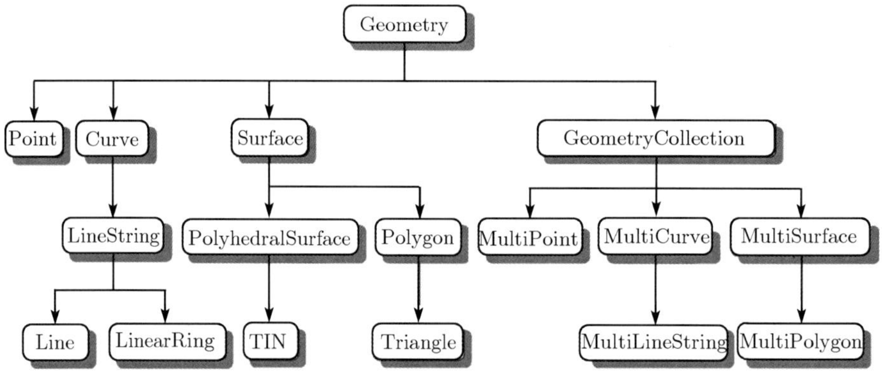

Figure 4 OGC simple feature specification[38]

Data visualization with Google Earth™

Created by Keyhole Inc. and originally named Earth Viewer, Google Earth™[39] maps the earth by superimposing satellite images and aerial photos on a virtual globe. Google Earth™ allows the user to interactively browse the globe in 3D and to zoom from space into street level views. Data is streamed from Google's Server upon request to the client computer. To date, most parts of the earth's surface are covered with images having a minimum resolution of 15 meters and a maximum age of 3 years. The aerial photographs are mapped onto a digital elevation model (DEM) provided by NASA. Google Earth™ uses a map projection called General Perspective to show the earth as it appears from space. This cartographic projection resembles an orthographic projection, but instead of an infinite point of perspective it has a finite point of perspective near the globe.

Google Earth^{TM} has the capability to combine the multi-resolution raster image representation of the planet with any kind of georeferenced data. The geobrowser allows overlaying a wide set of geographic features like streets, freeways, and county borders. The user can filter those data sets by space, time, and layer. Furthermore, Google Earth^{TM} supports an overlay of points of interests, placemarks as well as 3D buildings and structures on the surface. Layers containing geospatial data are managed through KML.

Results

This section reports the results of our estimation procedure for deriving building types from *UrbanSim* output and the implementation of the visualization using PostgreSQL and KML. *UrbanSim* baseline data on households is obtained from the Digital Phoenix project that has implemented *UrbanSim* for Maricopa County at both 1 mile and 150 meter resolutions. The households data for *UrbanSim* contain detailed information about each household in Maricopa County including the number of people, number of workers, number of children, racial and ethnic characteristics of head of household, income, and the number of cars. This data is synthesized from the 2000 Census of Population and Housing and 5 percent Public Use Microdata Samples(PUMS) using the household synthesizer provided as part of *UrbanSim*.

We derive the building types from the Maricopa County Assessors data for 2000, which is an extensive database providing detailed property information at the parcel level. The Assessor's parcel data file is intersected with the Census data at the grid cell level with the help of ArcGIS^{TM} for generating shapefiles. The Primary Use Codes in the Assessor's data provide an effective classification system and a filter. We use this code for defining single-family dwellings and apartments that form the two main residential building type categories in our analysis.

The single-family category is sub classified into lots with different sizes and the apartments are subdivided according to the number of housing units. Altogether, we obtain four different building type categories:

1 = single family, small lots

2 = single family, medium lots

3 = single family, large lots

4 = apartments

Figures 6 to 8 show aerial photographs of representative building types in the Phoenix metropolitan area.

Results of multinomial regression of household characteristics on building types

In our multinomial regression model for Maricopa County, we assign a dominant building type to each grid cell by identifying the dwelling type with most frequency counts normalized by all residential units in the grid cell.

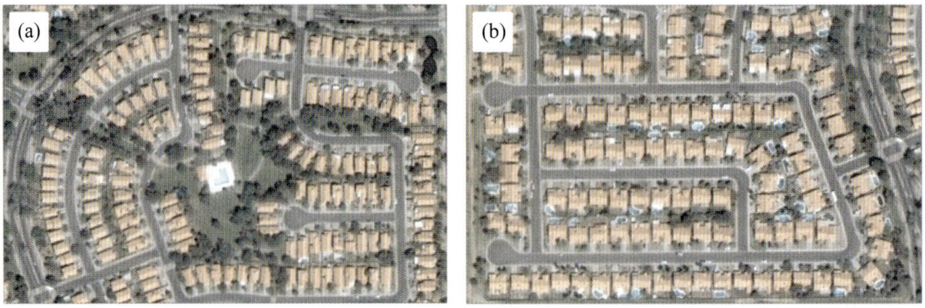

Figure 5 Single family dwellings, small lots

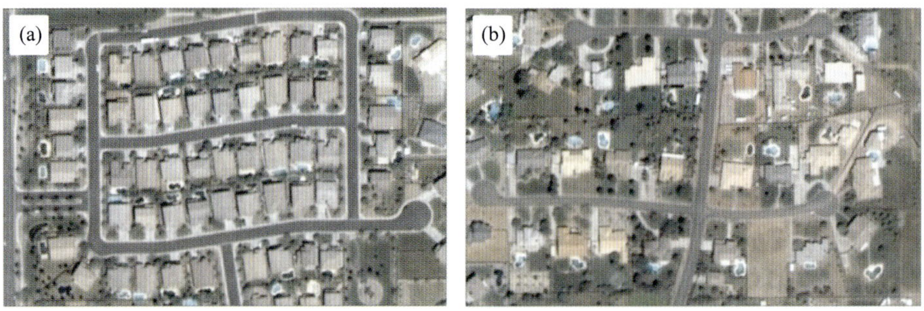

Figure 6 Single family dwellings, medium lots (a) and large lots (b)

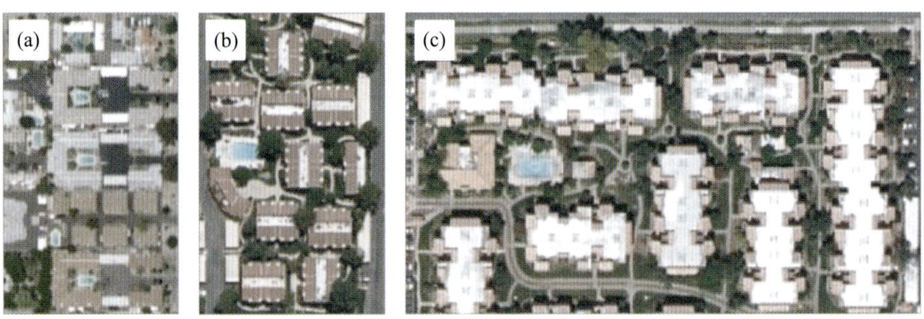

Figure 7 Apartments

Table 1 shows the number of grid cells for each building type, summing up to a total of 34,013 grid cells. The explanatory variables for our regression model are demographic household characteristics from Census. In order to parse Census data down to the level of

Table 1 Case processing summary

Building Type		*N*	Marginal Percentage
Category	1	6,339	18.6%
	2	3,872	11.4%
	3	22,704	66.8%
	4	1,098	3.2%
Total		34,013	100.0%

grid cells, we intersect the Maricopa County block group shape file with the *UrbanSim* grid cell file and assign Census demographics to each household in the corresponding grid cell.

Regression output Table 2 shows three pseudo R^2 estimates: Cox and Snell R^2, Nagelkerke's R^2, and McFadden's R^2. Larger values between 0 and 1 indicate a better explanation of the variation by the model and these results imply that our model performs reasonably well.

Table 2 Pseudo R^2

Cox and Snell	0.523
Nagelkerke	0.617
McFadden	0.393

The test statistic -2LL confirms the significance of the overall logistic regression model at a rounded 0.000 level (compare Table 3) and suggests that our model is well fitting. A Likelihood ratio test of individual model parameters reveals that all independent variables are linearly related to the log odds of the dependent variable also at a very high significance level (Table 4).

Table 3 Model fitting information

Model	−2 Log Likelihood(−2LL)	Likelihood Ratio Tests		
		Chi-Square	df	Sig.
Intercept Only	64,020.221			
Final	38,864.008	25,156.213	24	0.000

Table 4 Likelihood ratio statistics, reduced model

Effect	−2 Log Likelihood of Reduced Model	Likelihood Ratio Tests		
		Chi-Square	df	Sig.
Intercept	39,071.004	206.996	3	0.000
Population Density	50,861.503	11,997.495	3	0.000
Median Household Income	39,506.340	642.332	3	0.000
Minorities/%	39,534.047	670.040	3	0.000
Hispanics/%	39,093.025	229.017	3	0.000
Average Household Size	39,062.419	198.411	3	0.000
Distance to Highway	39,391.503	527.495	3	0.000
Children per Household	38,967.626	103.618	3	0.000
Cars per Household	38,987.171	123.164	3	0.000

A comparison of observed and predicted frequency values as displayed in Table 5 strongly supports the reasonably high predictive efficiency of the model. In general, single family dwellings (1, 2, 3) are rarely estimated as apartments (4), whereas apartments are mixed up with single family buildings on a small lot in about 50% of the cases. A possible explanation for this confusion might be a high similarity between incorporated demographic characteristics of households choosing to live in apartments and households living in small single -family dwellings.

Table 5 Classification table

Observed \ Building Type	Predicted				Percent Correct/%
	1	2	3	4	
1	3,924	240	2,100	75	61.9
2	1,451	457	1,960	4	11.8
3	700	43	21,961	0	96.7
4	548	7	37	506	46.1
Overall Percentage/%	19.5	2.2	76.6	1.7	78.9

A complete listing of model parameters for the multinomial logistic regression is summarized in Table 6. The estimated coefficients reflect the effect of our demographic variables

Table 6 Estimated coefficients

		Exp(β)	Std. Error	df	Sig.	Exp(β)
1	Intercept	1.196	0.177	1	0.0000	
	Population Density	−0.076	0.003	1	0.0000	0.927
	Median Household Income	0.000	0.000	1	0.0000	1.000
	Minorities/%	−0.057	0.005	1	0.0000	0.945
	Hispanics/%	0.030	0.007	1	0.0000	1.031
	Average Household Size	1.098	0.161	1	0.0000	2.999
	Distance to Highway	0.000	0.000	1	0.8894	1.000
	Children per Household	0.737	0.284	1	0.0095	2.089
	Cars per Household	−0.851	0.174	1	0.0000	0.427
2	Intercept	0.696	0.205	1	0.0007	
	Population Density	−0.114	0.003	1	0.0000	0.893
	Median Household Income	0.000	0.000	1	0.0000	1.000
	Minorities/%	−0.079	0.007	1	0.0000	0.924
	Hispanics/%	0.010	0.008	1	0.2558	1.010
	Average Household Size	2.165	0.176	1	0.0000	8.716
	Distance to Highway	0.000	0.000	1	0.9292	1.000
	Children per Household	−0.038	0.311	1	0.9037	0.963
	Cars per Household	−1.694	0.187	1	0.0000	0.184
3	Intercept	2.354	0.187	1	0.0000	
	Population Density	−0.283	0.004	1	0.0000	0.754
	Median Household Income	0.000	0.000	1	0.0000	1.000
	Minorities/%	−0.158	0.007	1	0.0000	0.854
	Hispanics/%	0.095	0.008	1	0.0000	1.100
	Average Household Size	1.650	0.171	1	0.0000	5.205
	Distance to Highway	0.000	0.000	1	0.0000	1.000
	Children per Household	1.567	0.303	1	0.0000	4.792
	Cars per Household	−1.272	0.188	1	0.0000	0.280

on the likelihood of households living in an apartment (reference category 4) relative to living in a single family dwelling (categories 1–3). In general, a positive sign of the coefficients indicates increased likelihood while a negative sign reduces the choice probability of the corresponding building type.

The explanatory variable population density has an overall negative effect. As population density decreases the likelihood of single-family dwellings with respect to apartments becomes higher. The fact that an increase in population density lowers the likelihood of single-family buildings is obvious, since building types are inherently related to population and housing density. The impact of the discussed explanatory variable is confirmed by the high significance of the corresponding coefficients. The variables income and distance to highway have a positive effect but it is smaller than 10^{-4}. Nevertheless, income is highly significant for the estimation at a 0.000 level whereas distance to highway lacks significance for most of the outcome categories.

The percentage of minorities has a negative effect on the model, suggesting that increasing proportion of minorities decreases the odds that the corresponding grid cells will include mostly single-family buildings. Findings also indicate that the demographic variable Hispanics has a slight positive effect on all associated alternatives. The effect's algebraic sign of the average number of children varies within single family building type categories. Furthermore, the explanatory variable children does not appear to be significant at all for choosing building type category 2 over apartments. A relatively large corresponding standard error indicates a lower precision with which the parameters are estimated.

Finally, we find the average number of cars and the average household size is important for building type prediction. Negative parameter estimates are associated with the variable cars in categories 1–3, whereas household size exerts a significant positive effect on the model. In summary, the high significance of the demographic explanatory variables in our model indicates their importance as determinants of building type choice. Population density is an especially important parameter for predicting different dwelling types.

The final model's predictive accuracy is reasonably good for all outcome categories, although the model has some weakness in distinguishing apartments from single family dwellings with small lots. Future research will be dedicated to finding variables that help differentiating between these categories. Moreover, further work is necessary to refine dwelling type categories. Additional building attributes should be taken into account such as the number of floors. At the time of our analysis, the number of floors was not available in parcel level data.

Geometry generation and visualization

We use a PostgreSQL geodatabase as basis for our multidimensional information visualization. *UrbanSim* input and output data and estimated building types are stored in this database and encoded as scalable 3D geometries. The geometries are then visualized in a geospatial context on top of Google Earth™. Geometric objects are created for every grid cell from geometric primitives. Geometry calculations are performed within PHP scripts. For each record in the geodatabase, modular scripts select the associated attributes to be visualized and generate user-specified geometries. At the same time, size and shape of the geometries are scaled according to attribute values.

The final visualization was accomplished by incorporating KML to display 3-dimensional geometries draped over the terrain visible in Google Earth™. First, the 150m×150m grid from *UrbanSim* was imported as layer into Google Earth™. Second, 3-dimensional geometries were developed for the estimated four building types. Third, three ranges of transparencies were applied to the building types to represent the probability of predicting the correct building type based on our empirical estimates. Fourth, five different density classes were established which could then be visualized by scaling the building footprint in relation to the expected density. Finally, a classification system was established to visualize other demographic attributes like access, population share, average number of cars, average number of children, and average income.

We provide three screenshots as examples of the flexible and interactive geometry layers designed to work in conjunction with Google Earth™. Figure 8 shows building types and densities together with average income in the grid cell. Figure 9 shows the distance to nearest highway as color-coded grid cells. Figure 10 displays a close-up of different building types, iconized as dwellings and distinguishable by different roof shapes.

The examples above demonstrate that our approach is generic and dynamic and offers excellent accessibility of the information through a web-based, intuitive interface. Each figure provides a means of visualizing multidimensional aspects of the data and offers an effective

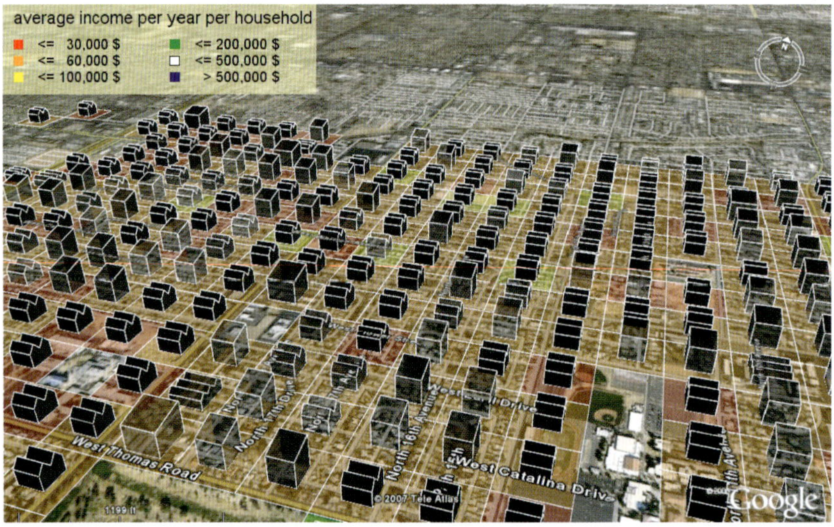

Figure 8 Visualization of (a) average income (color-coded grid cells), (b) building types (geometry), (c) population density (size of footprint), and (d) uncertainty of building type prediction (transparency)

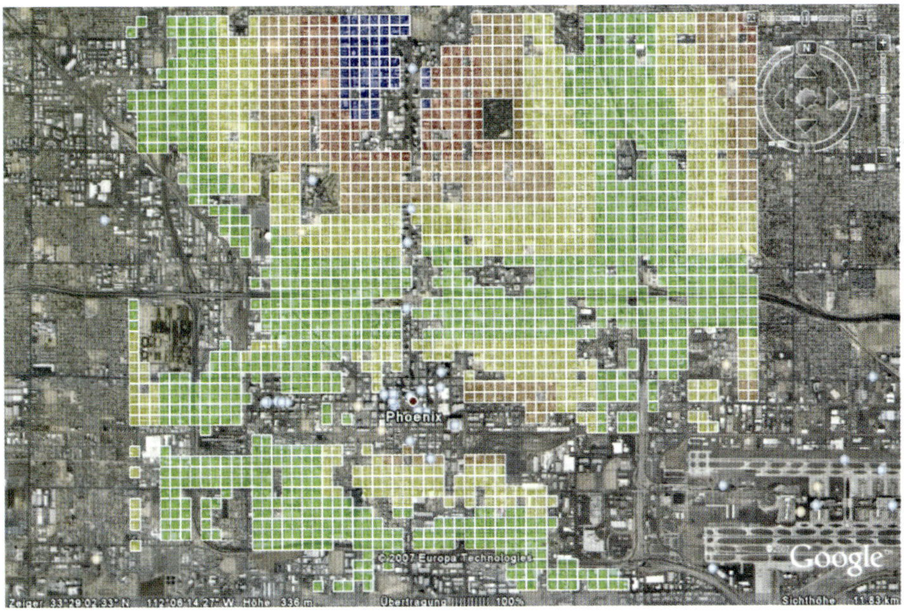

Figure 9 Phoenix Downtown, color-coded grid cells display distance to nearest highway

Figure 10 Close-up of icons for different building types

technique for examining relative patterns and relationships between different characteristics embedded in the information. Given that the procedure uses open geospatial data standards, it also offers an inexpensive yet powerful tool for spatial data visualization in three or more dimensions.

Conclusions

The motivation for this project was to develop an inexpensive, open-source visualization tool to communicate multidimensional attributes of a complex data set. The research also provided a means for visualizing future built environments and the characteristics of their inhabitants based on the textual output of the *UrbanSim* simulation package. *UrbanSim*, like most other urban futures projection models, offers very limited ability to visualize the output. Further, almost no land use change model offers effective means for visualizing information on built forms in three-dimensions. This chapter bridges this gap in the current visualization literature dealing with urban forms. It also provides a path beyond the rule-based methods of procedural modeling towards a more empirically-based framework for developing future environments.

Our visualization framework is not based on realism because realism is often unnecessary and inherently inaccurate for simulated future neighborhoods even when the underlying data has a reasonable degree of certainty. The presented framework incorporates abstraction methods from cartography and information visualization to amplify cognition. Demographic and building type data is communicated to users as abstract geometries, scaled by attribute values, in a 3D real context on Google Earth^{TM}. Our method provides realistic information without invoking a high degree of photorealism that can distract attention from the complexity of information that needs to be communicated. It provides a 3-dimensional representation of abstract data to amplify cognition and facilitate thinking, problem solving, and decision-making.

Acknowledgements This work was supported by the German Research Foundation (DFG, Grant No. 1131) as part of the International Graduate School (IRTG) in Kaiserslautern on "Visualization of Large and Unstructured Data Sets. Applications in Geospatial Planning, Modeling, and Engineering".

References

[1] Guhathakurta S (Editor). Integrated Urban and Environmental Models: A Survey of Current Applications and Research. Berlin, DE: Springer, 2003.

[2] Förstner W. 3D City Models: Automatic and Semiautomatic Acquisition Models, Photogrammetric Week'99. Heidelberg: Wichmann Verlag, 1999: 291–303.

[3] Hu J, You S, Neumann U. Approaches to Large-Scale Urban Modeling. IEEE Computer Graphics and Applications, 2003, 23: 62–69.

[4] Brenner C. Building Reconstruction from Images and Laser Scanning. International Journal of Applied Earth Observation and Geoinformation, 2005, 6: 187–198.

[5] Neidhart H and Sester M. Creating a Digital Thermal Map Using Laser Scanning and GIS. In: Proceedings of the 2004 District Heat and Cooling Symposium. Hannover, DE, 2004.

[6] Lechner T, Watson B A, Wilensky U, Felsen M. Procedural City Modeling. In: 1st Midwestern Graphics Conference. St. Louis, MO, 2003.

[7] Wonka P, Wimmer M, Sillion F, Ribarsky W. Instant Architecture. Transactions on Graph-

ics(SIGGRAPH 2003), 2003, 22: 669–677.

[8] Müller P, Zeng G, Wonka P, Gool L V. Image-based Procedural Modeling of Facades. In: Transactions on Graphics, SIGGRAPH 2007, 26.

[9] Müller P, Wonka P, Haegler S, Ulmer A, Gool L V. Procedural Modeling of Buildings. Transactions on Graphics, SIGGRAPH 2006, 25: 614–623.

[10] Parish Y I H and Müller P. Procedural Modeling of Cities. In: Proceedings of ACM SIGGRAPH 2001. New York: ACM Press, 2001: 301–308.

[11] da Silveira L G and Musse S R. Real-time Generation of Populated Virtual Cities. In: VRST'06: Proceedings of the ACM Symposium on Virtual Reality Software and Technolog. New York: ACM Press, 2006: 155–164.

[12] Slocum T, Blok C, Jiang B, Koussoulakou A, Montello D, Fuhrmann S, Hedley N. Cognitive and Usability Issues in Geovisualization. Cartography and Geographic Information Science, 2001, 28:61–75.

[13] Klosterman R E. The What If? In: Brail R K and Klosterman R E (Editors). Planning Support Systems: Integrating Geographic information Systems, Models, and Visualization Tools. Redlands, CA: ESRI Press, 2001.

[14] Kwartler M and Bernard R. CommunityViz: An Integrated Planning Support System. In: Brail R K and Klosterman R E (Editors). Planning Support Systems: Integrating Geographic information Systems, Models, and Visualization Tools. Redlands, CA: ESRI Press, 2001.

[15] Waddell P. UrbanSim: Modeling Urban Development for Land Use, Transportation and Environmental Planning. Journal of the American Planning Association, 2002, 68: 297–314.

[16] Joshi H, Guhathakurta S, Konjevod G, Crittenden J, Li K. Simulating Impact of Light Rail on Urban Growth in Phoenix: an Application of UrbanSim Modeling Environment. In: Proceedings of the 7th Annual International Conference on Digital Government Research, DG.O, 2006: 135–141.

[17] Chernoff H and Rizvi M H. Effect on Classification Error or Random Permutations of Features in Representing Multivariate Data by Faces. Journal of American Statistical Association, 1975, 70: 548–554.

[18] Viégas F B, Wattenberg M, van Ham F, Kriss J, McKeon M. Many Eyes: A Site for Visualization at Internet Scale. In: Proceedings of the IEEE Symposium on Information Visualization (InfoVIS 2007), Sacramento, CA, 2007.

[19] Bertin J. Semiology of Graphics. Milwaukee, WI: University of Wisconsin Press, 1967.

[20] Andrienko G L and Andrienko N V. Interactive Maps for Visual Data Exploration. International Journal Geographic Information Science, 1999, 13: 355–374.

[21] Skupin A and Fabrikant S I. Spatialization Methods: A Cartographic Research Agenda for Non-Geographic Information Visualization. Cartography and Geographic Information Science, 2003, 30: 95–115.

[22] Skupin A. From Metaphor to Method: Cartographic Perspectives on Information Visualization. In: Proceedings of the IEEE Symposium on Information Visualization (InfoVIS 2000), Salt Lake City, UT, 2000: 91–97.

[23] Fuchs G and Schumann H. Visualizing Abstract Data on Maps, IV'00, 2004, 00:139–144.

[24] MacEachren A M. Visualization–Cartography for the 21st Century. In: Proceedings of the 7th Annual Conference of Polish Spatial Information Association. Warsaw, Poland, 1998.

[25] Guo D, Gahegan M, MacEachren A M, Zhou B. Multivariate Analysis and Geovisualization with an Integrated Geographic Knowledge Discovery Approach. Cartography and Geographic Information Science, 2005, 32: 113–132.

[26] Skupin A and Hagelman R. Attribute Space Visualization of Demographic Change. In: 11th ACM GIS Symposium (GIS'03), New Orleans, Louisiana, 2003.

[27] MacEachren A M, Gahegan M, Pike W, Brewer I, Cai G, Lengerich E, Hardisty F. Geovisualization for Knowledge Construction and Decision Support. IEEE Computer Graphics and Applications, 2004, 24: 13–17.

[28] Kraak M-J. Visualization Viewpoints: Beyond Geovisualization. IEEE Computer Graphics and Applications, 2006, 26: 6–9.

[29] Pinnel L D, Dockrey M, Brush A J B, Borning A. Design of Visualizations for Urban Modeling. In: VisSym'00 Joint Eurographics: IEEE TCVC Symposium on Visualization, Amsterdam, The Netherlands, 2000.

[30] Schwartzman Y and Borning A. The Indicator Browser: A Web-Based Interface for Visualizing UrbanSim Simulation Results. In: Proceedings of the 40th Annual Hawaii International Conference on System Sciences (HICSS'07). Washington, DC: IEEE Computer Society, 2007:92.

[31] Tominski C, Schulze-Wollgast P, Schumann H. 3D Information Visualization for Time Dependent Data on Maps. In: Proceedings of the Ninth International Conference on Information Visualization (IV'05). Washington, DC: IEEE Computer Society, 2005:175–181.

[32] Borning A and Waddell P. Integrated Land Use, Transportation, and Environmental Simulation: UrbanSim Project Highlights. In: Proceedings of the 2004 Annual National Conference on Digital Government Research (DG.O'04), 2004:1–2.

[33] Borning A, Waddell P, Forster R. UrbanSim: Using Simulation to Inform Public Deliberation and Decision-Making. In: Digital Government: Advanced Research and Case Studies. New York: Springer-Verlag, 2006.

[34] Guhathakurta S, Holston J, Kobayashi Y, Lant T, Patel M. Digital Phoenix Project. http://www.digitalphoenix-asu.net [2007-11-21].

[35] US Census Bureau. Census 2000. http://www.census.gov [2007-11-06].

[36] PostgreSQL Global Development Group. PostgreSQL. http://www.postgresql.org [2007-10-23].

[37] Refractions Research. PostGIS. http://postgis.refractions.net [2007-10-23].

[38] Open Geospatial Consortium Inc. OGC Simple Feature Specification. http://www.opengeospatial.org/standards [2007-10-30].

[39] Google Inc. Google Earth. http://earth.google.com/ [2007-11-23].

Chapter 8 Evaluating Real-Time Landscape Visualization Techniques for Public Communication of Energy Crop Planting Scenarios

Andrew Lovett, Katy Appleton, Philip Paar and Lutz Ross

Introduction

The combination of developments in computer performance, software integration and data availability has transformed the capacity to generate 3D virtual environments or landscape views from GIS databases during the past decade. Applications of such geovisualization techniques have occurred in a wide variety of contexts, with a particular focus on the communication of information and facilitating stakeholder engagement in decision making[1, 2]. At present, the typical forms of visualization include photorealistic still images, animated sequences and real-time models where the user can interactively alter their viewing position or the environmental content[3, 4]. Arguably, the sense of control provided by a real-time model should enhance the viewing experience and so increase public engagement in decision making, but empirical evaluation of such potential benefits is currently quite limited[5–8]. There is also the issue that the creation of a real-time model usually involves some element of trade-off between interactivity (particularly smoothness of movement) and the detail with which features are represented. This is an important constraint because several studies have demonstrated that the level of detail, particularly for foreground features, is a key influence on how people relate to such computer-generated visualizations[9, 10].

As the graphics performance of PCs has improved so the degree of compromise required between interactivity and feature detail has declined. Nevertheless, it is still a real issue in certain contexts, one particular example being heavily-vegetated landscapes with many thousands of plants or trees that ideally should be represented as 3D objects rather than via terrain textures. Several methods exist for representing 3D vegetation in landscape visualizations including billboard images, geometric solids or hybrids of the two[11, 12]. There are also techniques for producing plant models through procedural geometry or rule- based methods and for generating realistic ecosystems of plants to cover defined geographical areas[13–15].

This chapter compares and evaluates several existing real-time landscape visualization systems in the context of representing the potential changes in an English rural landscape under increased planting of energy crops. These crops are a potentially important, if sometimes controversial, means by which countries can increase renewable energy generation and achieve targets to reduce greenhouse gas emissions[16, 17]. There are several types of energy crops, including those such as cereals, maize, oilseed rape and sugar beet which can be used to produce transport fuels and other forms of grasses and trees that supply biomass which can be processed to generate heat or electricity. This chapter focuses on one type of perennial biomass crop, miscanthus grass (see Figure 1), which is currently planted on around 15,000 ha of agricultural land in England. A recent government report estimates that the area of such crops in the UK could expand to 350,000 ha by 2020[18], a total which would probably translate to planting on around 10% of arable land in some regions. However, these perennial energy crops are physically different to most current rural land uses; they are in place for 7–25 years, harvest is normally early spring (February–March) and they are dense and tall (3–4m). These factors mean that expansion in planting has the potential to modify the rural landscape, with particular implications for visual appearance, cultural heritage, tourism, farm incomes, hydrology and biodiversity[19, 20]. Such issues are of interest to a range of government agencies and NGOs, and have stimulated discussion regarding the development of planning policies and tools to maximize the benefits of planting[21, 22].

Figure 1 Miscanthus crop in England

The social, economic and environmental implications of increased UK rural land under perennial energy crops are currently being investigated as part of the RELU-Biomass project (http://www.relu-biomass.org). One component of the project is examining public attitudes to energy crops and it is planned to use landscape visualizations to convey the potential impacts of different scales of planting to focus group audiences. An important preliminary stage of this work was to assess different means of generating such visualizations and for this purpose we compared three main software tools that can produce real-time landscape models. These were 1) the 3D Analyst extension of ArcGIS, 2) Visual Nature Studio (VNS) with Scene Express extension and the NatureView Express (NVE) and Virtual Terrain Project (VTP) Enviro viewers and 3) the Lenné3D Player (see Table 1 for further details). This selection represented a range from a general-purpose GIS, through a commercial landscape visualization product capable of generating a wide variety of outputs, to research software

Table 1 URLs for data sources and software

Data	URL
Bluesky Imagery	http://www.bluesky-world.com/
Flora3D plant models	http://www.lenne3d.com
NEXTMap Britain DEM	http://www.bluesky-world.com/dem-nextmap.html
OS MasterMap	http://www.ordnancesurvey.co.uk/osmastermap/

Software	URL
ArcGIS 9.2	http://www.esri.com/software/arcgis/index.html
ArcGIS 3D Analyst 9.2	http://www.esri.com/software/arcgis/extensions/3danalyst/index.html
Deep Exploration	http://www.righthemisphere.com/products/dexp/de_std.html
Hawth's Analysis Tools	http://www.spatialecology.com/htools/index.php
Lenné3D Player	http://www.lenne3d.com
ModelBuilder3D	http://www.presagis.com/products/multigen_paradigm/details/creator/
NatureView Express	http://3dnature.com/nv.html
oikPC	http://www.lenne3d.com
Polytrans	http://www.okino.com/conv
Scene Express	http://3dnature.com/scene.html
Visual Nature Studio	http://3dnature.com/vnsinfo.html
VTP Enviro	http://www.vterrain.org/
Xfrog	http://www.xfrog.com

designed specifically for real-time capabilities. The following section describes the selection of a pilot area, together with the data sources and methods used to produce visualizations. This is followed by presentation of some example views and an evaluation of the outputs achieved.

Idata Sources and Methods

Selection of pilot area for visualizations

Government grants to support the planting of miscanthus in England are available under the Energy Crops Scheme[23]. Information on planting agreements approved under the scheme from 2001 to mid 2006 was used to identify a number of possible locations for visualization and these were then visited. As a result of this exercise a site near the village of Dunholme in Lincolnshire (see Figure 2) was selected for visualization. Arable agriculture is the dominant land use in Lincolnshire and as Figure 2 shows the county is part of one of the major concentrations of miscanthus planting in England. Under the grant scheme, applications were approved to establish over 160 hectares of miscanthus during 2007 which would have covered many of the fields to the south and west of the village. In the event, site visits in July and November 2007 indicated that no planting had taken place (probably due to increases in prices for conventional arable crops) so there was an opportunity to collect information on

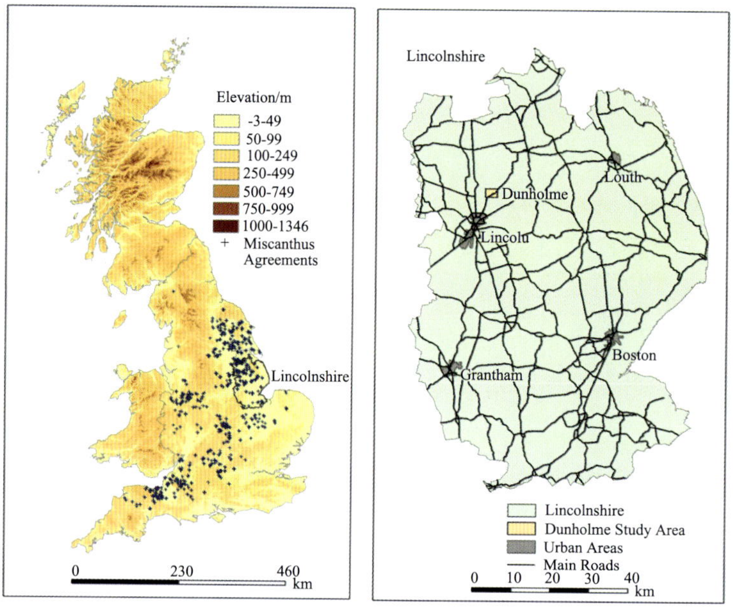

Figure 2 Location of miscanthus planting agreements and the Dunholme study area

the existing land uses and visualize how the area would change with different amounts of miscanthus being grown.

Data sources

An initial 3 by 2 km study area was defined to cover the village and fields extending to the south and west. Digital data assembled for this area included a 5 m resolution NEXTMap Britain DEM, a 25 cm Bluesky orthophoto (taken on 15th June 2003) and Ordnance Survey MasterMap vector mapping. Attributes from the MasterMap data were combined in ArcGIS 9.2 to produce a polygon land cover layer and information on the specific crops being grown in 2007 was added to this. Details of tree positions and hedgerows were digitized from the orthophoto and photos taken during site visits were used to create billboard and texture images for relevant vegetation. Other visits were made to sites where miscanthus is already grown to obtain photos of the crop.

A variety of 3D building models for the area were created by combining polygon footprints extracted from the MasterMap data with textures and height details from photos taken during field visits. The building models were generated in ModelBuilder3D (a subset of MultiGen-Paradigm Creator) and then converted from OpenFlight to 3D Studio format using Deep Exploration and Polytrans Further details of the source data and software used are available via the URLs in Table 1.

For the purposes of comparing the different real-time modeling tools it was decided to focus on creating visualizations for a smaller area covering 1,400 m by 1,400 m on the western side of Dunholme. Figure 3 shows this square against a background of the 2003 orthophoto. A hatched area represents a potential 10 ha field of miscanthus. It is common to establish miscanthus at a density of two plants per m^2 so a 10 ha field would be equivalent to 200,000 plants. Three viewpoints are also marked and these were used to generate sets of comparative images.

Figure 3 The Dunholme visualization area and viewpoints

Visualization methods

The real-time models were all created on PCs with at least a 1 GHz processor, 1 Gb RAM and a graphics card with 256 Mb memory. Several previous publications have discussed the general procedure involved in creating real-time models with the software tools used[3, 4, 24, 25] so this information will not be repeated here. Instead, the discussion below focuses mainly on issues associated with the representation of vegetation and certain problems that were encountered.

ArcGIS 3D Analyst

Three-dimensional real-time display capabilities in ArcGIS 9.2 are provided by ArcScene and ArcGlobe. Recent releases of the software have much improved capabilities to import 3D model formats and it is now possible to store many of these within a geodatabase. However, one of the limitations of using the software is that the 3D objects need to be assigned to defined points. As a consequence, hedges were created by transforming the digitized lines into points spaced every two meters and then using a random function to assign different bush symbols to these. For the miscanthus plot, a 3D symbol was created from a billboard image and saved in .3ds format. Hawth's Analysis Tools (a free ArcGIS extension) was then used to generate a regular point distribution within the field polygon and the miscanthus symbol was assigned to this pattern. It did not prove practical to do this at the planned planting density so the eventual field visualized contained some 11,200 plant models.

When it came to generating the final visualization output it was found that better results were obtained using ArcGlobe rather than ArcScene. The most likely reason for this is that ArcGlobe has an improved caching mechanism for the integration and display of large datasets. Even so, on a dual core PC with 2 Gb RAM it was found that when the rendering parameters were set to maximum quality and smooth shading the system response became sluggish and prone to freeze or crash when settings were changed.

Visual Nature Studio

VNS provides the capability to generate both highly-detailed still images and real-time models from the same underlying database. Real-time output was generated via the Scene Express extension, and was displayed using two free viewer tools, namely VNS Nature View Express and VTP Enviro. For the purposes of comparison, still images were rendered from several locations; these locations were included as saved viewpoints in the NVE and VTP output.

Representation of vegetation features followed similar procedures to those using ArcGlobe. However, the process was more efficient with VNS since the software includes Foliage Effects that can, for example, randomly assign billboard images to vertices along a line, and also allow the user to specify the relative proportions of each species/image. Similarly, the miscanthus plot could be automatically populated with a 'clump' billboard image at a specified density. Placements of this image, as well as variations in image height and orientation (X-flipping) were determined by VNS, avoiding unnaturally regular outcomes.

During creation of the real-time models it became apparent that the viewing tools, as run on the development machine, could not cope with a miscanthus field containing 200,000 plants so only an area of approximately 2.5 ha was eventually included. Both real-time formats allow some Level of Detail (LOD) control for foliage and building objects; experiments were carried out with foliage LODs, but the 'popping' effect of the miscanthus field as the viewer approached was found to be rather distracting. In any case, slowdown in response was most noticeable when the miscanthus field was closest to the viewer, and therefore the vegetation would have been switched on.

NVE can export multiple lights, which allowed the buildings to be lit more effectively; shading is flatter and less dark. VTP only allows one light to be exported, and shading varies across the triangulated polygons of the buildings. However, VTP does allow 'time of day' to be manipulated whilst viewing and the shadows can thereby be lessened to some degree for the viewpoint on display.

3D-Player

The Lenné3D system is a prototype for a GIS-data-based visualization tool for interactive exploration of three-dimensional landscapes[25, 26]. Three-dimensional scenes are generated using standard GIS data, *e.g.* land-use and biotope maps and a digital terrain model, an integrated vegetation model, and 3D objects, especially detailed plant models.

The oikPC software component calculates plant positions by means of heuristic-algorithmic models, on the basis of habitat data (*e.g.* biotope types), cropping practices, vegetation reference mappings and topographical site data. Habitats are broken down into various types of vegetation. The plants are then distributed randomly within the vegetation layers (strata) and units. Mapped plants (*e.g.* a solitary tree) can be positioned interactively or according to coordinates from a GIS point layer. This combination of facilities means that vegetation can be efficiently placed in simulated landscapes whilst avoiding unnaturally regular positions.

3D-Player is an interactive rendering program that interprets the scenic description and geo-data specified in a Lenné3D project (.eco). Plant models (.flora3d) are assigned to the distribution calculated by the oikPC component and positioned within the landscape model. These plant models are hybrids which contain a textured polygon mesh (generated through mathematical procedural modeling) for full resolution and billboards for mid and background representation. The full resolution models are created using the Xfrog software[27] and add-on programs. A tree model typically has around 100–300,000 polygons at full LOD resolution.

A key feature of the 3D-Player is LoD control for dynamic reduction of scene complexity. Efficient implementation of data structures and access meant that it was possible to include several million separate plant models in the Dunholme scene. However, the current data structures are quite static and do not work well with dynamically changing data. As a consequence, animations of plant growth are not possible and the user has limited scope to interactively change the landscape model. This is because the geometric description of many plants is combined to an overall representation of the scene to enable real-time rendering. Single objects can be added, but after changing substantial parts of the scene, pre-processing in GIS and oikPC must be performed again before display.

Evaluation Criteria

The performance of the different software tools was evaluated using four main criteria. Two of these concerned the ability to handle large numbers of plant models and the realism with which vegetation was represented. Ease of real-time navigation was the third criterion and the fourth concerned the extent of facilities to change displayed features on the fly or link in other data.

Results

Table 2 summarizes key characteristics of the real-time models generated by the different software tools. This table highlights a number of differences in resolutions and file sizes, as well as the capacity of the 3D-Player to handle a much larger number of plant models.

Figures 4 to 8 show a series of photos and visualizations of the study area. An aerial overview looking towards Dunholme is shown in Figure 4 and illustrates the relatively flat and open nature of the landscape. Figure 5 contains a panoramic ground level photo (approximately 135 degrees) looking from VP1 towards Dunholme. The photos constituting this panorama were taken in July 2007 and show a field of wheat between the viewpoint and the village. Under the visualized planting scenario the miscanthus would be grown at the far

Table 2 Comparison of real-time models

Item	Arc Globe	NVE	VTP	3D-Player
Terrain resolution/m	5	10.3	10.3	10.3
Texture resolution/m	0.25[3)]	1.3	1.3	1.3
Number of Plant models	12,917	46,589	46,589	5,233,669
Number of 3D objects[1)]	370	370	370	370
File size[2)]/Mb	300	10	10	182[4)] 100[5)]
Export time/min	–	16	16	5[6)]

1) Mostly buildings. 2) Including any subdirectories, image files, *etc.* 3) Input resolution, data are resampled automatically during viewing. 4) Scene file with 3D object positions. 5) 3D objects including 3D plant models. 6) For high-resolution renders of 5,330 by 3,990 pixels.

end of this field. Figure 6 contains another panoramic view (approximately 100 degrees) based on photos taken in November 2007 and looks south west from VP2 across the same field. The open nature of the landscape (it is actually possible to see the tower of Lincoln Cathedral over 10 km away) is again apparent.

Figure 4 Aerial view of the Dunholme area produced using the 3D-Player

Figure 5 Panoramic photo looking north east towards Dunholme from VP1, July 2007

Figure 6 Panoramic photo looking south west from VP2, November 2007

A set of example visualizations looking north from VP3 across the miscanthus crop is shown in Figure 7. The top left-hand image is a still rendered with VNS and was produced as a reference to illustrate the highest levels of feature detail that could be anticipated. Each of the other four images is a screenshot or rendered view from a real-time model and so allows comparison of their capabilities. Some of the key differences are as follows.

Left: Reference still image (Visual Nature Studio)

Below left: 3D-Player

Below right: ArcGlobe

Bottom left: VTP Enviro

Bottom right: Nature View Express

Figure 7 Example visualizations from VP3

- There are significant contrasts in visual appearance, highlighting the differences between the systems and the subjectivities involved in creating such visualizations.
- The ArcGlobe model is noticeably more artificial in appearance than the others, with

the 3D-Player and NVE displays being closest to the textures and feature realism in the still image.

- Both the VTP and NVE models were created from the same underlying data in VNS, but the former is substantially darker than the latter (due to limitations in lighting control).
- The ArcGlobe model has a lower miscanthus plant density, but does fill the 10 ha plot. In the VTP and NVE models the density is as required but only part of the plot is covered by miscanthus plants. The 3D-Player output was the only model to combine the required plot area and plant density.

Close-up visualizations of the miscanthus crop are presented in Figure 8. These images were generated from the same location as used for the photo in Figure 6 and so give a sense of how the view would change as a miscanthus crop became established. As with Figure 7, the top-left image is a still rendered from VNS and the remaining four are from the real-time models. With respect to the representation of vegetation, one clear difference is that the 3D-Player model is the only one that can display plants in the buffer strip around the miscanthus with similar detail to the still render. In the other three models this area has a rather bare appearance which varies according to how the ground texture image is handled.

There are also differences in how the miscanthus crop is shown. With ArcGlobe the density of plants is lower, the models are quite regularly spaced and sized, and there is a slight 'halo' effect around their edges which probably stems from how transparency information is stored in the alpha channel. The other real-time outputs do not show this problem and there is more realistic variation in the appearance and height of the miscanthus plants.

Real-time model navigation

The images in Figures 7 and 8 illustrate the appearance of output from the different visualization systems, but another important characteristic is the ease of actually interactively moving through each landscape model.

All four of the viewing environments provided a good range of navigation modes (see Table 3), but had problems with slower frame rates when moving in proximity to the miscanthus plants. For example, VTP achieved a rate of 32 frames per second (fps) with a distant view, an average of ~18–20 fps when approaching the miscanthus plot and ~2 fps when looking around at VP2. Comparable figures for NVE on the same computer were 12 fps, ~4 fps and ~2 fps respectively. At the latter speed control tends to become difficult as it is easy to issue multiple consecutive navigation commands before the first one has completed. One solution to this problem is to predefine sets of viewpoints and move between them on a tour sequence. Another option is provided by the 3D-Player which uses a preview facility to assist with navigation when frame rates are low.

Several other features deserve mention because of their potential value in decision-making contexts. For instance, one of the strengths of ArcGlobe is the integration of the display capability within a GIS which provides query/analysis facilities and a ready ability to change the displayed layers on the fly. Making such alterations is difficult in the 3D-Player and in NVE is essentially restricted to switching the display of entire classes of features (*e.g.* building or vegetation) on or off. The VTP viewer has a facility to move or rotate objects in real-time and development of these types of interactive or 'what if' tools will become important as real-time visualizations are used more widely to support public engagement in decision making.

Left: Reference still image (Visual Nature Studio)

Below left: 3D-Player

Below right: ArcGlobe

Bottom left: VTP Enviro

Bottom right: Nature View Express

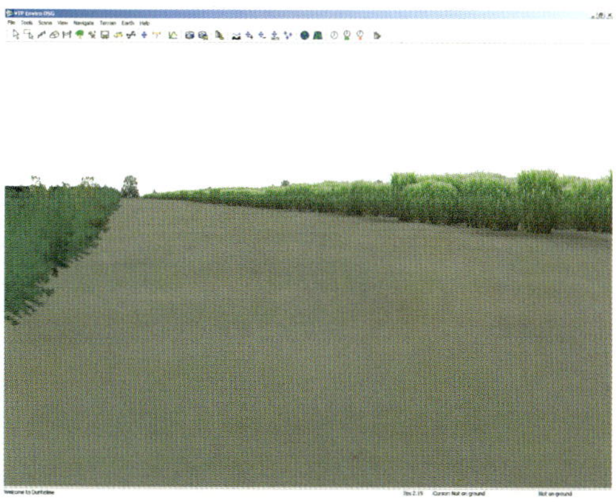

Figure 8 Example visualizations from VP2

Table 3 Characteristics of the real-time viewing environments

Viewer	Navigation Capabilities	Other Features and Observations
ArcGlobe	Wide range including fly and walk modes plus goto facility	Keyframes can be defined to make animations; Wide range of query facilities, hotlinking and ability to change layers on the fly
NVE	Four navigation modes, plus stop, goto and reset position Pre-defined viewpoints can be selected, toured either manually or automatically, and used for animations	Objects can be made clickable, linking to text, html, sounds, or other models. Logo or location map can be added; Classes of features can be switched on/off
VTP	Two mouse modes plus keyboard and joystick controls. Pre-set viewpoints can be selected, but no capability for tours	Objects can be moved and rotated in real- time; Moving vehicles can be added from a list
3D-Player	Fly and walk modes using mouse or keyboard; Pre-set viewpoints can be selected, toured and used for animations	Water shading, interactive water level control; Preview feature helps with navigation when frame rates are low

Conclusions

The examples presented in this chapter illustrate some of the challenges of creating landscape visualizations for heavily-vegetated environments and demonstrate that there are now real-time systems which can match the degree of feature detail provided by rendered still images. Nevertheless, it is also apparent that systems which excel in this respect may be limited in other aspects of interactivity (*e.g.* the ability of viewers to change the content of the environment they are examining). This emphasizes that at present there is no perfect visualization solution and reinforces the need for research to evaluate the importance of different capabilities in practical decision-making contexts. It is hoped that future work using visualization methods with focus groups as part of the RELU-Biomass project will contribute to addressing this challenge.

Acknowledgements The research described in this chapter was funded by the Rural Economy and Land Use (RELU) programme of the UK Research Councils and a British Council/German Academic Exchange Service ARC grant. We would also like to thank Luke Harmer of Natural England for supplying the information on miscanthus planting agreements and Gilla Sünnenberg of UEA for her help in creating GIS data layers and building models. The NEXTMap DEM data were supplied by the NERC Earth Observation Data Centre and the Ordnance Survey MasterMap data are © Crown Copyright/database right 2007. An Ordnance Survey/Digimap supplied service.

References

[1] Bishop I and Lange E (Editors). Visualization in Landscape and Environmental Planning. London: Taylor & Francis, 2005.

[2] Lovett A and Appleton K (Editors). GIS for Environmental Decision-Making. Boca Raton, FL: CRC Press, 2008.

[3] Appleton K, Lovett A, Sünnenberg G, Dockerty T. Rural Landscape Visualisation from GIS Databases: A Comparison of Approaches, Options and Problems. Computers, Environment and Urban Systems, 2002, 26: 141–162.

[4] Appleton K and Lovett A. Visualizing Rural Landscapes from GIS Databases in Real-Time—A Comparison of Software and Some Future Prospects. In: Madden M (Editor). The ASPRS Manual of GIS, American Society for Photogrammetry and Remote Sensing. Washington DC: MD, in press.

[5] Bishop I. Visualization for Participation—The Advantages of Real-Time? In: Buhmann E, Paar P, Bishop I, Lange E (Editors). Trends in Real-Time Landscape Visualization and Participation. Heidelberg, DE: Wichmann, 2005: 2–15.

[6] Paar P. Landscape Visualizations: Applications and Requirements of 3D Visualization Software for Environmental Planning. Computers, Environment and Urban Systems, 2006, 30: 815–839.

[7] Stock C, Bishop I, Green R. Exploring Landscape Changes Using an Envisioning System in Rural Community Workshops. Landscape and Urban Planning, 2007, 79: 229–239.

[8] Miller D, Morrice J, Coleby A, Messager P. Visualization Techniques To Support Planning Of Renewable Energy Developments. In: Lovett A and Appleton K (Editors). GIS for Environmental Decision-Making. Boca Raton, FL: CRC Press, 2008: 227–239.

[9] Lange E. The Limits of Realism: Perceptions of Virtual Landscapes. Landscape and Urban Planning, 2001, 54: 163–182.

[10] Appleton K and Lovett A. GIS-Based Visualisation of Rural Landscapes: Defining 'Sufficient' Realism for Environmental Decision-Making. Landscape and Urban Planning, 2003, 65: 117–131.

[11] Ervin S and Hasebrouck H. Landscape Modeling. New York: McGraw-Hill, 2001.

[12] Muhar A. Three-Dimensional Modelling and Visualisation of Vegetation for Landscape Simulation. Landscape and Urban Planning, 2001, 54: 5–17.

[13] Perrin L, Beauvais N, Puppo M. Procedural Landscape Modeling with Geographic Information: The IMAGIS Approach. Landscape and Urban Planning, 2001, 54: 33–47.

[14] Deussen O, Colditz C, Coconu L, Hege H-C. Efficient Modelling and Rendering of Landscapes, in I. Bishop I and Lange E (Editors). Visualization in Landscape and Environmental Planning. London: Taylor & Francis, 2005: 56–61.

[15] Röhricht W. Oik—Nulla Vita Sine Dispensatio—Vegetation Modelling for Landscape Planning. In Buhmann E, Paar P, Bishop I, Lange E (Editors). Trends in Real-Time Landscape Visualization and Participation. Heidelberg: Wichmann, 2005: 256–262.

[16] European Environment Agency. How Much Bioenergy Can Europe Produce without Harming the Environment. European Environment Agency, Copenhagen, Denmark, 2006.

[17] UN-Energy. Sustainable Bioenergy: A Framework for Decision Makers. United Nations, New York, 2007.

[18] Defra, DTI and DfT. UK Biomass Strategy. Department for Environment, Food and Rural Affairs, London, 2007.

[19] Upreti B. Conflict Over Biomass Energy Development in the United Kingdom: Some Observations and Lessons from England and Wales. Energy Policy, 2004, 32: 785–800.

[20] Rowe R, Street N, Taylor G. Identifying Potential Environmental Impacts of Large-Scale Deploy-

ment of Dedicated Energy Crops in the UK. Renewable and Sustainable Energy Reviews, 2009, 13: 271–290.

[21] English Heritage. Biomass Energy and the Historic Environment. London: English Heritage, 2006.

[22] Defra. Planting and Growing Miscanthus: Best Practice Guidelines for Applicants to Defra's Energy Crops Scheme. Department for Environment, Food and Rural Affairs, London, 2007.

[23] Natural England. Energy Crops Scheme-Establishment Grants Handbook. Sheffield, UK: Natural England, 2007.

[24] Appleton K and Lovett A. Display Methods for Real-Time Landscape Models—An Initial Comparison. In: Buhmann E, Paar P, Bishop I and Lange E (Editors). Trends in Real-Time Landscape Visualization and Participation. Heidelberg, DE: Wichmann, 2005: 246–253

[25] Paar P and Rekittke J. Lenné3D—Walk-Through Visualization of Planned Landscapes. In: Bishop I and Lange E (Editors). Visualization in Landscape and Environmental Planning. London: Taylor & Francis, 2005: 152–162.

[26] Werner A, Deussen O, Döllner J, Hege H-C, Paar P, and Rekittke J. Lenné3D—Walking through Landscape Plans. In: Buhmann E, Paar P, Bishop I and Lange E (Editors). Trends in Real-Time Landscape Visualization and Participation. Heidelberg, DE: Wichmann, 2005: 48–59.

[27] Lintermann B and Deussen O. A Modeling Method And User Interface for Creating Plants. Computer Graphics Forum, 1998, 17: 73–82.

Chapter 9 A GIS and CG Integrated System for the Automatic Generation of 3D Building Models

Kenichi Sugihara and Yoshitugu Hayashi

Introduction

Based on building polygons or building footprints on a digital map shown in Figure 1, we are proposing a GIS and CG integrated system that automatically generates 3D building models. A 3D urban model as shown in Figure 2 is an important information infrastructure that can be utilized in several fields such as landscape evaluation and urban planning, disaster prevention simulation, tourism, and many other business practices. In addition, disclosure of information about public projects such as city planning or historic buildings restoration in order to carry out public projects smoothly and to encourage their participation is a new application area where a 3D city model can be of great use in public policy making. In Japan, public agencies publicize urban planning and the results of building excavations by showing floor plans and side views and by explanations based on written papers and even paper plans that are difficult to understand. To facilitate public participation, 3D models simulating a real city or an ancient city using 3D CG can be of enormous help.

Figure 1 Building polygons on a digital map

Figure 2 An automatically generated 3D urban model

However, much time and labor has to be consumed to create these 3D models using modeling software such as 3ds Max or Sketchup. For example, when manually modeling a house with roofs by Constructive Solid Geometry (CSG), one must follow five laborious modeling steps involving: 1) Generation of primitives of appropriate size, such as box, prism or polyhedron that will form parts of a house; 2) Boolean operations among these primitives to form the shapes of parts of a house such as making holes in a building body for doors and windows; 3) Rotation of parts of a house; 4) Positioning of parts of a house; 5) Texture mapping onto these parts.

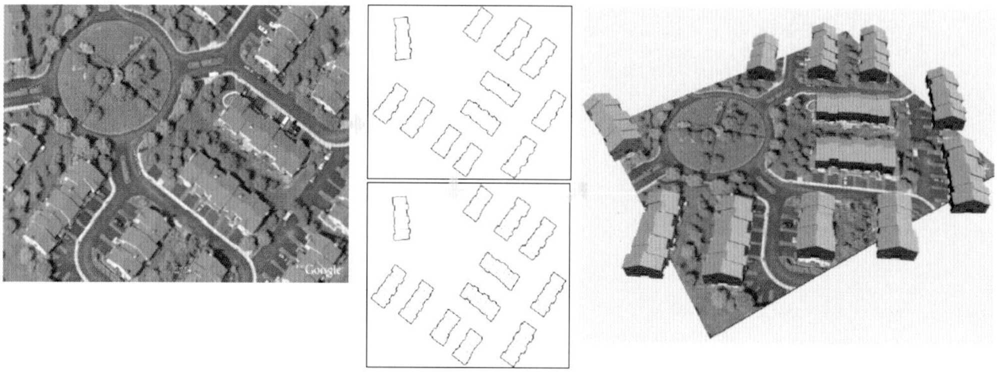

Figure 3　The process of automatic generation for 3D building models that are not created by straight skelton

Left: An ortho-image from Google Earth.
Upper middle: Building footprints or building polygons. Lower middle: Partitioning into a set of rectangles. Right: An automatically generated 3D building model.

In order to automate these steps, we propose a GIS and CG integrated system[1] that automatically generates 3D building models from building polygons on a digital map. As shown in Figure 1, most building polygons' edges meet at right angles (orthogonal polygon). An orthogonal polygon can be partitioned into a set of rectangles. Our integrated system partitions orthogonal building polygons into a set of rectangles and places rectangular roofs and box-shaped building bodies on these rectangles. In order to partition an orthogonal polygon, we proposed a useful polygon expression (RL expression) and a partitioning scheme that is used in deciding from which vertex a dividing line (DL) is drawn[2].

In this chapter, we propose a new scheme for partitioning building polygons and show the process of creating a basic gable roof model. We also present 3D models of a Japanese ancient temple and a pagoda, based on the generation process of a basic roof. Since 3D urban models provide important information infrastructure that can be utilized in several fields, the research into 3D urban models is in full swing. Procedural modeling is an effective technique to create 3D models from sets of rules such as L-systems, fractals, and generative modeling language.

Müller *et al.*[3, 4] have created an archaeological site of Pompeii and a suburbia model of Beverly Hills by using a shape grammar with production rules. They import data from a GIS database and try to classify imported mass models as basic shapes in their shape vocabulary. If this is not possible, they use a general extruded footprint together with a general roof obtained by a straight skeleton computation[5].

The straight skeleton can be used as the set of ridge lines of a building roof, based on walls in the form of the initial polygon[6]. However the roofs created by the straight skeleton are limited to a hipped roof or a gable roof with their ridges parallel to the longest edges of the rectangle into which a building polygon is partitioned. There are some roofs whose ridges are vertical to the longest edges of the rectangle, and these roofs are independently formed on individual rectangles as shown in Figure 3. Figure 3 also shows the process of automatic generation for such 3D roof models. For this reason, we have investigated partitioning orthogonal building polygons into sets of rectangles.

Laycock *et al.*[7] have combined the straight skeleton method and polygon partitioning in the following steps: 1) Partition the polygon into a set of rectangles by horizontal and vertical lines from all reflex vertices; 2) Construct the straight skeleton and grow an axis aligned rectangle (AAR) from each of the lines; 3) For each AAR, collect the rectangles which are interior to AAR and union them to obtain an exterior boundary; 4) Assign a roof model to each exterior boundary; and finally 5) Merge the roof models. This method seems effective in independently choosing a roof model for each rectangle and merging the roof models for polygons with a small number of vertices. However, for polygons with a large number of vertices, implementation of partitioning along all DLs (lines from all reflex vertices) often results in an unnecessarily large number of rectangles and collecting and merging steps becomes so cumbersome that they do not succeed in generating the required shapes.

Therefore, in our automatic generation system, we have tried to select a suitable DL for partitioning or a suitable separation of overlapping rectangles, depending on the RL expression of a polygon, the length of DLs and the edges of a polygon.

The Flow of the Automatic Generation Process

The automatic generation system consists of a GIS application (ArcGIS, ESRI Inc.), a GIS module and a CG module as shown in Figure 4. The source of a 3D urban model is a digital residential map that contains building polygons linked with attribute data such as the number of stories and the type of roof. The GIS module pre-processes building polygons on the digital map. Pre-processing includes filtering out a 'short edge' (a short edge that is between long edges of almost the same direction) and unnecessary vertices of a building polygon, partitioning orthogonal building polygons into sets of rectangles, generating inside contours

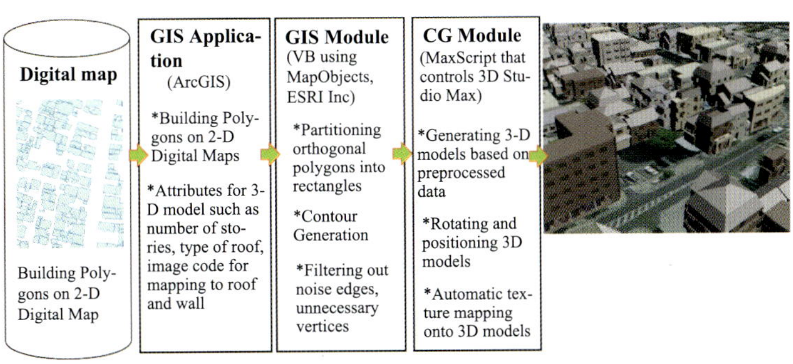

Figure 4 Flow of automatic generation system for 3D building models

for positioning walls and glasses of a building and exporting the coordinates of polygons vertices and attributes of buildings. The attributes data of buildings consist of the number of stories, the image code of the roof, wall and the type code of the roof (flat, gable roof, hipped roof, oblong gable roof, gambrel roof, Mansard roof and so forth). The GIS module has been developed using 2D GIS software components such as MapObjects from ESRI.

The CG module receives the pre-processed data that the GIS module exports, generating 3D building models. The CG module has been developed using Maxscript that controls 3D CG software (3ds Max, Autodesk Inc). In case of modeling a building with roofs, the CG module follows these steps: 1) Generation of primitives of appropriate size, such as boxes, prisms or polyhedra that will form the various parts of the house; 2) Boolean operation on these primitives to form the shapes of parts of the house, for examples, making holes in a building body for doors and windows; 3) Rotation of parts of the house; 4) Positioning of parts of the house; 5) Texture mapping onto these parts according to the attribute received; 6) Copying the 2nd floor to form the 3rd floor or more in case of building higher than 3 stories.

The Partitioning Scheme

Proposed polygon expression

At map production companies, technicians are drawing building polygons manually with a digitizer, using aerial photos or satellite imagery as shown in Figure 5. This aerial photo and digital map in Figure 6 also show that most building polygons are orthogonal polygons. An orthogonal polygon can be replaced by a combination of rectangles. When following edges of a polygon clockwise, an edge turns to the right or to the left by 90 degrees. Therefore, it is possible to assume that an orthogonal polygon can be expressed as a set of its edges' turning direction.

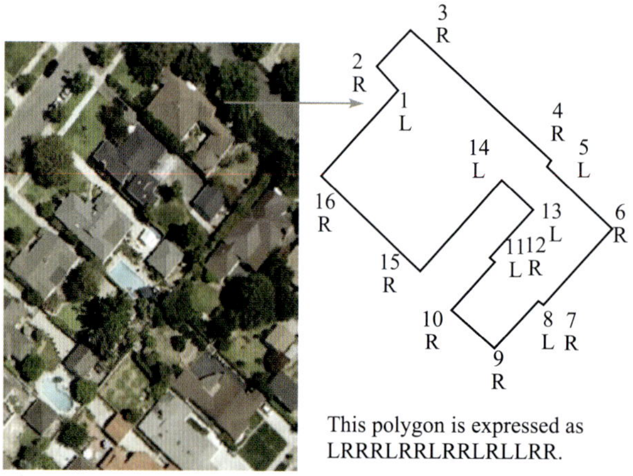

Figure 5 Satellite imagery and building polygon

Figure 6 Building polygons on a digital map

We propose a useful polygon expression (RL expression) in specifying the shape pattern of an orthogonal polygon[1]. For example, a polygon with 16 vertices shown in Figure 5 is expressed as a set of its edges' turning direction; LRRRLRRLRRLRLLRR where R and L mean a change of an edge's direction to the right and to the left, respectively. The number of shapes that a polygon can take depends on the number of vertices of a polygon. The advantage of this RL expression is as follows:

(1) RL expression specifies the shape pattern of an orthogonal polygon without regard to the length of its edges;

(2) This expression decides from which vertex a dividing line(DL) is drawn.

Partitioning scheme

The more vertices a polygon has, the more partitioning scheme a polygon has, since the interior angle of a 'L' vertex is 270 degrees and two DLs (dividing lines) can be drawn from a 'L' vertex. We propose the partitioning scheme that gives higher priority to the DLs that divide 'fat rectangles'. A 'fat rectangles' is a rectangle close to a square. Our proposed partitioning scheme is similar to Delaunay Triangulation in the sense that Delaunay Triangulation avoids narrow triangles and generates fat triangles. However, our proposal does not always result in generating plausible and probable 3D buildings model with roofs. For example, our proposed partitioning scheme shown at the right of Figure 7 indicates that a rectangle cut off by a DL is wider than a body rectangle, resulting in the branch roof being higher than a main roof, where a 'branch roof' is a roof that is cut off by a DL and extends to a main roof. A 'main roof' is a roof that is extended by a branch roof.

In our new proposal, among many possible DLs, the DL that satisfies the following conditions is selected for partitioning.

(1) A DL that cuts off 'one rectangle';

(2) Among two DLs from a same 'L' vertex, a shorter DL is selected to cut off a rectangle;

and

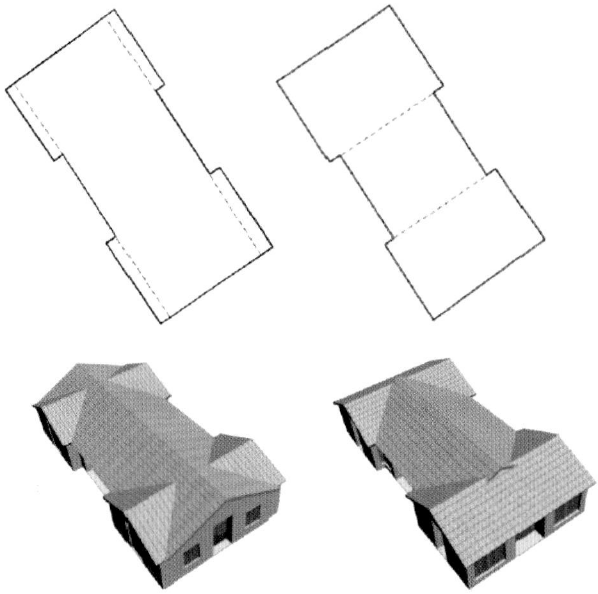

Figure 7 Polygons and 3D house models by different partitioning scheme

(3) A DL whose length is shorter than the width of a 'main roof' that a 'branch roof' is supposed to extend to.

Our newly proposed partitioning scheme as shown on the left of Figure 7 leads to a plausible and probable 3D house model with the main roof higher than a branch roof.

Figure 8 shows the partitioning process of an orthogonal building polygon into a set of rectangles. Stage 2 in Figure 8 shows an orthogonal polygon with all possible DLs shown as thin dotted lines and DLs that satisfy condition (1) shown as thick dotted lines. The example of a branch roof is shown as the rectangle formed by vertices 6,7,8,9 cut off by DL.

The reason we set up these conditions is that like breaking down a tree into a collection of branches, we will cut off along the 'thin' parts of branches of a polygon. Therefore, we propose a scheme of prioritizing the DL that cuts off a branch roof, based on the length of the DL. Since each roof has the same slope in most multiple-roofed buildings, a roof of longer width is higher than a roof of shorter width and 'probable multiple-roofed buildings' take the form of narrower branch roofs diverging from a wider and higher main roof. Narrower branch roofs are formed by dividing a polygon along a shorter DL and the width of a branch roof is equal to the length of the DL.

In the partitioning process as shown in Figure 8, the DLs that satisfy the mentioned conditions are selected for partitioning. By cutting off one rectangle, the number of the vertices of a body polygon is reduced by two or four. After partitioning branches, the edges' lengths and RL data are recalculated to find new branches. Partitioning continues until the number of the vertices of a body polygon is four. After being partitioned into a set of rectangles, the system places 3D building models on these rectangles.

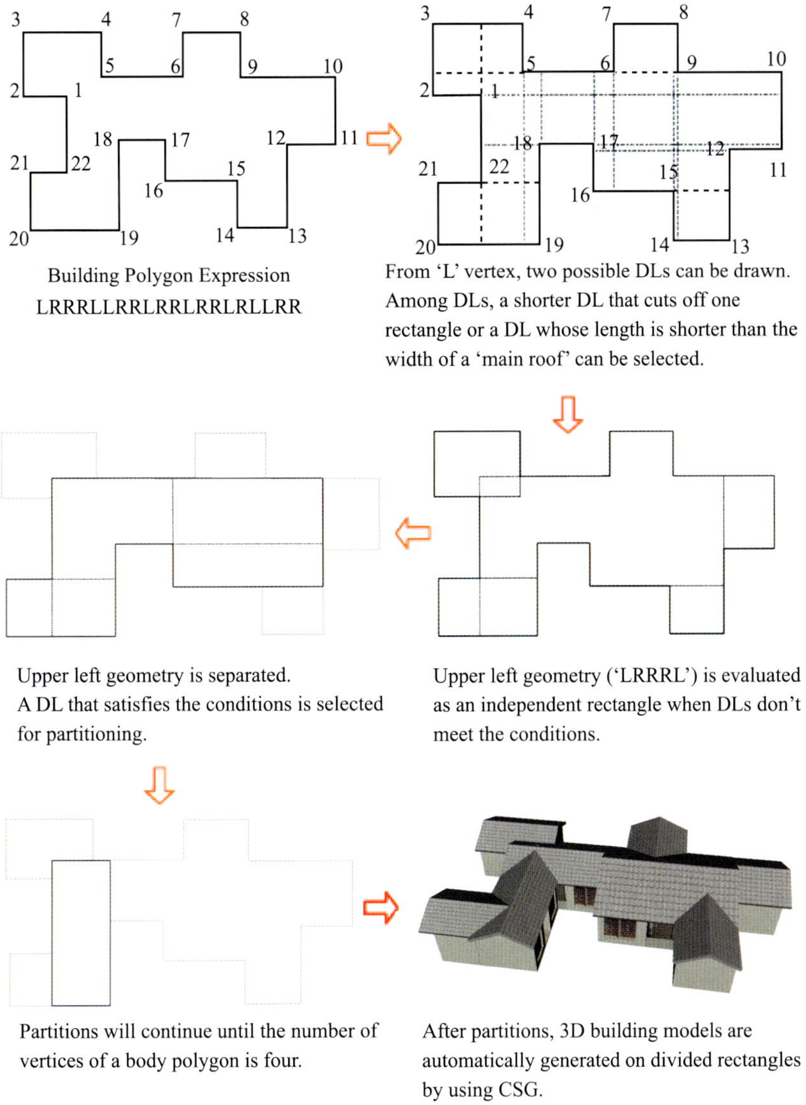

Figure 8 Partitioning process of an orthogonal building polygon into a set of rectangles

How to partition branches

The vertices of a polygon are numbered in clock-wise order as shown in Figure 8. Here, how the system is finding 'branches' is as follows. The system counts the number of consecutive 'R' vertices (= n_R) between 'L' vertices. If n_R is two or more, then it can be a branch. One or two DLs can be drawn from 'L' vertex in a clockwise or counterclockwise direction, depending on the length of the adjacent edges of 'L' vertex. Figure 9 shows three cases of drawing DL when n_R is two. The way of drawing DL depends on the comparison between Len(FCP) and Len(jpb). In Figure 9, FCP (Forward Cutting Point) is the point that precedes consecutive 'R' vertices and from which a DL can be drawn forward.

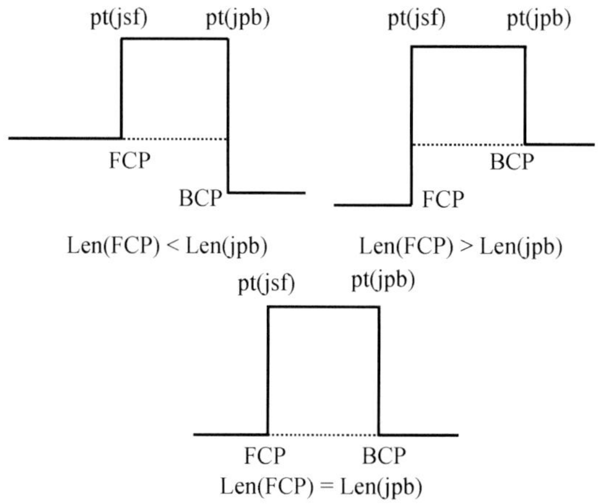

Figure 9 n_R=2, three cases of DL depending on the comparison between Len(FCP) and Len(jpb)

BCP (Backward Cutting Point) is the point that succeeds consecutive 'R' vertices and from which DL can be drawn backwards. 'jsf' is the index to specify the vertex that succeeds FCP by one vertex and 'jpb' is the index to specify the vertex that precedes BCP by one vertex. Len(FCP) means the length of an edge between FCP and jsf. Figure10 shows three cases of drawing DL when n_R is three. The way of drawing DL depends on the comparison between Len(FCP) and Len(j2sf) and the comparison between Len(jsf) and Len(jpb) where 'j2sf' is the index to specify the vertex that succeeds FCP by two vertex.

In the 3rd case of n_R=3 in Figure 10, the rectangle consisting of vertices pt(jsf) and pt(j2sf), pt(jpb), pt(A) is not partitioned but separated as an independent one when the system evaluates 'n_R=3' type as independent and prioritizes separating Figure 11 shows three cases of drawing DL when n_R=4. The way of drawing DL depends on the comparison between Len(jsf) and Len(j2pb) where 'j2pb' is the index to specify the vertex that precedes BCP by two vertex. The partition method of n_R=4 or more is the same as the method of n_R=3, except for the first case of n_R=4 where the vertices of pt(jsf) and pt(j2sf), pt(j2pb), pt(jpb) form one rectangle (1st case of Figure 11). This is because the system cut off 'one rectangle' only when n_R=2 or n_R=3. If n_R=4 or more, the branch that is formed by these 4 or more 'R' vertices will be self-intersecting.

Figure 12 shows a variety of shapes of orthogonal building polygons with DLs imple-

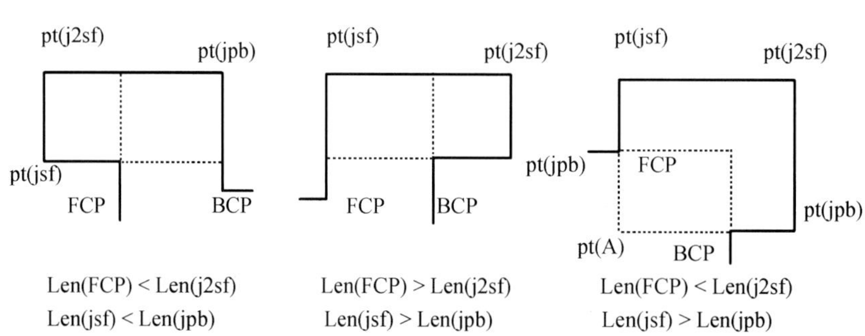

Figure 10 n_R=3, three cases of DL depending on the comparison between Len(FCP) and Len(j2sf) and the comparison between Len(jsf) and Len(jpb)

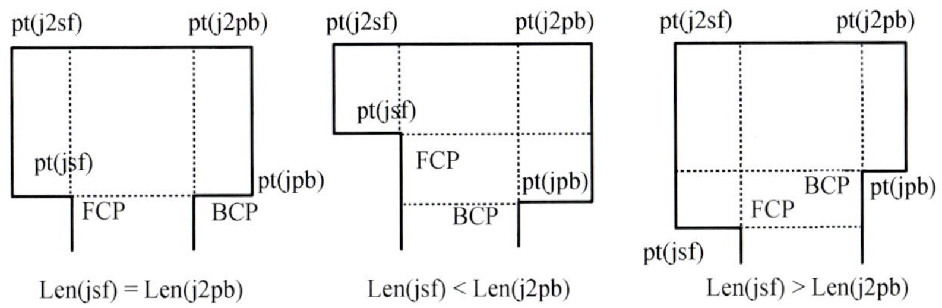

Figure 11 n_R=4: If Len(jsf)=Len(j2pb), then it forms rectangle; Otherwise, same as n_R=3

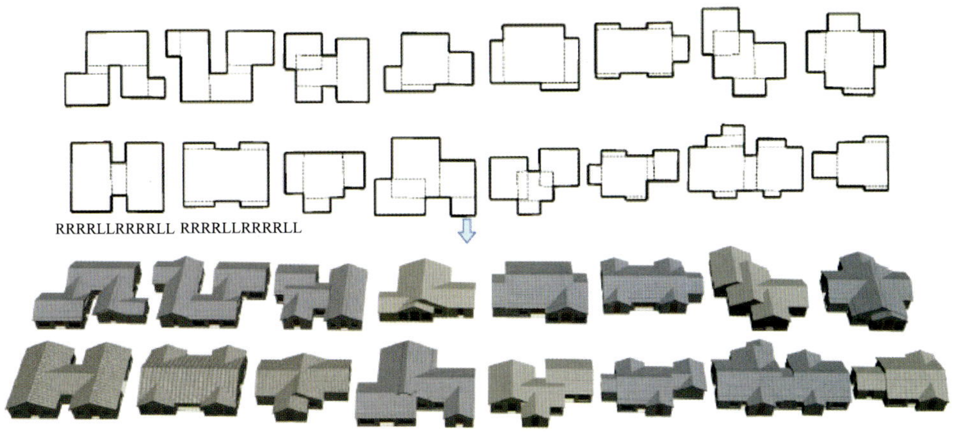

Figure 12 Complicated building orthogonal polygons with dividing lines and 3D building models automatically generated

mented and 3D building models automatically generated from partitioned building polygons. Two polygons at the lower left side of Figure 12 are expressed as the same "RRRRLLRRRRLL". But, shorter DLs are different in two polygons and different DLs are selected to cut off a rectangle by the proposed scheme. The rectangle partitioned is extended to a wider and higher main roof so that it will form a narrower and lower branch roof.

How to Create Roof Models

The integrated system partitions orthogonal building polygons into a set of rectangles and places rectangular roofs and box-shaped building bodies on these rectangles. The positioning of parts of a building is implemented in following steps. After measuring the length and the gradient of the edges of the partitioned rectangle, the edges are categorized into a long edge and a short edge. The vertices of the rectangle are numbered clockwise with the upper left vertex of a long edge being numbered 'pt1' as shown in Figure 13.

The control point ('cp') of a building part is placed at the point that divides edge12 and edge23 at a certain ratio. A 'cp' is different in each primitive. For example, a prism is used for the construction under roof boards. In 3ds Max which we use for the creation of 3D models, the 'cp' of a prism lies in one of the vertex of the base triangle in an upright position when a prism is newly created as shown in Figure 14. After rotating a prism by 90 degree around x-axis, the system rotates it around the z-axis by the degrees of the gradient of edge12. The top of a gable roof consists of two roof boards (two thin boxes). Since the 'cp' of a box lies in the center of the base, it is placed on the point that divides the line through $pt12$ and $pt34$ at the ratio shown in ground plan (Figure 13). The height of the 'cp's of two roof boards is shown in the front view of a gable roof (Figure 15).

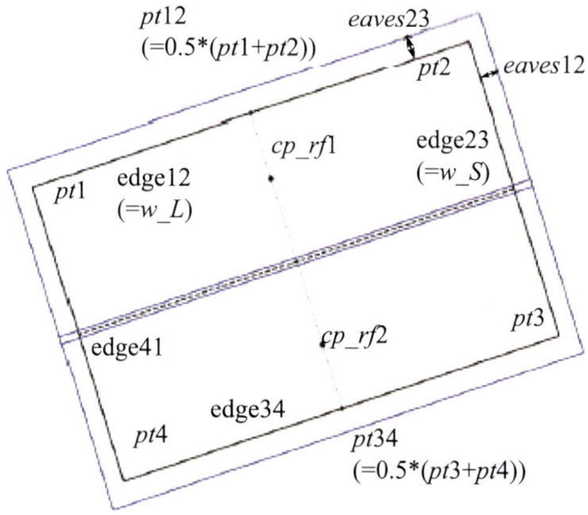

$$ratio\_s = 0.25 - \frac{0.5\times(eaves23\times\cos\theta + rf\_offs*\sin\theta)}{w\_S} + \frac{thick\_rf\times\sin\theta}{w\_S}$$

$$cp\_rf1 = (1.0 - ratio\_s)\times pt12 + ratio\_s \times pt34$$

$$cp\_rf2 = ratio\_s \times pt12 + (1.0 - ratio\_s)\times pt34$$

Here, $thick\_rf$ is a thickness of a roof board. $eaves23$ is the length of eaves in a direction of edge23, θ is an angle of a roof slope with a horizontal plane. rf_offs is the offset of a roof board from a prism as shown in Figure 15.

Figure 13 Ground plan of a gable roof and explanation of parameters of a gable roof

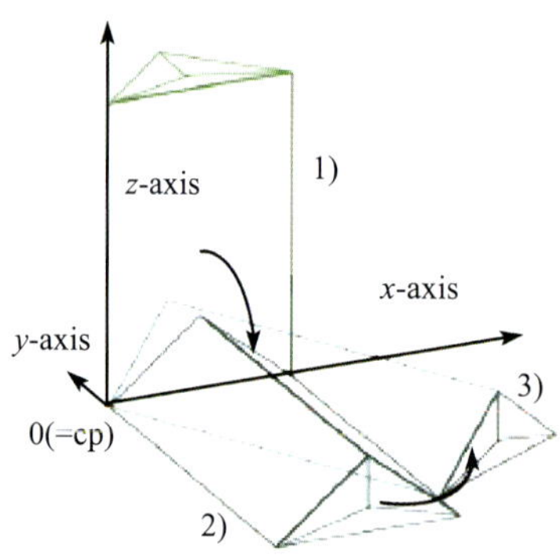

1) Upright position when newly created.
2) Rotate by 90 degree around x-axis.
3) Rotate the gradient of edge12 around z-axis.

Figure 14 How to place a prism

Also, the other complicated shapes of roofs, such as a temple roof, consist of many thin roof boards. Figures 20 and 21 show automatically generated 3D models with complicated shapes of roofs restoring a Japanese ancient temple and a pagoda. The 'cp's of thin roof boards consisting of a temple top roof are also placed on the point that divides the line through $pt12$ and $pt34$ at the ratio shown in the ground plan (Figure 16). The front view of a

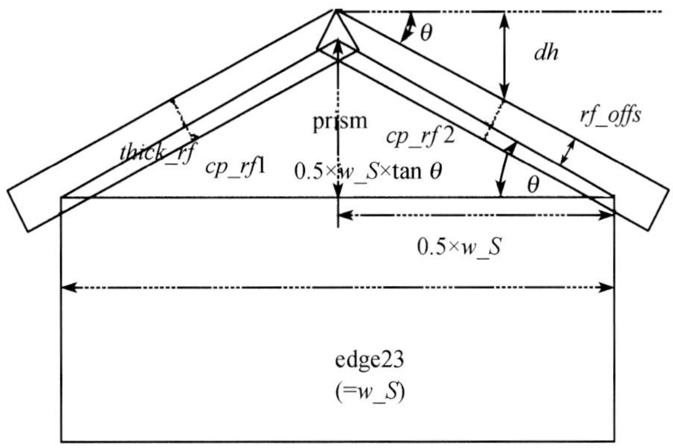

The width of a roof board is as follows.

$wid\_rfb = side23L + eaves23 + rf\_offs \times \tan\theta$

Here, $side23L = 0.5 \times w\_S \times \sqrt{1+\tan^2\theta}$

The height of a roof board is as follows.

$hei\_rf = st\_heit - 0.5 \times (side23L + eaves23 + rf\_offs \times \tan\theta) \times \sin\theta$

$-thick\_rf \times \cos\theta + rf\_offs/\cos\theta + 0.5 \times \tan\theta \times w\_S$

Here, st_heit is start height as follows.

st_heit = (floor-to-floor height) × (the number of stories)

Figure 15 Front view of a gable roof and explanation of parameters of a gable roof

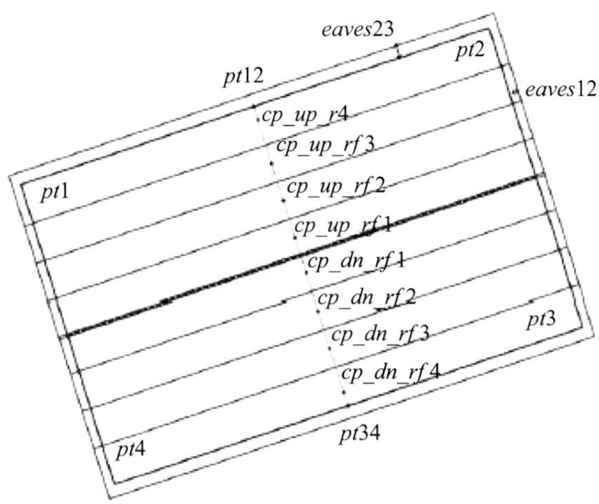

$$ratio(i) = \frac{n\_div + i}{2 \times n\_div} - \frac{0.5 \times (eaves23 \times \cos\theta i + rf\_offs * \sin\theta i)}{w\_S} + \frac{thick\_rf \times \sin\theta i}{w\_S}$$

$cp\_dn\_rf(i) = (1.0 - ratio(i)) \times pt12 + ratio(i) \times pt34$

$cp\_up\_rf(i) = (1.0 - ratio(i)) \times pt34 + ratio(i) \times pt12$

Here, n_div is the number of thin roof boards into which a roof is divided into.

$cp\_dn\_rf(i)$ and $cp\_up\_rf(i)$ are i-th control points of thin roof boards.

Figure 16 Ground plan of a temple top roof

temple top roof (Figure 17) reveals the height of the 'cp's of thin roof boards.

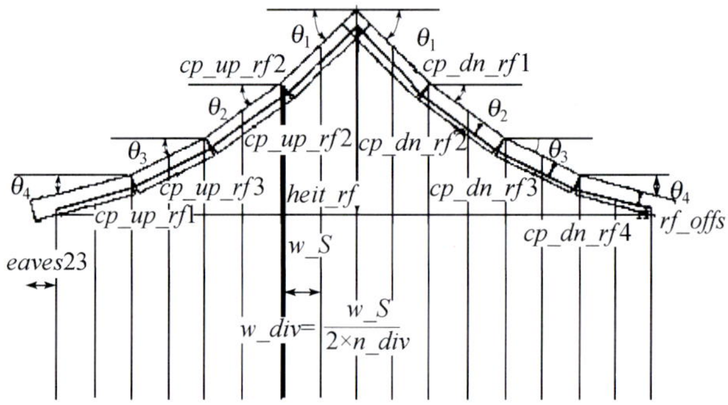

The gradient of each thin roof boards θ_i is

$$\theta_i = \arctan\frac{heit\_rf \times (rat\_heit(i) - rat\_heit(i-1))}{w\_S \div n\_div}$$

Here, *rat_heit*(*i*) is the *i*-th ratio to 'heit_rf' that is given by sampling catenary at a constant interval.

The height of *i*-th roof board (*hei_rfb*(*i*): *i*=1 to 3) is as follows. *i*=4; the width of the roof board is longer by *eaves*23.

$heit\_rfb(i)=0.5\times heit\_rf\times(rat\_heit(i)-rat\_heit(i-1))$
$+(rf\_offs-thick\_rf)\times\cos\theta_i$

$heit\_rfb(4)=heit\_rf\times rat\_heit(4)-0.5\times(w\_div\div\cos\theta_4+eaves23)$
$\times\sin\theta_4+(rf\_offs-thick\_rf)\times\cos\theta_4$

Figure 17　Front view of a temple top roof

As shown in Figure 18, a hipped roof is a type of roof with two triangular sides and two trapezoidal ones downwards to the walls. While the 'cp' of the trapezoid is placed on the point that divides the line through *pt*12 and *pt*34, the 'cp' of the triangle is placed on the point that divides the line through *pt*23 and *pt*41.

Figure 18　An automatically generated 3D urban model
(4 buildings with a hipped roof stand along a road)

Application and Conclusions

For residents, citizens or even students as well as the specialist of urban planning, a 3D urban model is effective in understanding what will be built, what the image of the town might be or what is the alternative plan to be realized. This model can act as a simulator to realize alternative ideas of urban planning virtually. Here are some examples of 3D models created by the automatic generation system. Figure 19 shows a 3D urban model simulating a typical Japanese town where a series of electric poles are standing along roads and the height of the buildings are not regulated. Figures 20 and 21 show 3D models restoring ancient temple called Mino Kokubunji and Kokubun-niji.

Figure 19 A 3D urban model simulating a typical Japanese town

Figure 20 An automatically generated 3D models restoring Japanese ancient temples called Mino Kokubunji

Excavation revealed that there used to be 7 and 5 story pagodas and a 2 story big gate, accommodations for monks, cereals storehouse surrounded by walls with small roofs. In this chapter, we proposed a new scheme for partitioning building polygons and show the process of creating a basic gable roof model. By applying a polygon partitioning algorithm, the system divides polygons along the thin parts of its branches. We also present 3D models for

Figure 21 An automatically generated 3D models restoring Japanese ancient gate and pagoda

restoring a Japanese ancient temple and a pagoda, based on the generation process of a basic gable roof.

Future work will be directed towards the development of methods for importing elevation data from DEM (digital elevation model) or some other source of remote sensing in stead of using the number of stories from GIS database, and then the 3D reconstruction algorithm to generate any objects in 3D urban model by using Computer Vision that defines the geometry of the correspondences between multiple stereo views and leads to 3D reconstruction.

References

[1] Sugihara K and Hayashi Y. Automatic Generation of 3-D Building Models From Building Polygons on GIS. In: Icccbe-Xi Proceedings of The 11th Icccbe. Montreal, Canada, 2006.

[2] Kenichi Sugihara. Generalized Building Polygon Partitioning for Automatic Generation of 3-D Building Models. ACM SIGGRAPH 2006, Poster Session, 2006.

[3] Müller P, Wonka P, Haegler S, Ulmer A, Andgool LV. Procedural Modeling of Buildings. ACM Transactions on Graphics, 2006, 25: 614–623.

[4] Parish Y I H and Müller P. Procedural Modeling of Cities. In: Proceedings of ACM SIGGRAPH 2001. New York: ACM Press, 2001: 301–308.

[5] Aichholzer O, Aurenhammer F, Alberts D, Gärtner B. A Novel Type of Skeleton for Polygons. Journal of Universal Computer Science, 1995, 1: 752–761.

[6] Aichholzer O and Aurenhammer F. Straight Skeletons for General Polygonal Figures in the Plane. In: Proceedings of the 2nd Annual International Conference on Computing and Combinatorics (Co-Coon' 96), 1996: 117–126.

[7] Laycock R G and Day A M. Automatically Generating Roof Models from Building Footprints. http://wscg.zcu.cz/WSCG2003/Papers_2003/G67.pdf [2003].

Part 3

User Interfaces, Public Participation and GeoVisualization

The user interface is a key component of a VGE in that such environments are primarily designed so that users can engage in interactive manipulation of the geographic content, be it for analytical professional purposes or for simply information that pertains to some more general interest. In short, user interfaces dominate VGEs. These range from the development of appropriate methods for embodying geographic knowledge in such a way that the analysis is better informed all the way through to more generic public participation in which users need to know content about specific problems that they are affected by. There are many varieties of user interface and in this part of the book, we examine two sides of this interface coin: the design of environments which extend the realism of the problem for the analyst, for the expert, and the design of environments where the prime use is by non-experts, by the public-at-large. Many of these issues of interface design which pertain to realism and context as well as access and interaction cut across the spectrum from the expert to the non-expert. But progress in this area comes from many sources and we will begin with three contributions that focus on the individual user's experience and they way such environments can be designed to extend our ability to interact with geographic data. Much of this work comes from detailed design within map-making and cartography although like many of the contributions in this book, there is a merging of ideas from many different sources.

In terms of developing appropriate interfaces, the objects of geographic study need to be considered with respect to how users interacting with them can perform various actions or inter-actions. This is the property of affordance as it has been developed in psychology and it is central to exploring ways in which a user makes an action which is reflective of the use in question. Context is also central to such design and information that establishes context is all important in developing good user interfaces. Once the tools are in place in spatial analysis within GIS, the task is to develop context adaptive geoinformation that responds to users in different situations. In fact many of these ideas are being implemented not on the internet or desktop but in terms of mobile displays which represents the cutting edge with respect to how maps are being fashioned for display in contexts that are difficult and highly variable. These kinds of interface can be established using various rule sets which merge the purpose of use with various spatial analytics in such a way that users find the experience salutary. The great challenge in this area is to improve human-computer interaction in such as way that this leads to better decision making and participation.

Meng first introduces the basic ideas that enable designers of software to produce good interactions with VGEs in terms of the best interfaces. From a technical point of view, it is possible for a visualization designer to embed as much information as possible in an open-ended VGE which allows interactive query and manipulation. However, the design practice hitherto has been more or less blind. In order to raise the awareness of this problem, she re-visits two concepts—affordance and reflex level which were separately developed for product design and image understanding. Affordance indicates the actionable properties of a thing, whereas the reflex level characterizes the mental processes in a human brain. The synergetic effect of these two concepts is elucidated and analyzed with examples from graphic user interfaces and visualization patterns in VGEs. With the objective of maximizing the information communication at minimal cognitive cost, she suggests an affordance-driven approach to geovisualization, and hence to virtual environments in general.

Cai on the other hand deals with context. He argues that the geovisualization research should go beyond the development of geovisual analytic toolbox, and pay attention to supporting human work across a broad activity scale. Future generations of VGEs are expected to be capable of monitoring and recognizing context changes, reasoning about implications, and adjusting their service behavior accordingly. This work advances the science of context-adaptive geoinformation services by suggesting an activity centered framework for structuring a diverse set of context factors which are given real form in mobile environments. Applications of this work to mobile GIS are then discussed using the example of designing a tourist guide.

Jahnke and his colleagues take the question of realism in VGEs on mobile devices much further. They suggest that the increasing amount of geographic data in urban environments and the technological advances in the domain of mobile devices have stimulated cartographers to try to improve the 3D visualization of urban information on small mobile displays. Their chapter is focused on one of the fundamental research questions: how can we design a mobile 3D virtual urban environment that allows its users cognitively adequate information processing? The prevailing photorealistic visualization techniques reveal some drawbacks with regard to mobile usage context. They argue that the non-photorealistic rendering (NPR) approach with its capability to unite the advantages of photorealistic presentation style and abstract 2D maps can better satisfy the design constraints for mobile applications. Starting from an outline of the development trends in virtual urban environments, they provide an overview of 3D NPR techniques along with the potential of 3D information communication to mobile users. Based on a general analysis of usability issues, they propose "cognitive adequacy" as the usability measure for the NPR approach. A set of applicable graphic variables and design options are presented by way of first results and in further stages of their work, two extensive virtual urban environments are to be implemented with NPR approach in accordance with the OGC format CityGML.

These contributions then shift to ways in which integrated VGEs can be designed for public participation. First Shen and his colleagues present their online design review tool which is used for gaining consensus on townscape rules among stakeholders. Their tools for modeling a 3D townscape by visualizing design guidelines using the virtual reality modeling language (VRML) technology has been developed as an online system. Using this tool, users across different locations can come together to create a virtual town by presenting, sharing and selecting different design elements, thereby enabling them to reach a consensus on the design guidelines of the townscape. They introduce a design-decision support system along with a visualization tool for members of a local planning committee to share common images about an entire townscape while also considering different design alternatives of the buildings to be reconstructed. Members can use the tool to check if the proposed plan and design conforms to the agreed townscape rules. Supporting a long-term planning committee for design review in an urban district, the tool is developed as an auxiliary measure to ensure that the townscape rules are followed where each reconstruction plan composed of different parcels is under review. Meanwhile, the residents of a town can use the online tool to access design reviews without any spatial or temporal limitations.

In the last chapter in this part, Portugali and his team outline the project O'Jerusalem which is a decision support system specifically designed for the city of Jerusalem and the region around it. Its aim is to provide decision makers and public discourse with accurate information in their debate over the fence/wall between Israel and the Palestinian territories in the West Bank. This is a controversial but a key issue that relates to the realism of the proposal. Many of the contributions to this book reflect the issues broached by Portugali and his team in this chapter. However specifically, they deal with two issues that need clarification at the outset: the first concerns the specific political situation related to the fence/wall while the second issue concerns the theory of cognitive maps. This chapter, although highly applied, points the way back to theory, tying these ideas about the most appropriate user interfaces for such participation back to psychological issues, to affordances and to the general context in which such participation takes place.

Chapter 10 Affordance and Reflex Level of Geovisualization

Liqiu Meng

Introduction

The growing availability of seamless digital worlds has largely facilitated the boom in virtual geographic environments (VGEs). Regardless of type and scope of a VGE, spatial modeling, geovisualization and human vision remain as its fundamental components. Spatial modeling is a necessary process to scale down the reality into an easy-to-handle digital model in favor of information storage, distribution, maintenance and retrieval. Human vision, on the other hand, relies on the collaboration of sensory-motor organs of a human body to attend to physical stimuli, pick up the invariants characterizing these stimuli and eventually associate them with meanings. The inherent gap between spatial modeling and human vision can be bridged by geovisualization through which the "lifeless" alphanumerical codes in the digital model are converted into visible, audible, or touchable stimuli. Geoinformation can thus intuitively flow into the human brain. As a constructive process, geovisualization can not and does not have to be a one-to-one graphic transcription of the digital model.

In current VGEs, geovisualization typically performs a two-fold task: 1) to create a display surface onto which a symbolized subset of digital geographic concepts are projected, and 2) to provide an interface for the user's access to secondary information that is either embedded in or related to the graphic symbols. In the former case, the communication of geoinformation starts with eye contact, *i.e.* seeing. The comprehension of what has been visually attended relies on an agreed visual literacy between the designer and the user. Developing such a visual literacy has been always a research focus of cartographic semiotics. In the latter case, however, geoinformation is essentially communicated and comprehended by means of user interactions which require seeing and doing at the same time.

The continuously evolving multimedia technologies and the steadily increasing computing performance during recent decades have led to a dramatic expansion of design freedom of geovisualization. Not only has the number of elementary graphic variables defined by Bertin[1] more than tripled[2–4], but many design patterns which are dedicated to satisfying constraints concerned with server-client architecture, storage capacity, rendering speed, display size, user's memory load *etc.* have been implemented[5]. In addition, the fact that a limited display surface can only accommodate a limited number of recognizable graphic symbols has forced the designer to adopt the "Swiss-knife" principle to hide much information beneath the display surface of individual graphic symbols. Apart from ethic concerns, without a smart information-hiding strategy the global platforms such as *Google Maps* and *Microsoft Live Search Maps* which allow the geotagging of micro-blogging services and photo-sharing services like *Flickr* may become too cluttered to be useful (as for example in Figure 1). However, information-hiding techniques need to be coupled with information -revealing techniques.

Figure 1 An information-rich but cluttered Google Map with geotagged photos and Wikipedia comments (©2008 Google and ©2008 Tele Atlas)

The Concepts of Affordance and Reflex Depth

Being aware of the typical problems an average user faces when using a complex VGE, we need to take a closer look at the nature of the graphic user interface and graphic symbols from the theoretical perspective of affordance and reflex level.

Affordance

The term affordance was initiated by Gibson[6, 7]. According to Gibson, an affordance is an actionable property of a thing (usually a tangible thing). As a part of nature, the affordance is objectively measurable and independent of the individual's ability to recognize it. It does not have to be visible, known or desirable. A thing can have many affordances. For instance, a graphic display surface affords to see, touch and further actions such as carry, damage *etc*. However, from a pragmatic point of view, an affordance makes sense only when the individual can be aware of it and act on it, depending on his capability and environmental constraints. Since the awareness of an affordance (perception and interpretation) is a prerequisite of visually guided behavior, Norman[8, 9] introduced an elaborated concept which he called "perceived affordance" in the context of product design. He believed that the designer cares much more about what the user perceives than what is actually true. "*A computer system, with its keyboard, display screen, pointing device* (*e.g., mouse*) *and selection buttons affords pointing, touching, looking, and clicking on every pixel of the display screen. Most of this affordance is of no value. Thus, if the display does not have a touch-sensitive screen, the screen still affords touching, but it has no result on the computer system*"[8].

Obviously, not all the real affordances are meaningful for the user. Moreover, the user may perceive wrong affordances. Therefore, the designer generally faces two challenges. On the one hand, he has to highlight meaningful affordances and inhibit irrelevant ones as far as possible. On the other hand, he needs a good insight into the relationship between each meaningful affordance and the user's eye-mind-body characteristics in order to ensure an efficient perception and correct interpretation of the affordance. In Figure 2(a), a user is interactively working with a screen display. Both affordances seeing and touching are meaningful. In Figure 2(b), however, the screen in a subway station affords to see but not touch because the display hangs too high for a human body to reach or because it is not allowed to cross the rail track between the screen and the observer. In addition, the perception efficiency and accuracy of an affordance is strongly dependent on how well the design fits the behavior pattern of the observer. It has been scientifically proved that there is a strong interplay between seeing and doing. A graspable object will draw more visual attention if it looks ready to be grasped[10]. This means, a good design allows not only a fast recognition of an affordance, but also makes its observer feel convenient to perform the action. Figure 3 shows two pictures of a chair from two different viewing angles. Obviously, the affordance of "sitting" is more apparent in the left one.

(a) (b)

Figure 2 (a) A screen display for both seeing and touching; (b) An information screen only for seeing

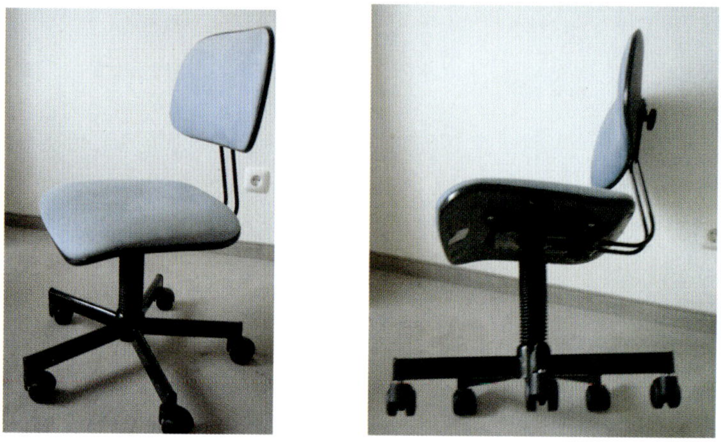

Figure 3 The left picture looks more comfortable for sitting than the right one

The affordance-driven development of a VGE is far more complex than these simple examples. Jordan *et al.*[11] reported one of the first applications. With the aim of improving the user-friendliness of a GIS, they modeled the places in the real world based on affordances instead of coordinates. This chapter tries to investigate the application potential of affordance in the interactive geovisualization process. In order to convey as much information as possible to the target user, the designer has to embed multiple interactive operations (affordances) in the graphics. However, being constrained by the limited display space, he can hardly make all the meaningful affordances simultaneously visible. Instead, he needs to define a ranking order for the affordances and then reveal them one after another. Whether such an order is suitably defined or not depends very much on how well it matches the reflex structure of the target user.

Reflex level

The concept of reflex level was explored by Rosner and Schmauks[12] in the context of image understanding. According to them, an image can present multiple meanings at one or more reflex levels arranged in a tree-like structure. Upon seeing an image, the observer will more or less unconsciously associate it with one or more primary meanings (first reflex level). If he continues to think or attend to the image, he may discover new meanings related to primary ones (second reflex level). In a similar way, a third reflex level may be triggered after a more intensive cognition. The cognition process stops when no further meanings can be discovered. The reflex depth of an image can thus be defined as the number of reflex levels. The number of meanings at each level, also termed as width, can be regarded as a measure of ambiguity. An example of such a tree-like reflex structure is illustrated in Figure 4. The relative depth of a special meaning is proportional to its mental distance to the observer. The shorter the mental distance, the less effortful a meaning occurs to the mind of the observer. Depending on the culture background, knowledge profile, gender, habit and many other factors, different reflex depths and widths of the same image can be revealed by different observers. According to Ellis[13], "*Seeing is an active, constructive process utilizing considerable non-visual information, some of which is in a Kantian sense a priori knowledge. Our knowledge of our environment provides a filter through which we see it.*" For this reason, a successful visualization must be based on an insight into the mental activities of its individual target users.

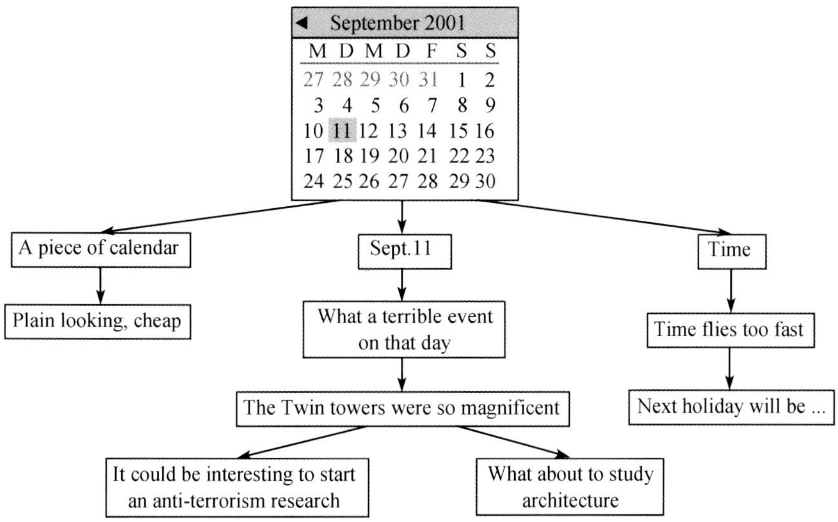

Figure 4 A reflex structure derived from a calendar picture on the top

In fine art, it is taken for granted that a valuable artwork is associated with multiple meanings and/or a large reflex depth, although what is interpreted does not necessarily correspond to what the artist has intended or anticipated. Misinterpretations could bring the artist more glory or more misfortune. The design of geovisualization, however, serves the purpose of conveying geoinformation in a controlled way. Mapping objects are systematically categorized or classified. Each category or class is then represented by a symbol. According to the motto "nothing should be left unexplained", a map legend is provided as an anchor point for the user to rely on. It guides the user to follow the designer's reflex structure. Since the primary meaning of each symbol is explicitly explained, different map users are either trimmed to the same and only reflex level or they start from the same lexical level to do further analysis and interpretation, *e.g.* identification of an individual object by associating its class symbol with its special location and name, estimation of distribution density of a certain object class, *etc*. This strategy works well for view-only maps because misinterpretations are kept within a pre-defined frame. However, in a VGE which is characterized by a highly interactive user interface, multimedia / hypermedia geovisualization and one or many underlying spatial databases, a map legend alone is no longer adequate because the perception task has been extended to include both seeing and doing. The interactions result in a mental map that contains not only a sensory part, but also a motor part[14]. This means, more advanced design methods are necessary to integrate affordances with reflex depths in a way that enables an easy detection of not only what kind of meanings the individual symbols bear, but also what kind of interactive operations are applicable on them and how they work.

Affordance and Reflex Level in Virtual Geographic Environments

When a VGE is started, the user typically sees a graphic user interface with help of which he can then bring a study area into view. Interactive tools are grouped into different folders based on their functional similarity. For example, tools for viewing, selecting and editing belong to three different folders. They are either simultaneously visible in menu bars or hierarchically accessible based on their calling frequency or relevance to a particular application. A legend of the visualized study area is not always presented due to a number of reasons. First, being different from printed maps where the main contents are framed within a sheet line, a VGE is seamless no matter how large the underlying real world can be. The screen space can hardly accommodate a complete explanation of all symbols. Second, many of the symbols explained in the legend would not appear in the subset within the viewing port. Third, objects of interest are often uniquely symbolized in the form of self-explaining infographics, pictograms, silhouettes *etc.* (see an example in Figure 5). It is neither possible nor necessary to explain them individually in the legend. Due to a missing legend, the user faces a high cognitive load to decide what to see and to do next, even if he has sufficient interaction experiences with computer systems.

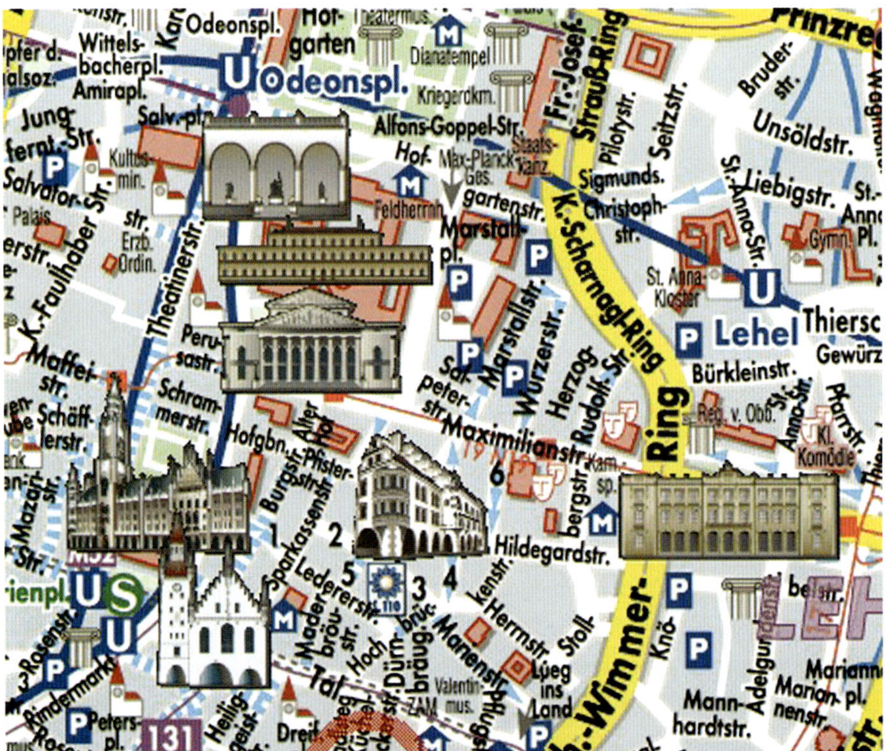

Figure 5 A viewing port showing the individuality of objects of interest (©Euro-Cities AG)

As illustrated in Figure 6, the visual stimuli of a VGE are mainly composed of two categories: 1) Small-sized icons arranged in a menu bar or toolkits superimposed on the display surface; and 2) Graphic symbols that represent geo-objects at different spatio-temporal scales. The first category contains the individual operations that have an impact on graphic symbols as well as their underlying geo-objects. Different user groups may have different access rights. That means each particular user may access a subset of all existing operations. The second category is connected with qualitative or quantitative meanings which can be directly retrieved or manipulated by means of operations from the first category.

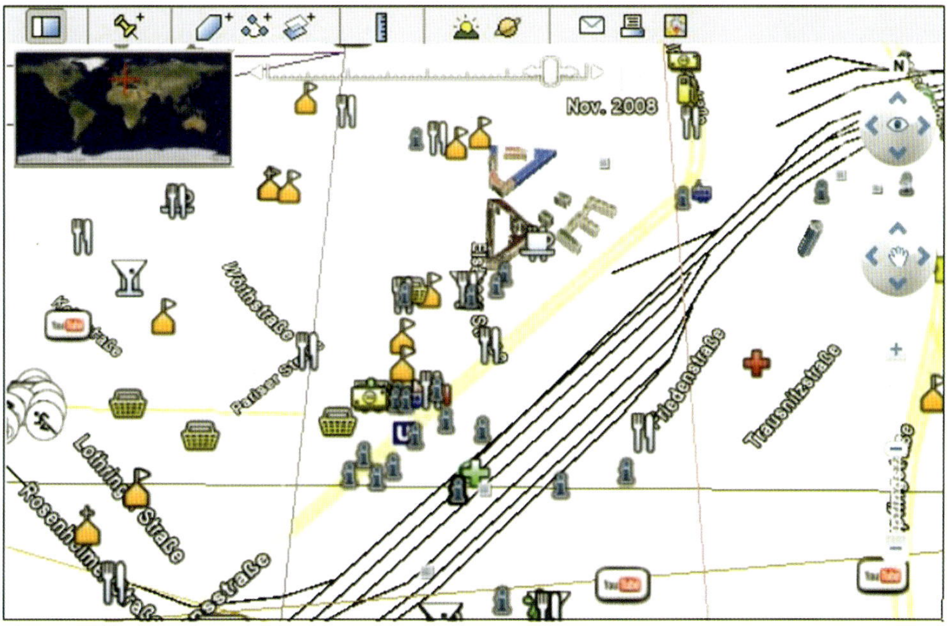

Figure 6 Visual stimuli in Google Earth V.4.3

Graphic user interface

Depending on the working state and the access rights of a particular user, a visualization system usually distinguishes applicable from inapplicable operations, *e.g.* by making their icons in dark grey and light grey respectively. In the case of small display devices, inapplicable operations may be removed from the menu bar. Since each operation corresponds to a dynamic procedure, its icon can metaphorically infer the form of a tool, the object on which the tool can be used or the states before and after the operation. Although standardization is very much desired by users for the purpose of minimizing the average learning effort and enhancing the interoperability among different systems, showing the identity of the system vendor is still a prevailing driving force in the practice of interface design. In other words, the designer expects the user to recognize the system vendor at first upon seeing the graphic interface and then begins to explore what the individual icons afford. Figures 7 and demonstrate some examples of vendor identities indicated by graphic user interfaces.

(a)

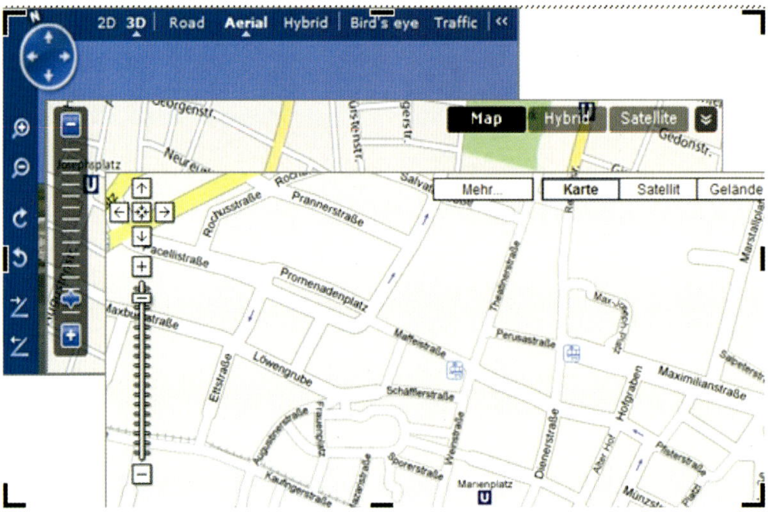

(b)

Figure 7 Vendor identities of (a) Commercial GIS—geomedia, MapInfo and arcview; (b) Map browsers—microsoft live search maps, Yahoo maps and google maps

As a result, it is the user who has to take the responsibility for performing cognitively burdensome tasks such as:

1. Perceive and interpret the meaning, *i.e.* the action indicated by each icon. The same action may be indicated by different icons in different systems.
2. Perceive and interpret multiple modes of an action. Many systems provide multimodal interactions, *e.g.* key stroke, voice, pointing device for the same input. This helps raise robustness because the system functionality may remain intact even if one of the

modes is defective. The user, however, needs to know what equivalent ways are connected to an action and how to switch between them.

3. Interpret the validation scope of an action. An action can be applied to a whole content layer (*e.g.* zooming), a window of a content layer (*e.g.* changing the display color), a subset of graphic symbols (*e.g.* listing the attribute values of the underlying geo-objects), a pair of graphic symbols (*e.g.* calculation of the overlap between two neighboring polygons) or a single symbol (*e.g.* editing the outline of a polygon).
4. Interpret the effect of an action. An action may 1) bring up an information box describing a graphic symbol or a relation between graphic symbols, 2) lead to changes of graphic symbols in terms of location, appearance *etc.* and 3) cause changes of geo-objects in the database in terms of location, geometric, topological and semantic attributes as well as relations. The changes include the creation, movement or elimination of graphic symbols and/or geo-objects.
5. Interpret the execution procedure of an action. Like verbs in sentences, the actions embedded in the icons do not stand alone. Instead they have various impacts on graphic symbols. In relation with its validation scope, an icon (*e.g.* zooming), once activated (*e.g.* by a mouse click), can have its lasting effect on the subsequent user actions till the same icon is deactivated (*e.g.* by a second mouse click). Sometimes, an icon will be automatically deactivated as soon as its corresponding action has been accomplished (*e.g.* drawing a circle). In other cases, an icon cannot be activated until one or many graphic symbols have been selected (*e.g.* buffering of a line).

Without sufficient visual cues or guides, icons are often difficult to interpret and remember. For example, a novice user may notice the two arrow signs in the upper left and upper right corner of the image map shown in Figure 8(a), but probably without knowing their meanings. If he sees Figure 8(b) at first instead, he may be able to imagine the effects of some arrow signs based on the available context information. Yet, it is still difficult for him to understand what the two capital letters K and H are about in the inset map and the meanings of other arrow signs. In complex VGEs which may involve a lot more user interactions and hardware devices, those visually obscure icons can be particularly error prone if they deviate too much from the user's habits and memory capacity. In order to keep pace with the cutting-edge technology, software vendors have been constantly upgrading their systems by inserting the newest functionality that can ever be realized on the high-end platform. These systems are potentially more powerful in terms of supporting the user to do more things. However, the upgrading would be in vain if many users are hardly aware of a substantial part of any accessible operations.

Graphic symbols

The graphic symbols in an interactive system can be subdivided into sensitive and non-sensitive symbols. The former category affords both to view and to click in various ways. To some extent, it can be regarded as synthesis of graphic symbols and interactions. The latter category is composed of view-only symbols. A salient graphic variable is required to make this distinction apparent to the target user. For example, it is hard to tell whether all the 3D building symbols shown in Figure 9 are sensitive or non-sensitive because they are designed using the same graphic variables. The distinction, however, can be well perceived in the design patterns demonstrated in Figure 10 where the sensitive symbols are highlighted using additional colors or depth cues. Obviously, the uplifted symbols in Figures 10(c)–(d) afford to click better than the 3D color symbols in Figure 10(a)–(b).

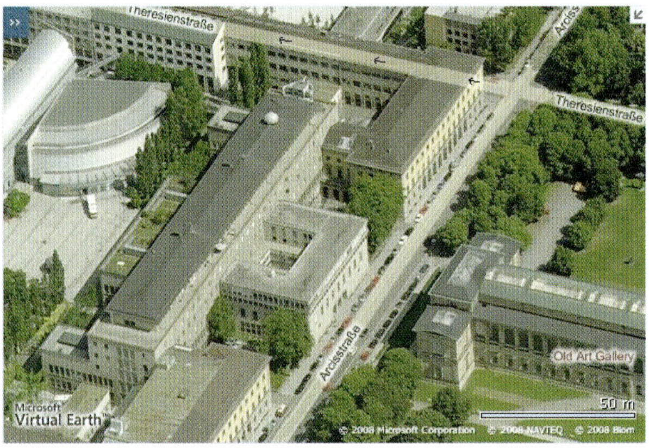

(a)

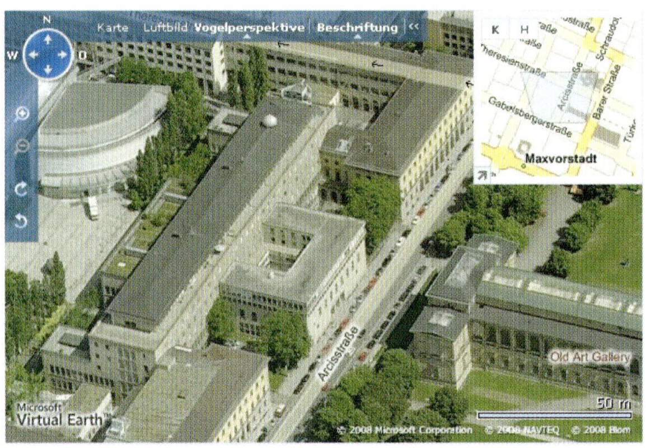

(b)

Figure 8 (a) Two different arrow signs in upper left and upper right corner of an image map in Microsoft Virtual Earth and (b) their effects intended by the designer

The design of view-only symbols is focused on the application of graphic variables and their combinations that can help embed multiple meanings and reflex levels in the symbols. The existing theories, methods and design patterns in traditional and multimedia semiotics have mostly considered the fundamental rules in Gestalt psychology. They provide a useful guide for the symbolization and understanding of geo-objects in nominal, ordinal, interval and ratio scales. Similar to the design of multimodal interactions, redundant variables can be applied to ensure the efficient recognition of significant variations in the data. As illustrated in Figure 11(a), both color hue and text are used to represent red, orange and green metro

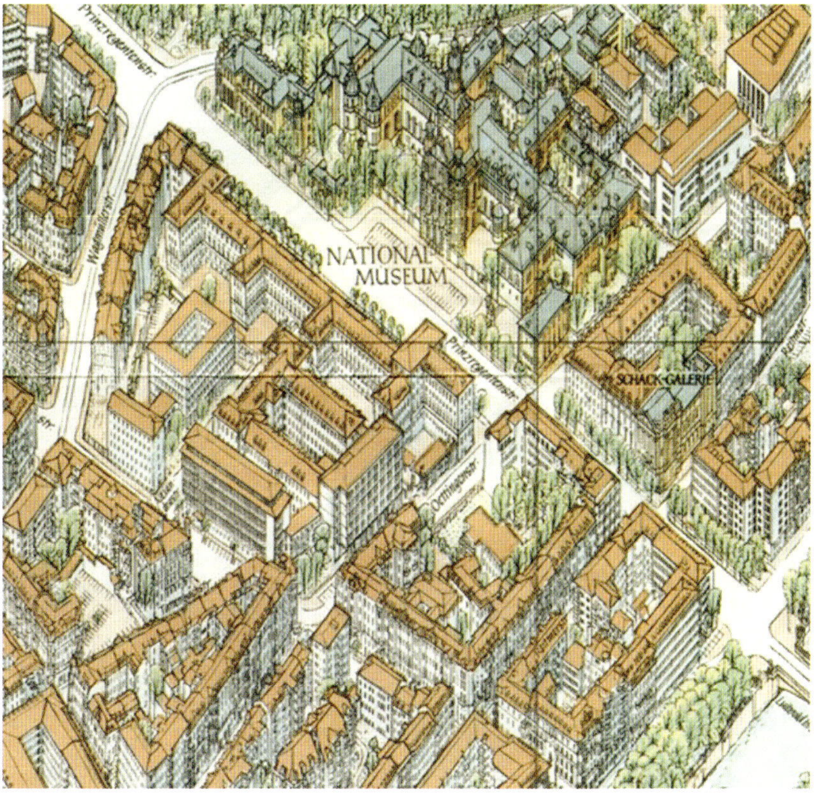

Figure 9 A section of Bollmann plan of Munich

lines in Washington DC. Though a colorblind user may not be able to perceive the variation in color hue, he may very well notice the text. The variable "shadow" of various polygon symbols in Figure 11(b) and the illustrative strokes within the circle symbols in Figure 11(c) contribute much to the attractiveness of the map, although they do not carry extra information.

The theories and methods for sensitive symbol design are still under development. The degree of sensitivity varies along a continuum from very insensitive to very sensitive. View-only symbols represent one extreme. At the other extreme, symbols are self-active and can be integrated in an agent-based VGE. After having collected sufficient evidence of user behavior (*e.g.* visited locations in the virtual environment), the agent is able to predict the next user actions and visually prompt the user to follow (see examples in Figures 12). Between the two extremes, a sensitive symbol is more or less responsive to user action. Generally speaking, interactive tools can help reveal the reflex levels of graphic symbols and support the user to navigate across geo-spatial dimensions, thematic layers and the scale space. Accordingly, the user expects sensitive symbols to incorporate both interaction operations (procedural information) and meanings of underlying geo-objects. For example, a sensitive symbol that represents a river should "invite" the user to associate it with a prototypical river object and to perform the query and analytical operations on it. If it additionally represents a

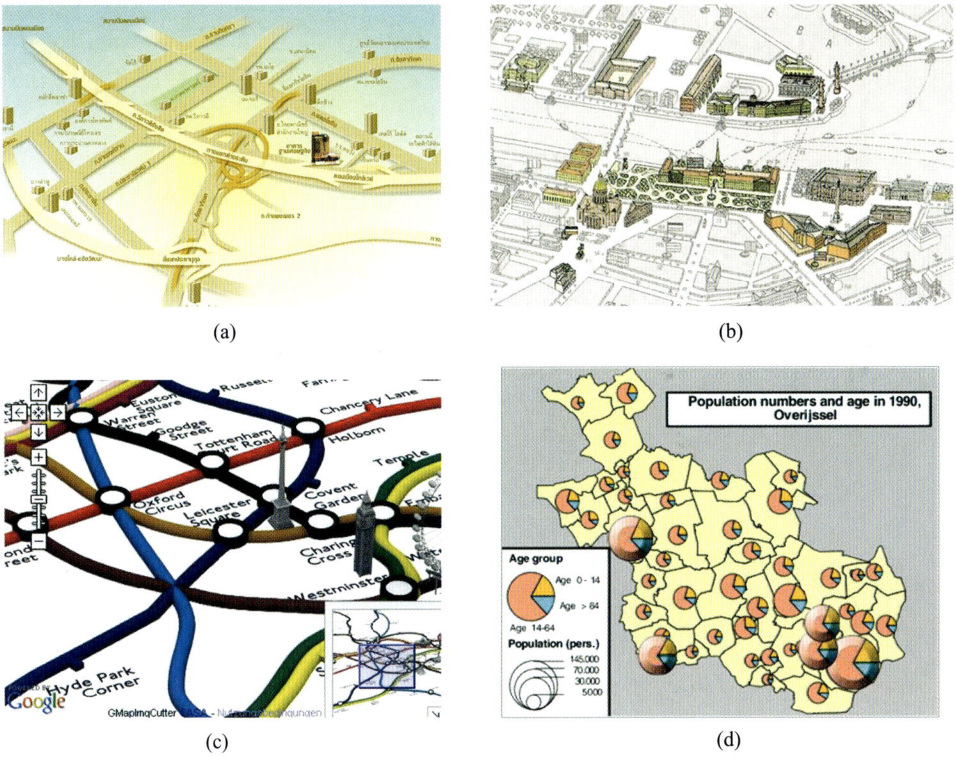

Figure 10 Sensitive symbols are highlighted with color in (a) Thansettakij Banquet in Bangkok[1], and (b) St. Petersburg architecture[2]; Sensitive symbols rise above the non-sensitive symbols in (c) London tube[3] and (d) A thematic map[4]

flood event, it should "invite" the user to follow the trajectory of the dynamic process. Further, if it also serves as a hyperlink, it should "invite" the user to click on it.

Figure 13 demonstrates some easily perceivable affordances such as clicking, rotating and scrolling. The following direct affordances on a sensitive symbol are widespread for computer literates:

- Touch or move a pointing device to its location (*e.g.* mouse-over effect)
- A single click
- A double click
- A short press
- A long press
- Press and hold
- Press and rotate
- Drag and drop

1) www.thannews.th.com/Banquets/img/MAP.jpg.
2) www.tcaup.umich.edu/stpetersburg/mapindex.html.
3) http://londonist.com/2007/08/3d_london_tube.php.
4) www.kartografie.nl/webcartography/webbook/ch07.

(a)

(b) (c)

Figure 11 (a) Redundant variables “color hue” and “text” in a section of Metrorail System Map of Washington DC[1]; (b) Variable “shadow” in a section of the city plan by ©Euro-Cities AG; (c) Illustrative strokes in circle symbols in a section of wall map at Moscow State University of Geodesy and Cartography (accessed 2007)

Often a direct affordance can either trigger further actions or reveal further meanings. For instance, a mouse click on a graphic symbol may lead to the adjustment of menu bar on

1) www.wmata.com/metrorail/systemmap.cfm.

Figure 12 Active arrows prompting users to follow in a car navigation system1) (left) and a pedestrian navigation system[15] (right)

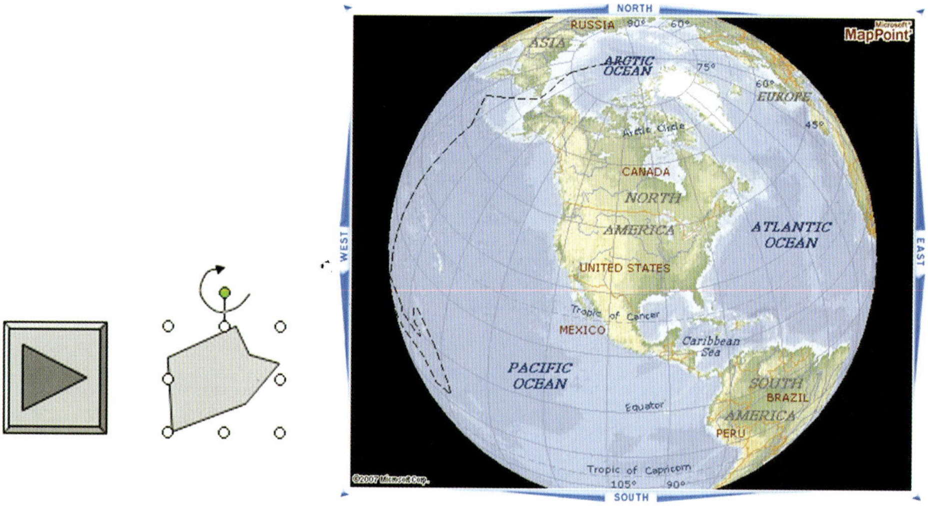

Figure 13 Symbols that afford to click (left), rotate (middle) and scroll (right)

which the applicable operations are highlighted or newly ranked, while other operations are either removed or they retreat into background. Similarly, a mouse click can lead to a new website, a new medium (*e.g.* video clip, photo) or a blown-up information box.

Ideally, at least one graphic variable must be reserved for each affordance or meaning so as to avoid confusions during interpretation. In current design practice, this constraint can hardly be satisfied. The "Swiss-knife" principle mentioned previously is associated with a dichotomized strategy. On the one hand, a sensitive symbol should bear the apparent meanings of the underlying geographic concept by virtue of graphic variables. On the other hand, it should provide its target users with subtle cues to trigger the applicable operations and/or (hyper) links, although it is technically possible to add to a display surface not only sensitivity, but also "thickness". In many cases, however, the designer does not bother to add an extra cue to make sensitive symbols outstanding, or he adds only one single cue to indicate

1) http://www.tomtom.com/.

all sorts of differences. In some systems, various mouse buttons are assigned to various operations. Many other systems, however, do not provide such distinctions. The same button may be used for a lot of different operations. On the other hand, the user is overwhelmed by interactive tools with which he is expected to see, see through and see beyond the graphic symbols whenever he starts the visualization system. Nevertheless, it is far from being straightforward for him to discover, either visually or with a mouse head as an extended arm, the cues pointing to the hidden information. Moreover, what would exactly happen after a press or mouse click on a symbol is often unpredictable. This makes the user feel uncertain. For instance, the user may hesitate to press the left symbol in Figure 13 if he is not sure whether he may trigger an action "go to next page", "play a video clip" or something else. As the result of these kinds of unpredictability, some useful affordances remain unused.

Discussion

Technical evolution has progressively increased the freedom for design and interactive information communication. However, the cognitive capacities of human users remain more or less the same as before while the sizes of the display spaces have not shown much variation. In a 2D or 3D VGE, each graphic symbol may bear not only meanings about itself and its underlying geo-objects, but also affordances of various sorts. This requires an extension of human perception from seeing to seeing and doing. The reflex structure of the human mind has to accommodate both the declarative information of what the individual symbols represent and the procedural information of how they work. For this reason, the designer faces two fundamental challenges of visualizing multiple meanings and affordances: How to couple the Swiss-knife principle for information-hiding with information-revealing techniques and how to match his design solutions with user's reflex structures.

By introducing the synergetic effect of affordance and reflex level, we have presented four themes based on: 1) raising the awareness of the blindness in technique-driven design of geovisualization; 2) providing a starting point for a more comprehensive study of user's mental processes during the interaction with a VGE; 3) appealing for sensitive symbol design that can intuitively "invite" the user to perform right actions; and 4) encouraging the research and development of affordance-driven design strategies that have potentials to improve the usability of geovisualization.

References

[1] Bertin J. Sémiologie Graphique: Les Diagrammes, Les Réseaux, Les Cartes. Paris, FR: Gauthier-Villars, 1967.

[2] MacEachren A M. How Maps Work—Representation, Visualization and Design. New York: The Guilford Press, 1995.

[3] Buziek G. Eine Konzeption Der Kartographischen Visualisierung. Hanover, DE: Habilitationsschrift,

Universität Hannover, 2001.

[4] Swienty O, Zhang M, Reichenbacher T, Meng L. Establishing a Neurocognition-Based Taxonomy of Graphical Variables for Attention-Guiding Geovisualisation. In: Proceedings of SPIE, Geoinformatics 2007. Nanking, 6751.

[5] Meng L. Egocentric Design of Map-Based Mobile Services. The Cartographic Journal, 2005, 42: 5–13.

[6] Gibson J J. The Theory of Affordances. In: R. Shaw, R. and Bransford J (Editors). Perceiving, Acting, and Knowing—Toward an Ecological Psychology. New York: Lawrence Erlbaum Associates, 1977.

[7] Gibson J J. The Ecological Approach to Visual Perception. New York: Lawrence Erlbaum Associates, 1979.

[8] Norman D. The Design of Everyday Things. New York: Basic Books, 1988.

[9] Norman D. Affordances, Conventions and Design. Interactions, 1999, 6: 38–43.

[10] Handy T C, Grafton S T, Shroff N M, Ketay S, Gazzaniga M S. Graspable Objects Grab Attention when the Potential for Action is Recognized. Nature Neuroscience, 2003, 6: 421–427.

[11] Jordan T, Raubal M, Gartrell B, Egenhofer M J. An Affordance-Based Model of Place in GIS. In: Poiker T and Chrisman N (Editors). 8th International Symposium on Spatial Data Handling, Vancouver, Canada, 1998: 98–109.

[12] Rosner R and Schmauks D. Die Reflektiertheit der Dinge und ihre Darstellung in Bildern. In: Homback K and Rehkämper K (Editors). Bild-Bildwahrnehmung—Bildverarbeitung. Sachs-Deutscher, Universitäts-Verlag, 2000: 15–31.

[13] Ellis S R. Pictorial Communication: Pictures and the Synthetic Universe. In: Kaiser M and Grunwald A J (Editors). Pictorial Communication in Virtual and Real Environments. London: Taylor & Francis, 1991.

[14] Bridgeman B. Separate Visual Representations for Perception and for Visually Guided Behaviour. In: Kaiser M and Grunwald A J (Editors). Pictorial Communication in Virtual and Real Environments. London: Taylor & Francis, 1991.

[15] Kolbe T. Augmented Videos and Panoramas for Pedestrian Navigation. In: Proceedings of the 2nd Symposium on Location-Based Services and Telecartography. TU Wien, Vienna, Austria, 2004.

Chapter 11 Modeling Contextual Knowledge for Adaptive Geographic Visualization*

Guoray Cai

Introduction

Geographic visualization (or Geovisualization) has played a key role in developing spatially-oriented virtual environments, particularly for stimulating visual thinking about geospatial patterns, relationships, and trends[1]. It is also one of the most active research areas within geographical information science (GI Science), dealing with challenges in human-GIS interactions, exploratory data analysis, and understanding human cognition[2]. Recently, geovisualization has gained new status as an enabling science and technology for supporting analytical reasoning with spatial data, and it has become part of visual analytics research[3, 4]. This is indicative of the fact that geovisualization is progressing into a science of action with methods and tools to support intelligence in decision- making.

As geovisualization is set to address the above new goals, there is a need to re-orient research efforts from traditionally data-centric perspectives towards more action-oriented and human-centered ones. In the recent past, significant advances were made in the development of theories, methods and tools for visually analyzing and exploring large spatial-temporal datasets. However, it is also clear to this scientific community that the problems of understanding complex social and environmental phenomena and their impact cannot be solved solely by maps[1]. We need interactive geovisual environments that can assist humans in the whole process of spatial decision-making where visualization and operational settings may be constantly in flux. However, existing interactive virtual environments are often designed for use in fixed environments for a single, well- understood task.

Due to increasing mobility and multi-tasking in human activities, the assumption of static and fixed context is no longer valid. The lack of awareness of the context (outside of the maps themselves) that users bring to the geovisual environment is a general problem that can partially explain the slowness of adopting geovisualization tools by many user communities[5, 6]. For geovirtual environments to be effective, designers must pay attention to the context in which maps were created, manipulated, and interpreted. The system should explicitly recognize and reason about the dynamics of context within which geovisualization tools operate. By incorporating context and context-adaptation in the core of a geovirtual environment, we hope to make such systems more responsive to changes in the operating

* Supported by the National Science Foundation (Grant No. EIA-0306845).

context as the user's activity unfolds.

As a step towards context-adaptive geographical information services, this chapter proposes to augment geovisualization environments with the ability to recognize, represent, and reason about contexts. Context adaptive services refer to those smart services that have the ability to make conscious adjustment of information delivery and presentation according to gathered knowledge about the context of use and its changes[7]. The fundamental difficulties of handling contexts in computer systems lie in two directions[8, 9]. First, contexts are *contingent* and *relative to the object of the main focus*. Second, contexts are *dynamic and emergent*. It is inherently difficult to differentiate contextual factors from non- contextual factors and to monitor changes. There has been a lack of workable models of context that can systematically answer the following three fundamental questions:

1. *How should a system understand and represent context*?
2. *If some aspects of the situation changed, how does the system decide if the change matters*?
3. *How does a system decide appropriate actions in response to the context change*?

The work in this chapter will focus on the theoretical aspects of context modeling for adaptive geovisualization, with an aim to address the above questions. The main ideas of this work can be summarized as follows. First, we believe that the way that a human activity is structured in a setting determines what factors of that setting are relevant. Therefore, the relationship between the geovisualization tools and its surrounding context must be understood through the structure of the activity to be supported. We use an activity as the fundamental, meaningful unit of analysis to understand the interacting relationships among human minds, artifacts, and the environment. Second, the target of the context-adaptation is to ensure the success of the entire activity as it unfolds in real settings. Therefore, context adaptation must explicitly reason about the dynamics of the ongoing activity. The most distinct aspect of this work is that adaptive behavior is achieved through late (run-time) binding of contexts to the ongoing activity according to how they contribute to the success of the activity. This is in direct contrast to traditional template-based approaches where behavior is determined by prescribed adaptation rules that directly bind one contextual factor with another.

Modeling Contexts of Geovirtual Environment

A model of context for a geovisual system refers to a systematic way of representing and reasoning about contextual factors and their effect on the behavior of the system. Context is a term that seems to be intuitive to everyone, but few can say clearly what exactly context is[10]. It is even trickier when we try to formalize it. Here we will provide a brief survey of relevant definitions and representational schemes, followed by our model of context.

Context

Given that contexts impact the behavior of the geovisual environment, the question arises naturally is: "what is context?" Most often, the definition is given in equally vague terms, such as "context is the current situation," "context is the setting," *etc.* Several workshops and conference series have generated extensive debate on this question, such as the *International and Interdisciplinary Conference on Modeling and Using Context* (context-07.ruc.dk). Unfortunately, researchers are far from reaching any agreement. A commonly adopted definition of context is "any information that can be used to characterize the situation of an entity"[11]. But such "all-inclusive" definition is too broad and foggy to be useful as a formal definition[12]. A more suitable definition was given by Chen and Kotz[13] who describe context as the set of situational states and settings that have determinate and predictive power over an applications' behavior. This definition allows us to focus on a subset of all possible factors. Others have tried to make the concept of context more concrete by differentiating several types of context (physical, technical, cognitive, social, *etc.*)[14]. We may have a better chance to make progress if we focus on a specific domain of applications. For the design of adaptive geovisualization systems, our definition of context is as follow:

Definition *A context is any identifiable configuration of physical, cognitive, social, and mission-related environment features that has predictive power for the behavior of geovisualization tools.*

This definition makes distinction between context and situation. A situation is an actual setting within which a tool is used, and potentially it can have a large (even infinite) number of identifiable features. Among all these features, there is a relatively small subset of features (and their configuration) that has impact on the behavior of the system. A context refers to a drastically reduced, relatively stable subset of features that are common to a large number of situations and that vary from one another along features of no significance. An aspect of the environment or situation becomes a relevant context when it defines (or helps to define) certain contextual states that have predictive power on how information services and presentations are provided.

Our definition emphasizes context as a special configuration of a set of features (rather than a simple list of features). In addition to the eight categories of contextual features (location, devices, purpose of use, time, physical surroundings and navigational history) defined by Sarjakoski and Nivala[15], our definition also includes cognitive features and social features as important dimensions of contexts. In particular, we argue that user's intention, cognitive representation of spatial problems, and beliefs on the intentions and capabilities of other collaborators strongly affect the choice of visual presentations and patterns of interactions.

Configuration of context factors

Context factors relate to each other in certain structures known as configuration. This configuration determines how changes in one of more context factors affect others. Every context-aware graphical information application has a model of how context factors are configured. Unfortunately, this is the least understood aspect of context, and applications often trivialize context to a list of attributes. There have been several well-known implementation efforts in the domain of mobile tour guides. Cyberguide[16] captures user's location and orientation (using GPS and infrared sensors). GUIDE[17] considered not only location but also the visitors' interests, the city's attractions, mobility constraints, available time, weather, and cost as context of using the mobile tour guide system. Dynamic Tour Guide (DTG)[18] makes use of a personal profile to capture personal interests. Each time the context changes, it recalculates tour routes to suggest to the user an adjusted tour. These implementations of context-aware applications address practical needs of context-adaptation, but their models of contexts tend to be overly simplistic. Typically, different context factors are enumerated in a fixed template as a one-dimensional structure. These template-based approaches suffer from a number of difficulties in designing context-aware behavior. First, they only work for routine activities, where contextual states and adaptation strategies can be determined a prior at the design time. Second, they tend to treat contextual factors as isolated variables that can be interpreted and adapted individually and independently. Third, they do not scale up with the number of contextual factors involved. As a consequence, template-based context adaptation often focuses on one type of context or a very small number of context factors. Last, existing context-awareness mobile applications almost exclusively focus on physical contexts and pay less or no attention to human cognitive contexts. Template-based context models are fairly effective for designing context-aware applications where the activity is simple, stable, and routine. But for complex and dynamic operational contexts, selection of context factors and adaptation targets is likely to miss the target without the knowledge on the configuration of context factors.

Towards a more principled way of modeling context, we must look at a picture larger than merely geographical information visualization and use. Tamminen and colleagues[19] argue that contexts are better understood by studying how the actions of everyday life are enabled by context-aware applications. Their studies revealed that user's mobile activities are highly dependent on context and vice versa. Reichenbacher[20] made a point that humans do not simply use geographical information, but they use it for the benefit of performing a meaningful activity. It is this activity that plays a central role in structuring context factors. An activity is associated with an objective and is composed of a set of actions that are part of a plan and are executed in sequence. As an activity unfolds, a plan of actions is adopted which requires certain agents, resources (information or physical), and tools. Similarly, Prekop and Burnett[21] argue that the context should be considered as surrounding the performance of an activity by an agent, *i.e.* the context will only come into existence when the activity is being performed. Kaenampornpan and O'Neill[22] also discussed the possibility of using Activity Theory to model contextual factors. They argued that context-aware applications have utilized only isolated subsets of their context and little research exploring the relationships between different elements of context.

All of these efforts attempt to identify an integrated context model from the activity-oriented perspective. However, none of them have made it clear how to do it formally and computationally. This is perhaps partially due to lack of methods to model activity itself.

As a way forward from the above idea, our work establishes a formal theory of context based on a computational model of activities. By modeling activity as an ever evolving and unfolding plan of actions, we are able to computationally determine how situational factors get selected into the activity's plan and become the ingredients of the ongoing activity. We will present a clear notion of activity in the next section. Before that, we reiterate the main argument made in this section.

Proposition 1 *The selection and configuration of situational factors into a context is ultimately determined by the on-going activity of the users.*

Activity and plan

An activity is a dynamic construct expressed over a period of time, and it consists of people, artifacts, an object or motive, socio-cultural rules, and roles. We subscribe to Activity Theory[23] in conceptualizing activity. From that angle, an *activity* is a form of doing directed to an objective. Whether an activity is conducted by a human and a computer or by multiple human and computational agents, it starts with an objective as motivations behind the activity. Multiple actions are performed to reach the overall objective. Each action is driven by a conscious intentional goal. Finally, actions are performed by operations that are unconscious, often routine actions carried out automatically as a service. An activity is the fundamental, meaningful unit of analysis to understand the interacting relationships among human minds, artifacts, and the environment.

Central to our view of activity is the role of plans in mediating the objective, actions, and situations. The theory of Situated Action[24] believes that an activity has an underlying plan that is prerequisite to and prescribe actions at every level of details. A plan is the product of agents actively reasoning about the course of actions and necessary conditions/resources in order to achieve the objectives of an activity. Agents collaboratively construct the plan and ensure mutual intelligibility of each others' actions. The significance and consequences of an action within an activity derive from its relationship to the plan.

Consistent with the above view of activity, we model an activity by a collaborative plan, following the computational theory of SharedPlans[25]. A collaborative plan is a moment-to-moment representation of an unfolding activity and includes not only a hierarchy of actions and recipes but also the knowledge and physical pre-conditions (modeled as parameters of recipes) as well as the set of mental states (beliefs, intentions, and commitments) that the participating agents have established towards the plan and its subplans (see Figure 1 for the structure of a collaborative plan). *Actions* in a plan may be basic or complex. Basic actions are performable at will, under certain conditions, while complex actions have associated recipes, their performance requiring the performance of each action in their recipe, under certain conditions. *Recipes* represent what agents know when they know a way of doing something. They represent information about the abstract performance of an action, and are composed of a set of constituent acts and associated constraints.

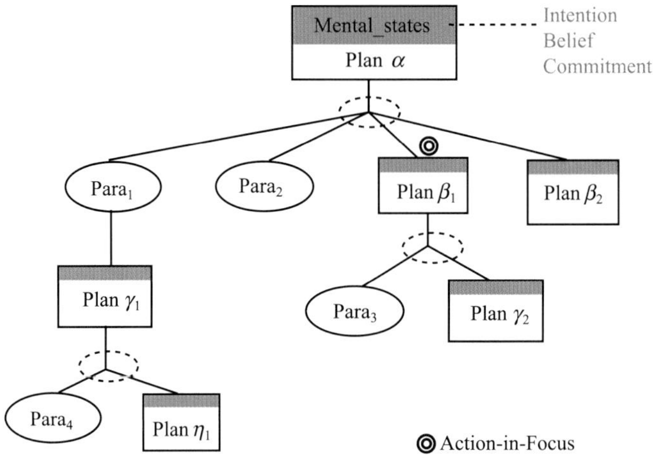

Figure 1 Structure of an activity represented as a SharedPlan[26–29]

As an example, Figure 1 shows a plan for α, which was elaborated with a recipe that has two parameters ($Para_1$ and $Para_2$) and two subactions β_1 and β_2. Parameters in a recipe are knowledge-preconditions which must be identified before any subactions can be worked on. Identifying a parameter can be a complex process and may need a plan (such as Plan γ_1). A SharedPlan of an activity explains why agents do what they did and is a model of the intentional structure of a collaborative activity.

An activity does not need to have a fully specified plan in order to be started. A plan can be partial (if there are un-instantiated parameters, un-executable leaf actions, or insufficient mental states), meaning that the activity is still ongoing and unfolding. Otherwise, a fully developed activity corresponds to a full collaborative plan. During the development of an activity, agents jointly evolve a collaborative plan from a partialplan towards a full plan. A group of agents GR is said to have a Full SharedPlan (FSP) on α when the state described in Table 1 is reached. Otherwise, the agents' plan will only be a *Partial Shared Plan* (PSP).

A plan represents the intentional structure of an activity. For a multileveled and complex activity, agents are likely to focus on a small part of the plan to elaborate and execute. The agents' focus of attention on the plan at any given moment is called the *attentional state* of the activity. For example, although a person's activity is to prepare a dinner, at any moment, the person will only focus on one or two items of the dinner. Agents attend to different parts of the plan as they make progress. The attentional state of an activity can be modeled as a stack[27].

Binding context to activity

When a collaborative plan is established for an activity, various contextual factors can be associated with the activity based on how such factors contribute to various components of the collaborative plan as it evolves towards a full plan. The black dots (•) in Figure 2 indicate where computational, physical, cognitive, and social contexts can be plugged into the plan of the activity. This figure places the activity (modeled as a collaborative plan) at the

Table 1 Specification of a Full Shared Plan (FSP)

(0) GR intends to perform α
(1) GR has a recipe R for α
(2) For each single-agent constituent action β_i of the recipe, there is an agent $G_{\beta_i} \in GR$, such that
(2a) G_{β_i} intends to perform β_i;
G_{β_i} believes that it can perform β_i;
G_{β_i} has an individual plan for β_i
(2b) The group GR mutually believes 2.a.
(2c) The group GR is committed to G_{β_i}'s success
(3) For each multi-agent constituent action β_i of the recipe, there is a sub-group of agents $GR_{\beta_i} \in GR$ such that
(3a) GR_{β_i} mutually believe that they can perform β_i
GR_{β_i} has a SharedPlan for β_i
(3b) The group GR mutually believes 3.a.
(3c) The group GR is committed to GR_{β_i}'s success

centre of everything, and attaches all the enabling conditions (social, cognitive, information environment, analytical tools, *etc.*) as contexts of the activity.

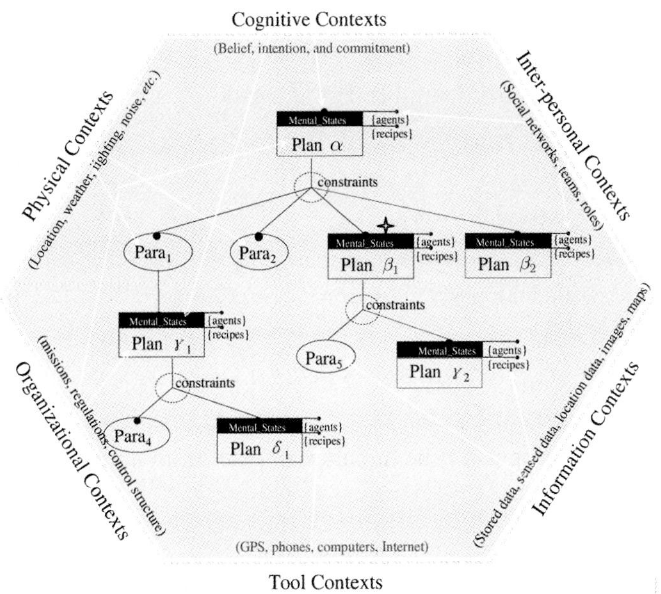

Figure 2 Binding situational factors to activity

The binding of context factors to activity can take the following forms:

Inter-personal contexts as 'agents' of a plan. The 'agents' attached to a plan node within a collaborative plan indicate who are the participants of that plan. The set of 'agents' for a plan is commonly negotiated within the social contexts of an activity. *Social contexts* include people, groups, organizations, and computer agents that have a degree of consciousness about the collective activity and their roles, responsibilities, and cooperative relationships.

Agents' intentions and beliefs as 'mental-states' of a plan. 'Agents' of a plan hold certain mental states towards the plan. For example, agent 1 may hold an intention to work on plan α, and believe that other agents will cooperate with it. At the same time, agent 1 may have committed to doing plan γ. Such cognitive states (intentions, beliefs, and commitments) exist internally in human minds, but they are important constituents of a collaborative plan.

Agents' knowledge provides recipes for plans. A plan on a complex action requires a piece of knowledge, called a "recipe", which specify one way of accomplishing the complex action. A recipe decomposes a complex action into subactions with (optional) knowledge pre-conditions (*i.e.* parameters). Recipe knowledge may come from the mind of human participants (*i.e. cognitive contexts*) or from the knowledge-base of a computer system (*i.e. computing contexts*).

Information contexts as parameters of the activity's plan. A plan on a complex action often involves the identification of one or more parameters as knowledge pre-conditions to the execution of the plan. Such parameters may be obtained by sensing the physical environment, by communicating with humans, or by consulting a knowledge-base. For example, user's physical location may be a parameter for the plan in a way-finding task. Similarly, time can be another parameter that could be read or sensed from the physical context.

Our model binds situational factors to the plan of an activity at run-time (while the activity is going-on). For the benefit of later discussion, we define the following notations:

$SF(A)$: all the Situational Factor of an activity A.

$CF(A)$: Context Factors of activity A are defined as the subset of $SF(A)$ that has been bound to the plan of the activity

$ACF(A, t)$: Attended Context Factors (ACF) of A at moment t. This is a subset of $CF(A)$ that is binding to the part of the plan under current attention.

Proposition 2 *For a complex and dynamic activity, we expect that* $ACF(A, t)$ *is significantly smaller than* $CF(A)$ *which is significantly smaller than* $SF(A)$.

Our activity-based model of context offers a principled way to sift through a potentially very large number of situational factors and narrow our attention down to only those that are relevant to the success of the activity.

Activity-Mediated Context Adaptation

Now, we turn our focus from the activity itself to the computational device that provides geographical information in support of the activity. Suppose that a GVE (geographic virtual environment) is available to an activity A. Then, the process of adapting the behavior of GVE to the context takes a number of steps.

First, we need the ability to detect if any situational factors have changed. For physical factors, like location, orientation, time, temperature, changes can be automatically detected by the proper use of sensors and instruments. However, changes to cognitive and social factors are not easily observable and require different detection/monitoring techniques. Work on conversational dialogue systems[26] offers one possible approach to allow communication of such changes from human knowledge to computers.

Second, we make inference on whether a change matters to the activity. Any change on a factor f matters to the activity if f is part of the $ACF(A, t)$ (*i.e.*, Attended Context Factors of A at moment t). If the change is minor such that nothing needs to be modified in the existing plan, then there is no need for GVE to adapt because the geographical information need will stay the same. If this change on f is significant enough such that previous plan is no longer viable, or an alternative (more efficient) plan is possible, it will trigger a replanning process. A plan of an activity is subject to replanning when one or more of the following three conditions happen: 1) There is a change of the intention of agents in terms of the objectives of the activity; 2) There are changes in some factors of the environment such that the current plan (or subplan) is no longer possible; 3) The execution of some part of the plan fails to achieve the expected effects. After replanning, a new pattern of binding between situational factors and the plan is likely to appear.

Third, given the updated knowledge on the plan of activity and $ACF(A, t)$, the GVE infers the geoinformation needs of the newly updated plan to decide if that differs from the current state of the device. This will be the basis for adapting geographical visualization behavior.

From the above three steps of the adaptation process, we may reach an important conclusion that our context-adaptation strategy is centrally mediated by the plan of the activity. We do not impose fixed rules that directly relate changes of one factor to the behavior of the GVE. Instead, the GVE only directly adapts to one thing, that is, the changes in the plan of the activity. There many novel adaptive behaviors that can be designed using the above principle. Here we mention two of them.

- *Shifting focus-of-attention of the activity*. Agents of an activity do not attend to all parts of the activity at the same time. Generally, there is always one action (plan node) that is the focus of the current activity at any given moment. Context factors that are considered relevant to the whole activity may not be relevant contexts at this moment, because only the contexts that contribute to the success of the action under focus are considered relevant. In other words, even in the case that the overall (physical) situation does not change, the shift of attention from one action to another will trigger an adaptation process.
- *Adapting to execution obstacles*. Change of contexts may sometimes cause the failure of executing a plan based on the currently selected recipe. For example, a device may normally expect receiving location data from a GPS sensor. In such cases, presenting a map of current location will be easy. However, if the GPS sensor network suddenly stops working, we still need to get the location data to ensure the success of the rest of

the activity. The activity will adapt to this contextual change by triggering a 're-plan' process to choose another recipe. A backup recipe is invoked which uses geo-referencing street addresses to identify the location. If the mobile device is aware of such context change, it will be able to adapt to the failure of the GPS event, but such adaptation is mediated through a conscious knowledge about what part of the activity has been re-planned and how such re-plan affects the mobile information service.

A Scenario

We will use a scenario to explain how the context-aware adaptation can be achieved. Suppose that a user, Tom, is interested in taking a tour in the city of Philadelphia. His planning of and executing the tour will be assisted by MapGuide—a mobile geographical information device for access to tourist information. Tom plans to visit a museum to catch a special exhibit at 1pm after touring an aquarium. If possible, he would like to have a lunch in a decent restaurant before visiting the museum. This activity includes several tasks that are interrelated both temporally and logically. The plan for this activity is shown in Figure 3.

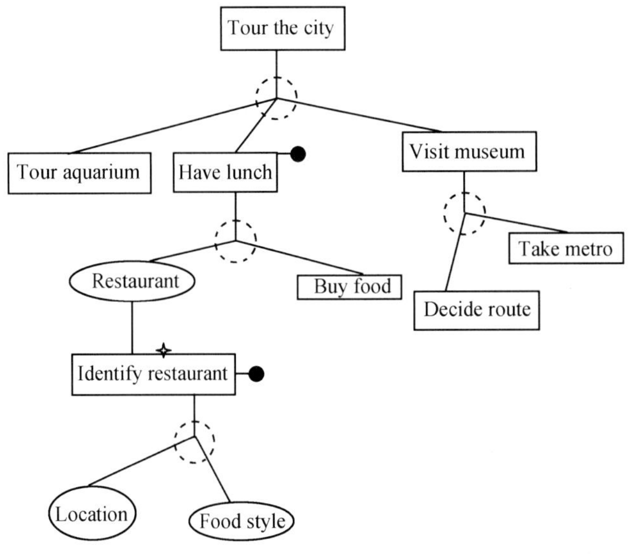

Figure 3 The collaborative plan of activity for a city tour (The action under focus is "Identify restaurant")

Suppose that Tom's visit to aquarium took a little longer than he had originally planned. Until he completed that part of the tour, the time was already 1:15pm. This forced Tom to redo the plan for the rest of the activity. Tom definitely wants to catch the 2pm exhibit at the museum, which means he had to cut time on the lunch. The question for Tom at that moment was: *how much time can he spend on lunch*? Given that he only had 45 minutes left until 2pm, Tom consulted MapGuide for informational help on the travel time to the museum. MapGuide responded with the following information: "you may find the nearest metro station in 5 minutes; it will take 25 minutes from the nearest metro station to the museum."

There was only less than 15 minutes left for Tom to have lunch. So Tom decided to modify his plan from a relaxed sit-in lunch to a simpler one (picking some fast food). This change in Tom's intention was communicated to MapGuide (perhaps through a dialogue). According to our context adaptation principle described earlier, Tom, with the help of the MapGuide, will re-plan the "have lunch" action. By performing means-end reasoning on this part of the plan, MapGuide inferred that it needs to come up with a recommendation to Tom on nearby restaurants that offer fast food. Based on the new plan for "have-lunch" action, MapGuide infers its behavioral change in terms of displayed map contents and their presentation. This modified provides proper support for Tom to choose a restaurant located on the way from his location to the nearest metro station.

From this scenario we can see that MapGuide maintains a model of plans about Tom's activity and actively reasons about the changes of the plan when other contextual factors change. The user's activity is the direct target that MapGuide adapts. For applications like MapGuide, the user's activity provides a comprehensive context for map adaptation.

Discussion and Conclusions

In visually-assisted geospatial activities, geographical information tools are an integral part of the activity that provides spatial knowledge and orientation to actions. This chapter analyzed the weakness of existing context models in dealing with design of map services for complex and dynamic activities. We have proposed an activity-based model of contexts that is able to address key computational issues in context-adaptive map services. Our definition of context includes not only external resources (people, artifacts, settings) but also internal mental processes (goals, intentions, and beliefs). We supported our model by applying the theoretical lenses of Activity Theory[23] and Situated Action[24]. Our model effectively addressed three difficult questions that have been troubling research in context-awareness applications. Because binding of contexts, activities and adaptation strategies is delayed till runtime, it is much easier and robust to determine what contexts are relevant and what appropriate actions to be taken.

Our model emphasizes the role of plans in mediating activity and context- adaptive behavior of devices. However, it is a tricky task if one wants to follow our method presented here, because it requires that a system has the ability to construct and update a model of the ongoing activity. This demands advanced techniques for not only sensing changes in physical environment, by also monitoring changes in intentional states and beliefs. This can potentially benefit from recent development in advanced human-computer dialogue systems[28, 29], but other approaches should be explored as well.

Acknowledgements The author would like to thank reviewers for their valuable comments on the draft of this chapter. This work is performed with partial support from the National Visualization and Analytics Center (NVAC), a US Department of Homeland Security

Program, under the auspices of the Northeast Regional Visualization and Analytics Center (NEVAC), USA.

References

[1] Kraak M-J. Beyond Geovisualization. IEEE Computer Graphics and Applications, 2006, 26: 6–9.

[2] Buckley A, Gahegan M, Clarke K. Geographic Visualization. University Consortium for Geographic Information Science, UC Santa Barbara, CA, 2000.

[3] Andrienko G, Andrienko N, Jankowski P, Keim D, Kraak M J, MacEachren A, Wrobel S. Geovisual Analytics for Spatial Decision Support: Setting the Research Agenda. International Journal of Geographical Information Science, 2007, 21: 839–857.

[4] Thomas J J and Cook K A. Illuminating the Path: The Research and Development Agenda for Visual Analytics. National Visualization and Analytics Center, http://nvac.pnl.gov/ [2005].

[5] Andrienko G, Andrienko N, Fischer R, Mues V, Schuck A. Reactions to Geovisualization: An Experience from a European Project. International Journal of Geographical Information Science, 2006, 20: 1149–1171.

[6] Lloyd D, Dykes J, Radburn R. Understanding Geovisualization Users and Their Requirements—A User-Centred Approach. http://kartoweb.itc.nl/icavis/2007/LloydDykesHelsinkiAbstractv5.pdf [2007].

[7] Oppermann R. Adaptive User Support: Ergonomic Design of Manually and Automatically Adaptable Software, Computers, Cognition, And Work. Hillsdale, NJ: Lawrence Erlbaum Associates, 1994.

[8] Dourish P. What We Talk About When We Talk About Context. Personal and Ubiquitous Computing, 2004, 8: 92–99.

[9] Greenberg S. Context as a Dynamic Construct. Human-Computer Interaction, 2001, 16: 257–268.

[10] Dey A K. Understanding and Using Context. Personal and Ubiquitous Computing, 2001, 5: 4–7.

[11] Salber D, Dey A, Abowd G D. The Context Toolkit: Aiding the Development of Context-Enabled Applications. In: Proceedings of CHI 99, 1999: 434–441.

[12] Dransch D. Activity and Context—A Conceptual Framework for Mobile Geoservices. In: Meng L, Zipf A, Reichenbacher T (Editors). Map-based Mobile Services. Berlin, DE: Springer, 2005: 31–42.

[13] Chen G and Kotz D. A Survey of Context-aware Mobile Computing Research, TR2000-381, Dartmouth College, Computer Science, Hanover, NH, 2000.

[14] Dix A, Rodden T, Davies N, Trevor J, Friday A, Palfreyman K. Exploiting Space and Location as a Design Framework for Interactive Mobile Systems. ACM Transactions on Computer-Human Interaction (TOCHI), 2000, 7: 285–321.

[15] Sarjakoski L and Nivala A -M. Adaptation to Context—A Way to Improve the Usability of Mobile Maps. In: Meng L, Zipf A and Reichenbacher T (Editors). Map-based Mobile Services. Berlin, DE: Springer, 2005: 107–123.

[16] Abowd G D, Atkeson C G, Hong J, Long S, Kooper R, Pinkerton M. Cyberguide: A Mobile Context-Aware Tour Guide. ACM Wireless Networks, 1997, 3: 421–433.

[17] Cheverst K, Davies N, Mitchell K, Friday A, Efstratiou C. Developing a Context-aware Electronic

Tourist Guide: Some Issues and Experiences. In: Proceedings of the ACM CHI 2000. The Hague, Netherlands, 2000.

[18] Kramer R, Modsching M, Schulze J, ten Hagen K. Context-Aware Adaptation in a Mobile Tour Guide. Lecture Notes in Computer Science, 2005, 3554: 210–224.

[19] Tamminen S, Oulasvirta A, Toiskallio K, Kankainen A. Understanding Mobile Contexts. Personal and Ubiquitous Computing, 2004, 8: 135–143.

[20] Reichenbacher T. Adaptive Egocentric Maps for Mobile Users. In: Meng L, Zipf A and Reichenbacher T (Editors). Map-based Mobile Services. Berlin, DE: Springer, 2005: 141–158.

[21] Prekop P and Burnett M. Activities, Context and Ubiquitous Computing. Computer Communications, 2003, 262: 1168–1176.

[22] Kaenampornpan M and O'Neill E. Modelling Context: An Activity Theory: Approach. Lecture Notes in Computer Science, 2004, 3295: 367–374.

[23] Nardi B A. Context and Consciousness: Activity Theory and Human-computer Interaction. Cambridge, MA: The MIT Press, 1996.

[24] Suchman L. Plans and Situated Actions: The Problem of Human-Machine Communication. New York: Cambridge University Press, 1987.

[25] Grosz B J and Kraus S. Collaborative Plans for Complex Group Action. Artificial Intelligence, 1996, 86: 269–357.

[26] Cai G, Wang H, MacEachren A M, Fuhrmann S. Natural Conversational Interfaces to Geospatial Databases. Transactions in GIS, 2005, 9: 199–221.

[27] Lochbaum K E. A Collaborative Planning Model of Intentional Structure. Computational Linguistics, 1998, 24: 525–572.

[28] Allen J, Byron D, Dzikovska M, Ferguson G, Galescu L, Stent A. Towards Conversational Human-Computer Interaction. AI Magazine, 2001, 22: 27–37.

[29] McTear M F. Spoken Dialogue Technology: Enabling the Conversational User Interface. ACM Computing Surveys, 2002, 34: 90–169.

Chapter 12 Non-photorealistic Visualizations on Mobile Devices and Usability Concerns*

Mathias Jahnke, Liqiu Meng, Jan Eric Kyprianidis and Jürgen Döllner

Introduction

The rapid development of photogrammetric and remote sensing technologies has led to the increasing availability of highly detailed 3D digital city models that typically consist of buildings with roof and façade elements, trees with various branch types, step structures that connect buildings with terrain forms and streets with associated navigation facilities *etc.*[1]. Being draped with simulated textures or photos from the real world, such a city model would give an immersive photorealistic impression if suitably rendered on a graphic display medium. Due to its ability of allowing an intuitive perception of spatial relations, 3D photorealistic visualization has prevailed within desktop mapping systems for nearly two decades. Their graphic quality has been progressively improved according to the motto "making things truer than the truth". To implement a photorealistic city model on miniaturized displays like smart phones and PDAs is therefore a very natural result of their technical evolution. However, the designer faces two research challenges: How far is it possible to transmit and render the detailed geometries in real-time on a mobile device which has limited storage and computing capacity? And is a photorealistic duplicate the optimal way of communicating 3D urban information? While the first question strongly depends on technical progress in terms of data structures, rendering algorithms, hardware properties and wireless communication bandwidth, the second one—the focus of our approach—is both a technical and a human issue because the user is an inseparable part of the usability engineering task. General principles of usability engineering were described by Nielsen[2] and Shneiderman and Plaisant[3]. Based on ISO-standards, Hermann and Peissner[4] have developed a specialized usability approach for 3D virtual urban environments.

Shneiderman[5] states that "Some designers dream about building interfaces that approach the richness of 3D reality. They believe that the closer an interface resembles the real world, the easier the usage. They strive for resolution that matches film—with rapid camera movement and lively animated objects. This is a dubious proposition since user studies show that disorienting navigation, complex user actions, and annoying occlusions can slow down performance in the real world as well as 3D interfaces." (page 12). MacEachren and Kraak[6]

* Supported by the German Research Foundation (DFG, Grant No. ME 1846/4-1, DO 697/5-1).

assume that the " ... approaches to desktop interface design will probably not work when used with ... immersive Virtual Environments or on small mobile PDAs ..." (page 5). They point out the two basic constraints of graphic design for mobile devices: limited display space and limited interactivity. Users in a mobile environment are constantly influenced by external visual, auditive and haptic stimuli beyond the mobile display[7]. These stimuli may, on the one hand, distract and hinder the attentional performance, *e.g.* by deviating gazes from the display, but on the other hand, they may also contribute to overall information acquisition and understanding. The inherently strong interaction between the virtual and the real world requires a smooth switch from one to another, hence, the immediate usability of the virtual environment. With this regard, a photorealistic 3D city model suffers from three drawbacks:

1. A photorealistic city model with typically unclassified geometric details reveals a low abstraction level. The user faces a high cognitive workload of information processing[8]. S/he has to rely very much on cognitive skillsof visual attention, *i.e.* the ability of sifting through the rather crowded graphics in order to notice the relevant information for applications such as pedestrian navigation.
2. Due to the limited interactivity and reduced vision field, the recognition of depth cues embedded in a photorealistic city model is less straightforward and cognitively more demanding on small displays than on desktop PCs.
3. Users' attention may be entirely locked on the appearance so that less effort will be made to look for the information behind the surface. In addition, users may be likely to beconfused by details that are not up-to-date.

Keeping these considerations in mind, we adopt the 3D Non-photorealistic Rendering (NPR) approach with the objective to create various visualization alternatives that possess the following useful characteristics:

1. They have a reduced visual complexity as compared with photorealistic visualization, but still they give a naturalistic impression of the underlying city model, thus allowing an intuitive and fast identification of the individual urban objects.
2. They reveal a higher abstraction level and allow the accommodation of more semantic information than a photorealistic visualization would do.
3. By means of simplification, exaggeration or any other artistic treatments of 3D objects they stimulate users' attention and curiosity for querying further information either from the underlying database or from the mobile environment.

NPR represents a new genre in 3D computer graphics. It addresses illustrative, expressive and artistic computer graphic techniques[9]. Various possible graphic variables and their combinations can be used as metaphors to highlight geometric and semantic characteristics of 3D objects.

3D NPR techniques

Among the typical NPR techniques used in 3D computer graphics are 'Cartoon Rendering',

'Sketchy Rendering' and 'Artistic Rendering'[10]. Similar to photorealistic techniques, they are realized through computer graphic hardware along the rendering pipeline. As shown in Figure 1, a rendering pipeline consists of basic steps of transforming a 3D data set (*e.g.* geometric elements of an urban environment) into a 3D model, then projecting the model onto a 2D display surface along with the necessary depth cues so that the user gains a 3D sensation. To achieve this goal, a geometric transformation model, an illumination model and a shading model are implemented on the graphic board. Unlike the photorealistic rendering that strives for the unique "truth", these three models can be manipulated in various ways so as to give the user different illustrative 3D impressions. A first promising approach for the usage of NPR techniques to visualize 3D city models was demonstrated by Döllner and Walther[11]. An example is given in Figure 2.

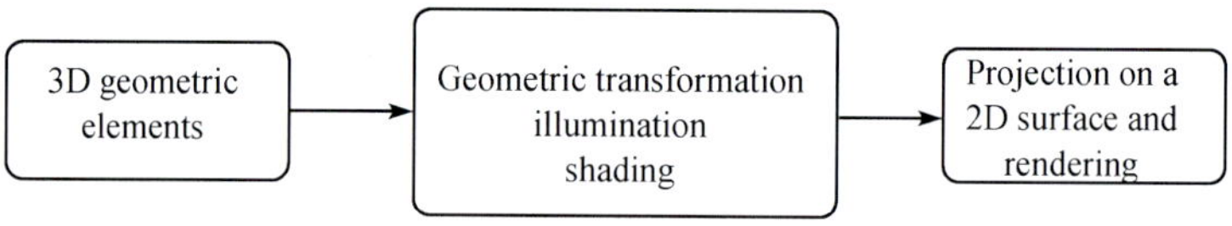

Figure 1 The rendering pipeline

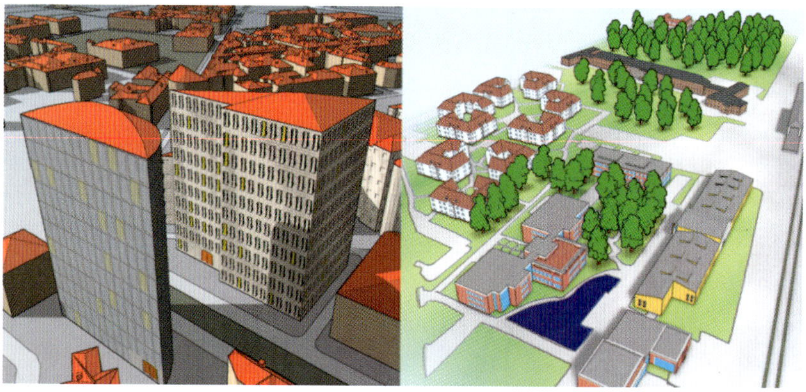

Figure 2 Non-photorealistic visualization of a 3D city model of Potsdam[1]

Existing NPR techniques have been almost exclusively developed for desktop computers which have much higher computing capabilities and graphic resolution than small display devices. Adapting these techniques to mobile devices and usage context is therefore by no means a simple transplantation task. Since there is still too little knowledge about how users get along with complex virtual environments (brain processes) and what kind of impact user strategies would have on the design[12], we take a first step to identifying, adapting or developing graphic variables that are feasible for NPR of 3D virtual urban environments on mobile displays.

Our approach adopts CityGML which is an OGC standard file format for 3D city models. Being based on the Extensible Markup Language (XML), CityGML is implemented as an application schema for the Geography Markup Language (GML). With CityGML, it is possible to process graphical and geometrical models as well as topological and semantic information for 3D applications. The 3D visualization software LandXplorer which can process the CityGML file format is used as our experimentation platform[13].

From 2D maps to 3D non-photorealistic graphic

Traditionally, 2D maps serve as highly abstract scientific documents which are made possible by precise map projection, symbolization and generalization. Map symbols on the legend usually represent object categories and classes. The individuality of a particular object is only implicitly indicated by its symbol bound to a particular location and its name. If for example a city with a population between 50,000 and 100,000 is represented by a circle symbol of a certain size, this symbol could remain up-to-date for a rather long time as long as the population does not grow or shrink beyond the class boundaries. For this reason, the abstraction ability of 2D maps also proportionally reflects their cost-benefit efficiency in terms of both production and maintenance expense. On the other hand, however, abstraction makes the map use a cognitively burdensome task that needs intensive training. With the democratization of mapmaking, the widespread use of internet maps and changed map use behavior, the traditional look of 2D maps has long been extended to include other design styles with similar cost-benefit efficiency. City plans, for example, can have both classified abstract symbols and highly individualized landmarks. Figure 3 briefly illustrates a con-

Figure 3 From self-explanatory rendering to abstract symbolization

tinuum of symbolization from abstraction to self-explanatory realism.

Today, many visualization tools allow an easy combination of a 2D plan with air photos or satellite images. The resulted image map gives its user a more or less immersive impression of seamless coverage of the mapped region. However, it leaves little space to accommodate additional thematic information such as population density in urban environments. 3D city models are comparable with image maps because they attempt to resemble the environment, *e.g.* with orthophotos projected on roofs and façades. 3D city models have gained a growing popularity among decision makers and normal citizens during recent years, especially since the Google Earth revolution in 2006. However, due to insufficient semantic information, the majority of existing 3D city models could only be used as photorealistic show pieces instead of urban information systems. With the rapid development of geoinformation technologies and Web 2.0 as a global platform for sharing collective intelligence[14], it is now increasingly possible to integrate more and more semantic information with geometric details. Correspondingly, there is an increasing demand for the visualization of information-rich virtual urban environments, which should serve as a driving force to the development of the so-called informed society. The approach of 3D photorealistic visualization will doubtlessly be further elaborated together with high-end computer graphic technologies. With regard to the diversified applications of 3D urban information, the expanding range of display sizes as well as the growing awareness of usability issues, the potential of NPR approach on mobile devices has also gained increasing attention in the research community[15]. As a whole, there has been a technical evolution from 2D map to non-photorealistic graphics as we show in Figure 4.

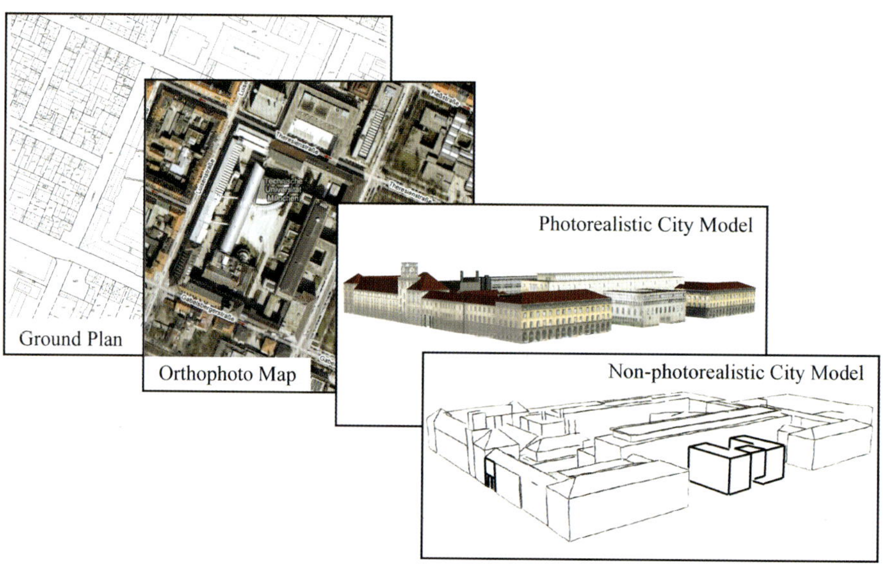

Figure 4 From 2D abstract graphics to 3D non-photorealistic graphics

Virtual Urban Environments on Mobile Devices

Design alternatives

Depending on the adopted symbol style, 3D virtual urban environments can be realized in various ways, ranging from schematic topogramms to immersive photorealism. Schematic topogramms are favorable for providing overviews of urban areas which are characterized by street networks and locations of Points of Interest (POIs). However, they are less efficient for mobile users to explore the local environments around their current position due to over-large geometric distortions on the graphics. 2D city maps are favorable for visualizing ground plans, geographic objects, and corresponding semantic information. But again the large visual discrepancy between the map and the reality does not allow an immediate understanding of the mobile environments. 3D photorealistic city models allow users a straightforward interpretation of the geographic space, but their inherently high graphic complexity makes them less suitable for mobile applications due to the drawbacks mentioned in Section 1. The NPR approach has the potential to unite the advantages of the above design styles and avoid their disadvantages. Since it does not have to duplicate the reality, there is a plenty of design freedom left to the developers. It allows not only the reduction of overall graphic complexity, but also a demand-driven visual highlighting of certain depth cues or important objects. Technically, non-photorealistic city models require less frequent updates than photorealistic city models. In a first user test of NPR approach, Plesa and Cartwright[15] confirmed that non-photorealistic city models are generally better accepted and understood by users due to their clarity, legibility and aesthetic appeal as shown in Figure 5.

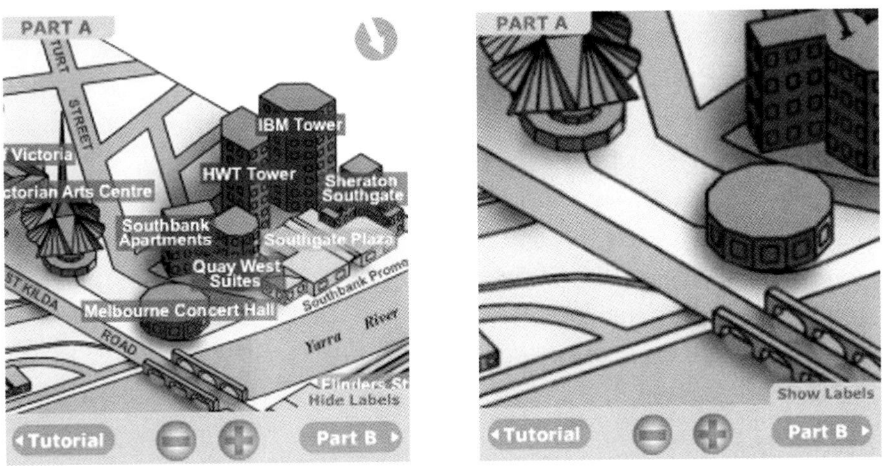

Figure 5 The NPR prototype developed by Plesa and Cartwright (2008) (left) and an enlargement (right)

However, whether an NPR approach will be finally accepted or not relies also heavily on the art of design. Since there may be multiple technically equivalent NPR design solutions for the same task and target user, the confirmed usability of one NPR solution may not hold for other solutions. In order to gain an insight into the potential and general usability of NPR approach, we need to take a closer look at the cognitive capacities of mobile users and their expectations from a non-photorealistic virtual urban environment.

Cognition-based design constraints

According to MacEachren and Kraak[16], the usage of a geovisualization system can be divided into data exploration, analysis, synthesis and presentation. Visualization on mobile devices is mainly intended for the communication of known geographic information to a private person. Reichenbacher[17] mentioned that orientation, localization, navigation, search, identification and event check are the elementary user tasks in a mobile environment. While the majority of existing mobile map services have been developed to support mobility-relevant tasks such as orientation, localization and navigation, future development will have to meet the growing user need for getting informed of instances from information-rich urban environments through mobile devices. Accordingly, mobile services should not only support mobility, but also facilitate information communication including search, query, identification and event checking. The following cognition-based design constraints and associated research questions are involved in the development of NPR:

(1) *Information package.* How do we package the large amount of information into spatially separable units that can be easily attended and interpreted? To answer this question, there is a need to investigate the working memory of mobile users. The "golden" rule of 7 ± 2 information units that fits the memory capacity for stationary perception environment needs to be adapted to the mobile context.

(2) *Interaction between the virtual environment and the real world.* What is the desired degree of attention to the small display size (6 cm×8 cm) of mobile devices? The mobile user typically has his/her primary interaction with the real world, while taking the mobile device as a supporting tool. Therefore, the visualization should not draw too much visual attention to the users. Instead, it should allow immediate identification or re-visiting relevant information , thus facilitating easy recall.

(3) *3D sensation.* How do we preserve the 3D sensation on the simplified graphics? Since the 3D sensation positively influences the emotional state of users, it helps to enhance the efficiency of orientation and navigation. There is a need to strike a balance between reduction of graphic complexity (information hiding) and exaggeration of important objects or landmarks (information highlighting).

Usability Concerns

Usability measure

According to Nielsen[2], the usefulness of a system is reflected by two fundamental aspects—utility and usability as in Figure 6. While utility describes the fitness of a system for a certain task which can be relatively easily quantified, usability defines how well users are

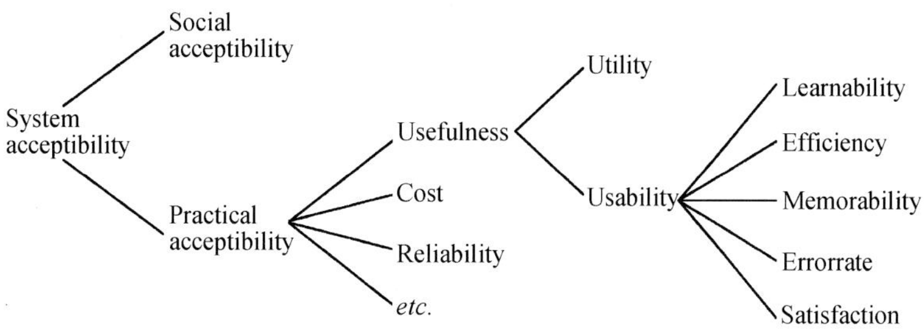

Figure 6 Attributes of the system acceptability[2]

supported by the system. It is difficult, if not impossible, to quantify usability because it involves human-related parameters such as learnability, efficiency, memorability, error rate, and degree of satisfaction.

Currently available visualization services on mobile devices have been mainly focused on their technical feasibility. Although more and more general ergonomic characteristics are being considered during the design[4], the mobile user remains an unknown or imprecise parameter. Consequently, there is a mismatch between what the visualization service offers and what an average mobile user can quickly get or between what is technically possible and what makes sense from a user's perspective. During visual information processing, users make use of existing knowledge or information from the mobile environment to get an idea of *what* they are searching for using a top-down strategy. Additionally, the sensory input from a cognitive adequate NPR solution will guide users to attentively know *where* relevant information is located from the bottom-up. The interaction of bottom-up and top-down processes controls *where*, *how*, and *what* users pay attention to[18], hence selective information acquisition. Being aware of the iterative information process and the existing mismatch, we define *cognitive adequacy* as a usability measure for NPR solutions on mobile devices. Concretely, cognitive adequacy can be expressed as a weighted mean from four submeasures: 1) The relative speed at which user's visual attention (gaze movements) is directed or re-directed to objects of interest; 2) The degree of visual concentration on the objects of interest (gaze fixations); 3) The degree of easy remembering of perceived objects of interest; and 4) Task-driven cognitive efficiency such as the relative speed at which the objects on the display are matched with their counterparts in the real world (identification), or the relative speed at which certain information is interactively acquired (event-check). The first three submeasures reflect the capability of NPR solutions to guide users' attention to the relevant information and remember it, while the fourth one emphasizes the impact of the mobile task on users' selection of information. Among many technically equivalent NPR solutions, the one with the highest cognitive adequacy must be regarded as the optimal solution from a usability perspective.

Graphic variables

In the pre-computer era, graphic design was realized by a taxonomy that essentially consists of six graphic variables "shape", "size", "orientation", "color hue", "value", "spacing" and "texture" and their combinations[19]. The development of multimedia technologies during the last two decades has substantially extended this taxonomy to include not only the further graphic variables such as color saturation, sharpness, resolution and transparency, but also auditive and haptic variables[20–24]. Figure 7 illustrates a set of graphic variables that are applicable for 3D non-photorealistic design. Due to a lack of systematic study of the visual effects of new variables, however, the developers have had to rely on their personal experiences and preferences in their design practice. Sometimes, the improper use or abuse of graphic variables may cause stimulus saturation in perception[24].

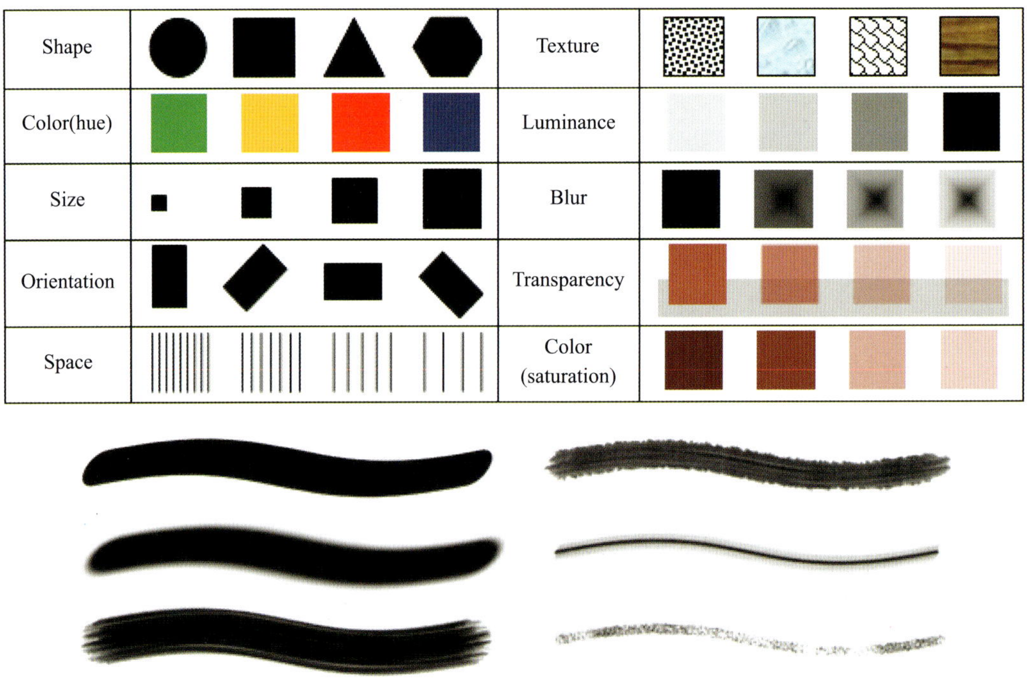

Figure 7　Applicable graphic variables for 3D non-photorealistic design

The 3D NPR approach for mobile devices needs to consider several constraints in selecting the suitable graphic variables and their combination. First, design freedom is constrained by the 3D structure of urban objects. That is, the graphic variables have to be bound to basic geometric elements—faces, edges or nodes of 3D objects as in Figure 8. Second, due to the perspective distortion, some graphic variables may cause perceptual confusion among users. For example, a constantly spaced surface polygon may be interpreted as variably spaced in a 3D visualization. Third, the unpredictable illumination conditions in a mobile environment may restrict or even prohibit the use of certain color hues. On the other hand, graphic variables such as shadow, shading, occlusion and figure-ground contrast can be flexibly combined to emphasize depth cues or eye catchers in a 3D virtual environment[25].

According to our first investigations, "value" can be applicable to all three geometric elements. "Size" could be embedded in edges or nodes to highlight important information or

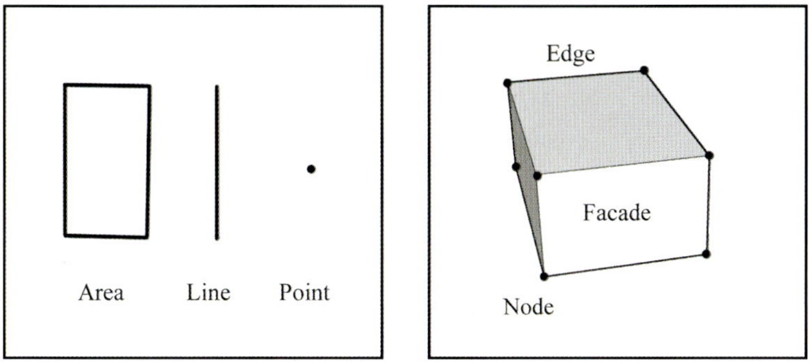

Figure 8 Basic geometric elements of 3D objects

enhance depth cues and texture can be embedded in facades. “Sharpness” or “blur” could help to emphasize the distinction between important and less important objects[26]. Figure 9 shows a number of NPR alternatives based on the combination of applicable graphic variables.

Figure 9 NPR alternatives of a photorealistic church

It is worthwhile mentioning the versatile usage of texture as a graphic variable when dealing with virtual urban environments. In many cases, photographed textures from real world buildings are mapped back onto facades to render a naturalistic look. Textures have the advantage that they can be easily modified with image processing algorithms to give a model an abstract appearance.

Using Modified Texture for Visualization

Due to a lack of highly detailed 3D urban environments in a vector-based file format like CityGML texture mapping is often used as a practical compensation for city model visualizations[27, 28]. In combination with other graphic variables such as value, size, sharpness and blur, the texture can be modified to give the city model an abstract or illustrative representation. There are two alternative ways to modify textures. First, a certain degree of abstraction can be reached by means of different methods from computer graphics[29] and image processing methods. Second, cartographic principles can be applied to generalize and adjust textures. We applied segmentation and morphological algorithms to a building texture which are described in Jähne[30]. We used the method on a texture which is photographed from a building of our test site. To reduce geometrical distortions of the original photograph, we have adopted rectification to get a geometrical correct image of our facade for further processing.

Figure 10A shows a section of the original texture from our test object whereas Figures 10B and C point out two abstraction degrees. From Figure 10A to C the amount of perceptible information decreases consequently, the user's cognitive workload will also decrease. The facade details are reduced by grouping the pixels into distinctive regions and assigning each region an average hue.

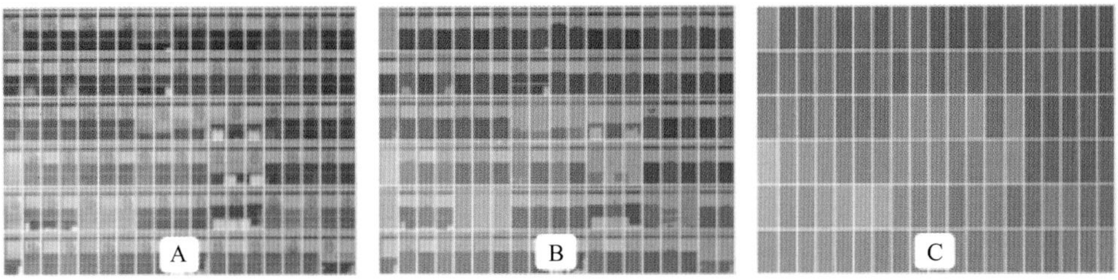

Figure 10 Non-photorealistic alternatives of the graphic variable "texture"
A: original image. B: 1st degree of abstraction. C: 2nd degree of abstraction

An associated effect of abstraction is the reduction of storage for textures and models. Figure 11 shows the relationship between the storage capacity and the abstraction degree of the textured building. The generalized textures require much less storage capacity than the original one. The reduced file size would fit mobile devices better and could accelerate the information transmission. Consequently, users would get faster wireless mobile services.

To approve the advantage of abstraction which is an essential component of an NPR approach, we conducted a very first "survey" on the visualization shown in Figure 12. The group of participants consisted of 22 males and 5 females with an age ranging from 22 to 27 years. The survey revealed a good acceptance and understanding of the 3D visualizations with modified textures.

In the next steps of our design practice, the constraints associated with small mobile devices and their limited computing capabilities will be considered with the aim of identifying an acceptable compromise among a number of cognitively equivalent solutions.

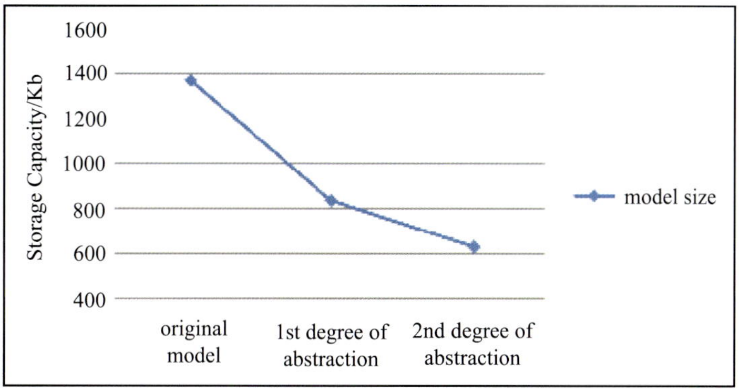

Figure 11 The relationship between the storage capacity and the abstraction degree of texture

Figure 12 The visualization in Google Earth?

A: the original model with draped façade photo. B and C: two successively derived abstraction degrees of the texture

Outlook

This chapter has introduced the NPR approach, its potential for the 3D virtual urban environments on mobile displays and feasibility of implementation with the OGC standard CityGML. The first results of our work are presented. A small survey has confirmed the acceptability of a non-photorealistic visualization based on texture modification. More extensive investigations concerned with the usability issue and user-centered design are necessary. According to our plan for the next stages, a number of design options for two extensive 3D virtual urban environments (Munich and Potsdam) will be implemented with NPR techniques on PDAs. Subsequently, usability tests will be conducted with the aim of identifying the cognitively most adequate solution.

Although we are still in the beginning stage of an on-going joint research project between the Technical University of Munich and the Hasso-Plattner-Institut, we wish to share our early thoughts with researchers and gain further inspirations, especially in aspects of design constraints and usability.

References

[1] Meng L, Heipke C, Mayer H, Chen J (Editors). Interoperation of 3D Urban Information. Photogrammetrie, Fernerkundung, Geoinformation, 2007, 6: 401–404.

[2] Nielsen J. Usability Engineering. New York: Academic Press, 1993.

[3] Shneiderman B and Plaisant C. Designing the User Interface. Reading, MA: Addison Wesley, 2005.

[4] Hermann F and Peissner M. Usability Engineering für kartographische Visualisierungen: Methoden und Verfahren. Kartographische Nachrichten, 2003, 6: 260–265.

[5] Shneiderman B. Why Not Make Interfaces Better than 3D Reality? IEEE Computer Graphics and Applications, 2003, 23: 12–15.

[6] MacEachren A M and Kraak M-J. Research Challenge. Geovisualization, Cartography and Geographic Information Science, 2001, 28: 2–13.

[7] Reichenbacher T and Swienty O. Relevanz und Kognition in der mobile. Geovisualisierung, Kartographische Schriften, 2006, 10: 83–130.

[8] Bunch R L and Lloyd R E. The Cognitive Load of Geographic Information. The Professional Geographer, 2006, 58: 209–220.

[9] Strothotte T and Schlechtweg S. Non-Photorealistic Computer Graphics: Modeling, Rendering and Animation. New York: Morgan Kaufmann, 2002.

[10] Gooch B and Gooch A. Non-Photorealistic Rendering. Berlin, DE: A K Peters, 2001.

[11] Döllner J and Walther M. Real-time expressive rendering of city models.In: Proceedings of the 7th International Conference on Information Visualization (IV'03). Seattle, WA, 2003: 245–250.

[12] Slocum T A, Blok C, Jiang B, Koussoulakou A, Montello D R, Fuhrmann S, Hedley N R. Cognitive and Usability Issues in Geovisualization. http://www.geovista.psu.edu/sites/icavis/agenda/PDF/SlocumLong.pdf [2001].

[13] Döllner J, Baumann K, Kersting O. L and Explorer—ein System für interaktive 3D Karten. Kartographische Schriften, 2003, L7: 67–76.

[14] O'Reilly T. Design Patterns and Business Models for the Next Generation of Software. http://www.oreilly.de/artikel/web20.html [2005].

[15] Plesa M A and Cartwright W. Evaluating the Effectiveness on Non-Realistic 3D Maps for Navigation with Mobile Devices. In: Meng L, Zipf A and Winter S(Editors). Map-based Mobile Services—Design, Interaction and Usability. Berlin, DE: Springer, 2008: 80–104.

[16] MacEachren A M and Kraak M-J. Exploratory Cartographic Visualization: Advancing thc Agcnda. Computers & Geosciences, 1997, 23: 335–344.

[17] Reichenbacher T. Adaptive Egocentric Maps for Mobile Users.In: Meng L, Zipf A and Winter S (Editors). Map-based Mobile Services–Design, Interaction and Usability. Berlin DE: Springer, 2008: 107–121.

[18] Corbetta M and Shulman G L. Control of Goal-Directed and Stimulus-Driven Attention in the Brain. Nature Reviews Neuroscience, 2002, 3: 201–215.

[19] Bertin J. Sémiologie Graphique: Les Diagrammes, Les Réseaux, Les Cartes. Paris, FR: Mouton/Gauthier-Villars, 1967.

[20] Mackinlay J. Automating The Design Of Graphical Presentations of Relational Information. ACM Transactions on Graphics (TOG), 1986, 5: 110–141.

[21] DiBiase D, MacEachren A-M, Krygier J B, Reeves C. Animation and the Role of Map Design in Scientific Visualization. Cartography and Geographic Information Systems, 1992, 19: 201–214.

[22] Krygier J. Sound and Geographic Visualization. In: MacEachren A M and Taylor D R F (Editors). Visualization in Modern Cartography. Oxford, UK: Pergamon, 1994: 149–166.

[23] MacEachren A M. How Maps Work. Representation, Visualization and Design. New York: The Guilford Press, 1995.

[24] Buziek G. Eine Konzeption der Kartographischen Visualisierung. University of Hannover, Habilitationsschrift, Hannover, DE, 2001.

[25] Meng L. Missing Theories and Methods in Digital Cartography. In: Proceeding of the 21st International Cartographic Conference. http://129.187.175.5/lfkwebsite/index.php?id=25 [2003].

[26] Kosara R, Mikisch S, Hauser H. Focus+Context Taken Literally. IEEE Computer Graphics and its Applications, 2002, 22: 22–29.

[27] Frueh C, Sammon R, Zakhor A. Automated Texture Mapping of 3D City Models with Oblique Aerial Imagery. In: Proceedings of the 2nd international Symposium on 3D Data Processing, Visualization, and Transmission. http://ieeexplore.ieee.org/xpl/freeabs_all.jsp?arnumber=1335266.

[28] Kada M, Klinec D, Haala N. Facade Texturing for Rendering 3D City Models. In: ASPRS 2005 Annual Conference Geospatial Goes Global: From Your Neighborhood to the Whole Planet. http://www. ifp.uni-stuttgart.de/publications/2005/kada05_asprs.pdf [2005].

[29] Santella A and DeCarlo D. Visual Interest and NPR: An Evaluation and Manifesto. In: Proceedings of the 3rd international symposium on Non- photorealistic animation and rendering. http://portal.acm.org/citation.cfm?id=987669 [2004].

[30] Jähne B. Digitale Bildverarbeitung. Berlin, DE: Springer, 2002.

Chapter 13 An Online Design Review Tool: Agreeing Townscape Rules Among Stakeholders on the Internet *

Zhenjiang Shen, Mitsuhiko Kawakami and Masayasu Tsunekawa

Introduction

The visualization of urban projects on a human scale in a virtual environment is helpful for stakeholders of the city to understand how their urban environment will be transformed after construction and development. This will help in achieving a consensus on design regulations that should precede the construction process. In this chapter, we attempt to visualize the townscape design guidelines for components of the physical environment such as buildings, street furniture and roads. After a consensus among the stakeholders on the design guidelines of an urban district, the design guidelines can be passed on to the local planning committee for reviewing the design of buildings and street furniture. In this chapter, we will discuss how to utilize computer technology to present design guidelines. This will help in the design coordination of different design alternatives of a 3D townscape in a virtual world, thereby enabling members of local committees to participate in a meaningful discussion on townscape design.

Current research

Research concerning the applicability of visualization tools in urban planning and design has recently been conducted around the world in the context of system development and planning practice. In order to develop the visualization of design guidelines for townscape design on the Internet, we first review some of these existing projects on visualization in the field of urban planning and design.

Basically, visualization for urban planning and design can be conducted in two ways. One involves the development of 2D or 3D aggregated spatial patterns as geographic information for presenting the results of planning analysis. The other involves the creation of a 3D human-scale virtual landscape, for presenting the urban environment usually using the Web. Hence, visualization is helpful in enabling stakeholders to share a common image of the future, thereby allowing them to reach a consensus on the design guidelines of their town

---

* Supported by the Japan Society for the Promotion of Science (Scientific Research Project (C) Grant No. 19560613).

according to the results of simulation that are based on different scenarios. CommunityVIZ[1)] employs the above two methods for visualization of urban spatial patterns and virtual landscapes. What If ?[2)] and INDEX[3)] can present different spatial patterns as results from planning and design. Recently, there have been many research reports pertaining to the simulation of land use patterns based on economic, demographic, employment and environmental policies. Moreover, the form of a city gradually changes due to different kinds of urban projects and it is helpful to employ urban land use transport models such as UrbanSim[4)] for visualizing the influence of development projects on communities and the city as a whole.

Comparatively speaking, in the case of local residents living in urban districts and who are concerned about what the state of their properties will be like in the future, the visualization of townscape designs at the human scale is an important issue. Stakeholders must also try to understand how various aspects of urban districts such as the road network, lots, blocks and building facades could be transformed in the future. Meanwhile, the types of street furniture and public facilities that are necessary for daily life along with the design preferences of the city's residents are important topics in formulating the design guidelines of the townscape. For example, in the PP-SDSS public housing project reported by Barton, Plume *et al.*[1]and the Woodberry regeneration project reported by Hudson-Smith, Evans *et al.*[2] the possibility of using visualization tools for generating alternatives regarding for a building layout at the planned site was discussed. Caneparo[3] has presented a shared virtual reality (SVR) system that was applied to a large-scale project — a new railway junction and the surrounding urban area in the city centre of Turin, Italy. SVR is bundled with communication tools to improve collaboration and supports a distributed environment to process information across networks. Since the execution of these tools and our own tools are at a human-scale virtual environment over the Internet, they are different from CommunityVIZ, UrbanSim and INDEX.

To improve public participation in our project, spatial multimedia employed for visualization is an important theme that supports decision-making of townscape design guidelines in planning practice. With the spread of the Internet and development of computer technology, visualization allows people to participate in urban planning right from the stage of how to make the planning information public. Previously, various types of spatial multimedia including text, video, photography and virtual reality along with WebGIS were employed to enable public participation in townscape planning, thereby helping stakeholders acquire a better understanding of the planning and design process[4, 5]. Bulmer[6] argued in favor of digital democracy in the case of public involvement in the planning process through visualization tools such as CAD, 3D-GIS and VRML Visualization tools employing VRML and WebGIS technologies can be hosted on the Internet through the httpd service and can be

---

1) www.communityviz.org.
2) www.what-if-pss. Com.
3) www.crit.com.
4) www.urbansim.org.

freely accessed by local communities; this cost-effective aspect of the tool allows for further public participation. There also exist research reports on visualization using Cave Automatic Virtual Environment (CAVE) and augmented reality[7] which are expensive in terms of public involvement. In our project, the documents of planning information are made public through the Internet; this allows a greater number of stakeholders to freely access information without spatial and temporal limitations. In this chapter, we intend to discuss how to visualize design guidelines for townscape design using VRML; hence we pay more attention to the development and implementation of a visualization tool according to the requirements of committee members in a local city of Japan.

Online design tools that will enable design coordination will facilitate the process of collecting planning and design proposals of different stakeholders. Many projects have been undertaken for the development of such online design tools. A visualization tool for the design of a public park was introduced as an online design collaboration system[8] for collecting participants' designs on the Internet. In this system, participants can create new ground surfaces and arrange the design of or select elements at the planning site in the VRML world. A similar work based on the visualization of planning and design alternatives through public involvement is proposed for public park design by Pettit and Nelson *et al.*[9]. With regard to townscape design, there exist research reports on the Dazaifu project[10] in Japan that discussed the potential and limitations of a visualization tool in encouraging public involvement in the design process for the purpose of reaching a consensus on the design guidelines of townscape rules at a district level. However, in this project, a virtual reality (VR) representation of the town was only shown in an exhibition hall for residents. Further investigation revealed that owners of private properties would be apprehensive and could get offended if the planning information of their properties were to be made public. These online design tools are developed for users to remotely design the layout of site plans according to their preferences; our project is a multi-user tool for stakeholders in different locations to remotely discuss various alternatives of design guidelines with each other while an operator coordinates different design options and presents a design guideline according to the requirements of the members in committee meetings. With respect to the coordination requirements of online design, a prototype of a multi-user tool for human or artificial design agents in a distributed network[11] is presented for discussing the possibility of generating a coordinated design. The aim of our project is to show how a visualization tool can be employed for planning practice.

For assistance in bottom-up decision-making, an online learning system to help users gain a better understanding of planning issues has been researched extensively. It is also important to help stakeholders utilize the visualization tool and learn from their experience in the virtual world. Hamilton, Trodd *et al.*[12] related aspects of learning systems to the features of the Internet, VR and GIS technologies, which are used to create models; they argued about the potential of such systems as agents for change in attitudes towards decision making in the urban scenario. Ieronutti and Chittaro[13] were involved in research on virtual education illustrating the use of virtual humans in environments (EVEs) and presented a software architecture that allows Web3D content creators to integrate interactive virtual humans who can play the role of teachers, assistants or guides in these environments. In this chapter, the implementation of a visualization tool is discussed according to the planning practice in Japan; a knowledge-base system accompanied with a visualization tool, however, will constitute future research.

As described above, present computer technologies such as virtual reality and WebGIS on the Internet can be employed by users across locations to come together and create a virtual town by presenting, sharing and selecting different design elements. There are reports on the opening up planning information, online design coordination and learning systems to assist users in different types of urban development projects. Meeting the requirements obtained from planning practice in the real world is an important step for developing such a tool. In our project, we consider a multi-user online visualization tool that allows a facilitator to combine different design options and present a design guideline that conforms to the requirements of the members in a committee meeting. By using this tool, it will be possible for members to reach a consensus on the design guidelines.

Research objectives-visualizing townscape

There exist design guidelines for different kinds of urban projects such as main street projects, residential improvements, industrial park developments, central commercial business districts redevelopment and shopping centre developments. Visualization tools will enable the public to gain a better understanding of the design guidelines. The design elements in the guidelines include the form, color and style of street furniture, reconstruction buildings and site layout in urban space. These elements are necessary for creating a reasonable composition of urban space that caters to urban activities in business, commercial or residential spaces. For this purpose, stakeholders should conform to the construction behaviors prescribed in the guidelines.

Moreover, if the project is related to public spaces like parks, it is easy to make planning information public in a planning committee because there is no private information; however, there are privacy issues associated with making design guidelines of privately owned properties public. It is impossible to make decisions on private ownership properties in a public planning committee[1]. In planning practice, even though a consensus is reached among stakeholders on the basic design guidelines of privately owned properties in urban districts, there are no concrete conclusions about each private building. Therefore, a design review process is necessary for checking the plan and design of buildings in the planning site, which will ensure that the owners of private properties adhere to the guidelines.

In this chapter, we focus on how to visualize design guidelines for a main street project in a local city centre in Japan. The local organization strives for 'Matsudukuri', in which the objective is to propose a design guideline for improving the residential environment and increasing the attractiveness of the shopping environment. Local residents attend public workshops for exchanging opinions regarding architectural style and usage of public space. Finally, residents will have to reach a consensus over a unified design style for buildings and public spaces.

Visualization alternatives of public spaces-consensus for design guideline of townscape

In order to arrive at a consensus on the design principles for townscape design, stakeholders, administrative officers and developers conduct workshops with local residents living at the planning site according to the planning process described in the urban law. At this stage, the workshop is conducted every few months over a short-time period. When the design committee reaches a consensus, the rules for townscape design should be opened to the public, who in turn, are under obligation to adhere to the rules.

A report presented by Tanoue and Arima[10] reveals that private residents are offended when information about their properties is discussed in a public committee. Owing to this, a consensus can be reached on only abstract design principles. Therefore, visualizing alternatives about street furniture in public spaces is effective in a public committee. In opposite fashion, the design guidelines for private buildings are difficult to review carefully in a public forum.

Visualization alternatives of private buildings-design review during formative proess of townscape

The design and construction of a townscape is a long and formative process undertaken by local owners and developers. Local citizens and developers must understand the design guidelines and make a note of the key ones. For this purpose, the design of buildings to be reconstructed or new buildings should be checked carefully so that design guidelines are effective. Therefore, a committee is set up for design review after this deliberation and participation. In the design review process, visualizing design alternatives of buildings is helpful to check whether or not the design guidelines are adhered to. Precedents such as these will help new developers conform to the design guidelines. Accordingly, the buildings that pass the design review can be added to the database and opened to the public for reference.

Recently, most visualization tools have been developed for gaining consensus, that is, a short-term project for making planning information public and gaining consensus on the design guidelines; a few of them are employed as a design review system for long-term management. Accordingly, the visualization tools for design review also provide a new attempt in planning support systems.

The Research Approach: Visualization Tools on the Internet

To improve the visualization tool in planning practice, it is important to understand the actual requirements. For this, we consider an urban project carried out in Nanao city, Japan as an example for systems analysis and we develop a visualization tool according to the requirements of planning practice. In this chapter, we focus on how to develop a tool for visualizing a townscape in the case of a town in which a road extension project is being carried out.

It is important to know the key requirements of the visualization tool for the purpose of designing a townscape. As stated above, there are two phases for achieving consensus on design guidelines. The first is the visualization of design alternatives of public spaces for gaining consensus on design guidelines before executing these urban projects; the second is the visualization of design alternatives of private buildings for design review during a long formative process of townscape design in an urban district. Accordingly, we suggest that three processes should be employed during visualization. They are as follows: the design consensus for presenting options about street furniture in a public space, the design review process for reconstruction of private buildings, and the long-term formation process that involves opening up different aspects of the virtual world to the public after the design review.

Furthermore, even though electronic media such as digital photography and film can be employed for visualizing planning alternatives, they do not provide the kind of interactive environment that is required. For developing the visualization tool, aspects of virtual reality which the Internet offers, is taken into account for the review process or for the purpose of reaching a consensus; VRML technology is helpful to formulate a common image between members of committee meetings for obtaining a better understanding of the proposed guideline. In the VRML world, images of different design elements are visually perceptible and inter-calibration between different design elements is possible. The VRML world can be distributed on the Internet, whereby participants can access information they require without spatial and temporal limitations. For these reasons, we employed VRML for the development of the visualization tools.

Planning Practice in Townscape Design in Japan

The process of reaching a consensus on the design guidelines of the town is called 'Machidukuri' in Japan. Recently, local residents have begun their Machitsukuri projects on a local level as a process of bottom-up decision-making, as more and more people are unwilling to depend on the central government to kick start their projects.

Traditional methods in planning practice

There exist research reports[14] on the process of reaching a consensus over design guidelines with the help of traditional methods such as field surveys, questionnaires, votes and other means that allow effective deliberation and consensus on design guidelines. With regard to making proposals for the design review process in a traditional planning committee for public participation, participants can present their concepts by preparing documents, drawing illustrations and models. However, if only the draft drawings and documents are submitted, the degree of consensus achieved during the deliberation process might be limited to some extent as stakeholders would be unable to share a common image of the plan proposed; this is difficult without a visualization tool.

A real model[15] is used for sharing images with committee members and a CCD camera is used to capture the real model on TV so that the committee members can access the model from different viewpoints. However, the model needs a permanent space for exhibition and storage and in order to show different design alternatives, model materials will need to be replaced.

In our projects, even though a new visualization tool is employed, the draft drawings and documents are also available as basic information for the plan alternatives. They are prepared as HTML documents and hosted on the Internet for free access to everyone without any temporal or spatial limitation. On the other hand, traditional planning committee meetings can only be held at predetermined times and places.

Townscape design with the road extension project in Nanao City

As shown in Figure 1 and Figure 2, the improvement of the townscape design in the central area of Nanao City (Nanao Project) was accompanied by a road extension project that was implemented by the local regional government from 2004. Local officers encouraged residents living in the central area to adjust their private land parcels and reconstruct their buildings in order to finalize the road extension project and thus improve the townscape. The government's budget for road extension only includes road extension and street furniture; reconstruction of private houses has to be carried out by the respective owners. In other words, road extension is undertaken by the local government whose planning and design is the job of planners and designers. The adjustment of land lots and reconstruction of private houses is a task that has to be carried out by the owners. However, there is a common space that exists between the road and the private building. This space consists of the pedestrian road and the street furniture on it along with the façade of private buildings; these elements are topics of deliberation and a consensus is required on their design guidelines. Therefore,

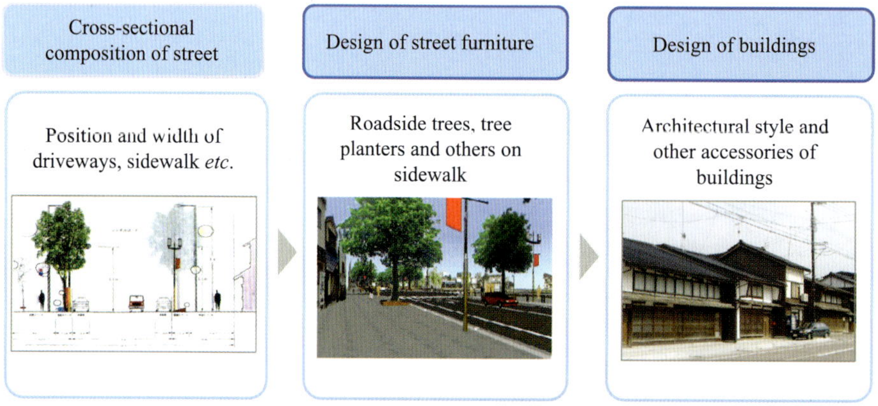

Figure 1 The three steps of urban project in Nanao City

Figure 2 VRML world of Nanao project area based on GIS database

the design guideline of townscape includes two sectors—public space and private space. In this case, except for the cross-sectional design of the road, which has already been constructed, the design guidelines for reconstruction of private houses, street furniture are awaiting consensus in the open committee.

Consensus for design guideline regarding street furniture in Nanao Project

In the case of the Nanao road extension project, a local committee has been set up for reaching a consensus on the design guideline, as shown in Figure 3. A facilitator is chosen from an NPO and 10 residents are invited to be the members of the committee. The officers of the local government and planners of the consultant company are a part of a working group to present design alternatives about street furniture and building design codes. There are also some experts and professors who attend the committee meetings and comment on various alternatives. The local committee meetings were held once a month from September 2004 to December 2004.

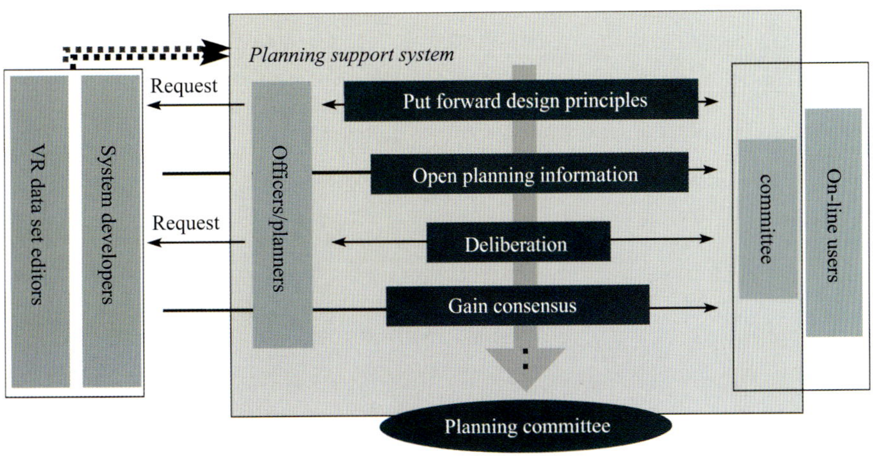

Figure 3 Committee meetings, system developers and VR data editors

As an outcome of the committee meeting, one of the design alternatives of the street furniture was chosen and the design guidelines called design codes of private buildings arrived at a consensus-based agreement.

Design review for private buildings in Nanao Project

A committee appointed for reviewing the design is established after a consensus on the design guideline is reached in order to review whether or not the reconstruction is in accordance with the design guideline. With respect to the review process of private properties, architects usually apply for construction permissions on behalf of their owners in a local city hall. When an architect applies for construction permission, officers transfer the necessary documents to an after-committee for design review according to the design guideline before formal authorization as designated by the law. This is required in order to make sure that the construction plan is not inconsistent with the design codes. Moreover, tools on the Internet can also be utilized for remote accesses in order to hold committee meetings without spatial and temporal limitations. The visualization tools not only visualize design alternatives, but also act as a platform to allow stakeholders to share common images of townscape design in a virtual environment. The tools can also make planning information public and allow local residents to understand the kind of design that can be accepted by the design review committee for reconstruction.

Visualization Tools Implemented in the Nanao Project

System framework using VR platform on the Internet

The project involving Alpha World by Smith, Dodge *et al.*[16] has proven that visualization technology developed for a multi-user environment opens up possibilities for each user to build his/her own space and communicate in a virtual world. The Alpha World Project was not a real construction project but it can be considered as an initial step for supporting planning practice. As a follow-up to this project, the Ryoanji Project[17], which was developed as a game for assigning the positioning of stones in a garden, adopted a similar remote coordinated design approach.

In the Nanao Project, a web application is considered as the core part of the tool, as shown in Figure 4, similar to the Ryoanji Project. An application server of the virtual reality platform[1)] based on the HTTP server is employed for users to share and coordinate different design alternatives on the Internet. Through this, the application server committee members can discuss alternatives using the 'sharevent' function of the Blaxxun platform in the virtual world; free remote access is also available for all residents in the planning site. Similar VR platforms such as Second Life and Active Worlds also exist. However, Blaxxun Inc. provides a free license for researchers to employ the Blaxxun community server as a VR plat-

---

1) www.blaxxun.com.

form for pure research and social experimentation. With this platform, virtual communities can be created by users across locations and shared for presenting and coordinating design alternatives for the townscape.

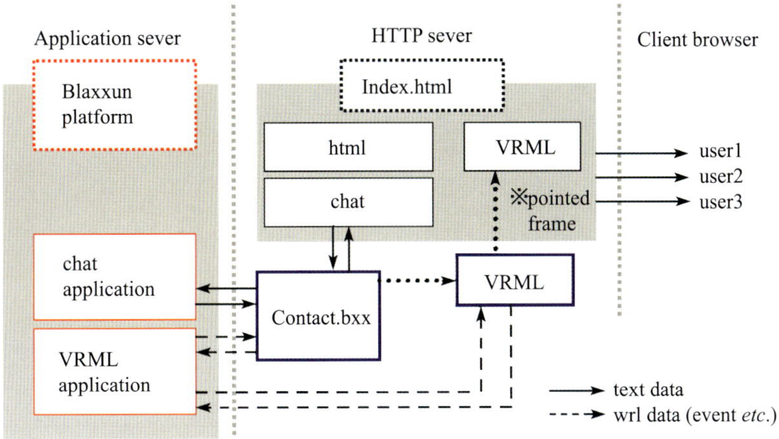

Figure 4 Core part of the system framework (with Blaaxun platform)

Consensus on design elements regarding street furniture using the tool

A visualization tool is essentially an enhancement of the traditional methods; however, it cannot replace all of these. For example, a field survey is necessary regardless of whether or not a new visualization tool is employed in the workshop. The tool can only be used to visualize design alternatives presented by the working group of the local planning committee.

In practice, the committee is composed of officers, planners, professors and local residents discussing design alternatives about street furniture in a public space and about the design principles for private buildings. Table 1 shows the design elements discussed in the committee; the visualization tool allows for deliberation of different alternatives in the committee. This tool is developed as a web application on the Internet which can be accessed by the committee from a remote site allowing them to experience these alternate worlds without spatial and temporal limitations. Figure 5 shows the interface for remote access and Figure 6 shows the combination of different design elements.

The working group put forward a design guideline before the committee, in which the design elements were separators between the sidewalk and the driveway, separators between the sidewalk and the park, roadside trees, and many other design elements such as the window style, roof style, wall color and so on. In the first meeting, the members checked the design of tree planters, roadsides trees carefully along with the separators through experiencing the virtual world in the next meeting. In the second meeting, the members finally reached a consensus on the separators between the sidewalk, park and driveway; meanwhile, the members wanted to check the position of the guide boards along the main street in the

Table 1 Design elements for street furniture in public space

Classification	Rules	Alternatives
Street furniture	Road side trees	
	Tree planters	
	Separators in sidewalk/driveway	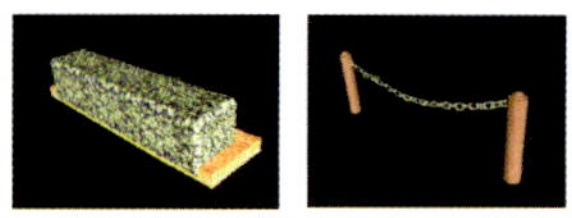 Separators in sidewalk/park 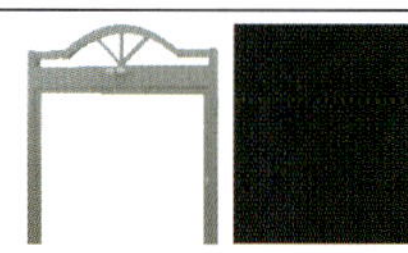
	Flowerbeds	

Figure 5 One specific alternatives for the public space

next meeting. In the third meeting, a consensus was reached on the design of all street furniture along the main street according to the design guidelines. The positioning of the guide boards along the main street was also checked in the virtual world. In other words, the VR dataset was exchanged every time in order to maintain consistency with the requirements of members in the committee meetings.

Even though a consensus was reached regarding the design elements to be used in public places, a consensus regarding the design guidelines of private buildings was deliberated at an abstract level as shown in Table 2. Design alternatives for private buildings with different architectural styles were also expressed in the committee; however, stakeholders refrained from deliberation with respect to their private ownership properties. As a result, the current private buildings were visualized as the background for deliberating on street furniture in the

A. Japanese Judas tree (spring) B. Japanese Judas tree (autumn)

C. Chestnut (spring) D. Chestnut (autumn)

Figure 6 An array of alternatives for the public space

Table 2 Design codes of private buildings

		Design review codes (Authorization of local planning committee is necessary before construction)	Necessity	Consideration in design
A	A1	Roof shape	Yes	
		1) Sloping roofs at a prescribed angle	Yes	
		2) Parallel to the road		
		Front façade		Yes
		1) Traditional painting and traditional windows design		Yes
		2) Natural materials (wood and brick)		
	A2	Color of exterior wall		Yes
		1) Harmony with the colors of natural materials	Yes	
		2) The brown, beige color or white and black design if no natural materials are used	Yes	
		3) Only black used for roofs		
B	B1	Green spaces		Yes
	B2	Shopping store on the 1st floor		Yes
		Show windows facing the road		Yes
		Ease with which tenants can open stores if owners do not open stores	Yes	
		No adult entertainment shop	Yes	
	B3	Advertisement considering townscape	Yes	
		1) No other advertisement beside owners' advertisement	Yes	
		2) Advertisement stand under the eaves		
		3) Design for covering air-conditioning and other machines	Yes	
		4) No parking spaces occupying pedestrians path	Yes	

public space. The committee was still faced with an issue, that is, of formulating design guidelines for private buildings which needed to be reviewed case by case.

Design review regarding private buildings using the tool

After reaching a consensus on the design guidelines regarding private buildings, a committee to review these designs was set up from 2005. There are three cases of design review already in practice. In the design review process, an architect who designs the reconstruction plans for local owners sends the design alternatives for review. The architect can provide the drawings or blueprint; thus, the task of the committee is to recreate the drawings in the virtual world. If we use the prints submitted to the committee as is, the reconstructed building will appear as a low-quality images in the virtual world. Certainly, if the designer's drawings are created and rendered in CAD or 3D-CG, they will be more helpful in creating a more photorealistic virtual world.

When design alternatives are presented in the virtual world, the committee reviews the design from the viewpoint of which alternative is in harmony with the townscape and whether the alternative conforms to the design guidelines. A verdict on the conformation of the design principles is actually easy using the traditional method which involves checking each design code individually. However, whether or not the design is in harmony with the townscape is very difficult to determine by using the traditional method. Thus, the visualization tool for visualizing private buildings is helpful in the design review process. After the first meeting, committee members determine what should be reviewed. The reviewed elements are always elements such as the flowerbeds, wall colors and parking spaces with lawns in front of private properties as shown in Figure 7; if necessary, the design styles of the windows, walls or doors in the facade of private buildings should be also reviewed as shown in Figure 8.

Figure 7 A tool for design elements relative to public spaces in front of a private building for design review

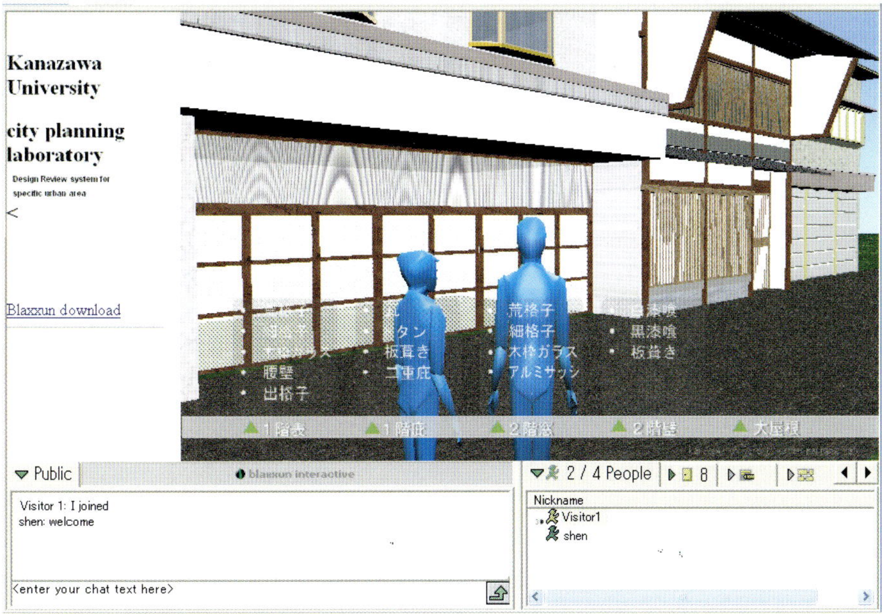

Figure 8 Design review for the facade of a private building

Long-term formative process and database maintenance

The construction of a townscape is a gradual process that requires long-term planning and management in order to control the design of private buildings that are to be reconstructed.

Figure 9 Open to the public for participation and engagement after the design review

After the process of design review, the reconstruction of private buildings will be carried out. The buildings that passed the design review can be added to the VR database for reference as shown in Figure 9. Therefore, the maintenance of the VR database is necessary for long-term management of the townscape.

Stakeholders' Evaluation Regarding the Effectiveness of the Tool

As described above, a consensus on the design guidelines was reached among stakeholders in the committee meeting scheduled in December 2004. When the design guidelines were employed for the design review from 2005, a committee set up comprising the same members as those in the committee meeting (shown in Table 3) was convened for reviewing the design guidelines for private buildings that were to be reconstructed in the urban district. However, the design review for private buildings in the urban district is a long process and it is still difficult to evaluate the tool for design review.

For evaluating whether the visualization tool is effective for presenting the design guidelines and reviewing the reconstruction alternatives of private buildings, a questionnaire was administered by the committee members during the meeting. The questionnaire comprised questions on VR representation and design coordination during the committee meeting.

Table 3 Commission members (from September 2004)

	Committee members	Head count
Facilitator	NPO	1
Experts	Landscape architects, color coordinator, professor	4
Residents	Land owners and leaseholders	10
Executive office	Officers in Nanao city, Ishikawa province, planners in consultant companies	8

Tables 4 and 5 show the results of the questionnaire. Seventeen out of 23 people in the committee said that that the VR dataset represents the alternatives well because the operator using the visualization tool, can coordinate different design options freely and present the design guidelines according to the requirements of committee members. Otherwise, a new VR data set with different alternatives was created according to the requirements from the last meeting and presented to the next meeting. Eleven people had asked the operator for a walkthrough around different places in the virtual world in order to review each design element carefully. In discussion, while the planners explained design alternatives with the help of documents and visuals, eleven people said that they had tried to understand some points through the VR representation. For example, they felt it difficult to understand an oral explanation of whether the color of streetlights was in harmony with the color of separators between the driveway and sideway. In deliberation, eight people admitted that they were motivated to ask planners to explain some of the points in detail. However, only four people believed in the possibilities of the virtual world in encouraging a meaningful discussion concerning urban planning. The remaining felt that merely a shared common image of the virtual world is not enough to reach a consensus on the design guidelines of the town. In addition, lack of expertise regarding product performance evaluations of each design element that would be constructed on the planning site was cited as another reason that may prevent the wide-scale adoption of the tool.

Table 4 Effectiveness of VR representation

	Respondents 23 (100.0%)
Do you agree that virtual reality can represent the alternatives well?	17(73.9%)
Do you walk though the virtual world to check out the design elements from different viewpoints?	11(47.8%)
Do you understand some of the topics through VR that are difficult to explain in words?	11(47.8%)
Do you think the VR tool motivated you to get involved in the deliberation of the alternatives?	8(34.8%)
Is it possible to engage in a meaningful discussion regarding urban planning in virtual reality?	4(17.4%)

Furthermore, as shown in Table 5, 89.5% of the respondents stated that formulating a common image between stakeholders was effective. When asked whether the visualization tool helped them work out the concept of design alternatives, 89.5% of them responded positively. 73.7% of the respondents believed that the tool was helpful in improving the deliberation process in the meetings. In addition, 84.2% confirmed that the tool was helpful in pointing to the improved elements of the alternatives and 73.7% said that the tool helped them make an effective evaluation of the design alternatives. It can be seen that on the whole, the tool received positive responses. This indicates that the tool had a positive effect in the design review stage as well as during comparison and discussion of design alternatives. Thus, the visualization tool is effective for obtaining a better understanding of design guidelines presented by the planners in a virtual environment. The visualization tool is also helpful in formulating a common image between stakeholders regarding townscape design because alternatives about the design guidelines are represented in a comprehensible manner. It also motivates the public to participate and deliberate on various aspects of planning and design during committee meetings.

Table 5 Evaluation about the visualization tool in the meetings

	Fully effective	Effective	Intermediate	Less effective
Helpful to formulate a common image between stakeholders	13(68.4%)	4(21.1%)	2(10.5%)	
Helpful to work out the entire concept of design alternatives	6(31.6%)	11(57.9%)	2(10.5%)	
Helpful to improve deliberation in meetings	5(26.3%)	9(47.4%)	3(15.8%)	2(10.5%)
Helpful to confirm the improved points of the alternatives	8(42.1%)	8(42.1%)	2(10.5%)	1(5.3%)
Helpful to make an evaluation of the alternatives	5(26.3%)	9(47.4%)	5(26.3%)	

Note: No respondents to "No effective".

Conclusion

Virtual reality and the Internet are advanced technologies, and employing them for tasks like urban planning and design can be a challenge. The technologies of Second Life and Active Worlds are very useful for visualizing planning and design and the use of these technologies for planning support is gaining popularity. In this chapter, we employ the Blaxxun platform for developing a visualization tool for representing a 3D townscape and discuss its possibility in encouraging public participation and its role in achieving a consensus on design guidelines for public spaces and private buildings.

There are many case studies on the application of VRML technology to urban planning and design; most of them were developed as a set of design alternatives for deliberation in public meetings. It is easy to reach a consensus over the street furniture in public spaces in an opening committee; however, it is difficult to reach a consensus on the design guidelines of a private building in a public forum with an open committee. Therefore, a VR dataset should be prepared at two parts including both public spaces and private buildings for visualizing townscape. We suggested that three processes be employed for visualization of planning practice. They are as follows: 1) The process of design guideline consensus for presenting options of street furniture for public spaces, 2) The design review process for reconstruction of private buildings and 3) The long-term formation process while opening up different aspects of the virtual world to the public after the design review.

In order to make planning committee meetings more interactive, we deliberated on how to prepare the VR dataset and employ the visualization tool according to the requirements of planning practice. For this project, we developed an online visualization tool in which the VR dataset can exchange in real time in response to the requirements in the meetings. Different design options can be coordinated by a facilitator in a multi-user environment over the Internet in order to present design guidelines that conform to the requirements of committee members and help them in reaching a consensus on the design guidelines for the town. Otherwise, the visualization tool in our project has to be developed according to the committee agenda. We prepared the VR dataset according to the requirements of the last committee meeting and presented them in the next committee meeting. Therefore, the VR representations matched the requirements of the committee well.

A questionnaire was administered to the committee members for evaluating whether the visualization tool is effective in representing alternatives of design guideline and for reaching a consensus through the Internet. As a result, it was useful for representing alternatives for design guidelines of street furniture and reviewing reconstruction alternatives of private buildings during the long and formative process of townscape planning.

This project is limited to an urban district; however, if the whole city can be simulated in a virtual environment, all the planning and design information of city planning committee meetings can be hosted on the Internet for the public. Even though it is difficult for private information to be open to public, its VR representation after design review can be made public.

It is important to discuss how these visualization tools can support different types of urban projects. However, it is not feasible to do so for each project as there are many types of urban development projects that have different parameters and requirements based on the geographical region and type of urban system that is being simulated. It is therefore necessary to determine different requirements from different urban projects and organize them as similar system components for a general purpose analysis of the planning process. Otherwise, a knowledge-based database system will help participants gain a basic understanding of the design elements in public spaces and functions. Similar experiences in other regions including spatial utilizations, management and maintenance are also helpful if there exists a knowledge-based database system. Therefore, an on-line ontology-based data retrieval tool will be helpful for gaining general knowledge of the design elements.

References

[1] Barton J, Plume J, Parolin B. Public Participation in a Spatial Decision Support System for Public Housing. Computers, Environment and Urban Systems, 2005, 29: 630–652.

[2] Hudson-Smith A, Evans S, Batty M, Batty S. Online Participation: The Woodberry Down Experiment. Centre for Advanced Spatial Analysis (CASA), University College London, Working Paper Series, http://www.casa.ucl.ac.uk/working_papers/paper60.pdf [2002].

[3] Caneparo L. Shared Virtual Reality for Design and Management: The Porta Susa Project. Automation in Construction, 2001, 10: 217–228.

[4] Shiffer M J. Spatial Multimedia for Planning Support. In: Brail R K and Klosterman R E (Editors). Planning Support Systems. Redlands, CA: ESRI Press, 2001: 309–385.

[5] Moon T H. Development of Web-Based Public Participation and Collaborative Planning System.In: The 8th International Conference on Computers in Urban Planning and Urban Management (CUPUM), Sendai, Japan, 2003.

[6] Bulmer D. How Can Computer Simulated Visualizations of the Built Environment Facilitate Better Public Participation in the Planning Process? http://www.casa.ucl.ac.uk/planning/articles.htm [2001].

[7] Penn A, Mottram C, Fatah gen Schieck A, Wittkaemper M, Stoerring M, Romell O, Strothmann A, and Aish F. Augmented Reality Meeting Table: A Novel Multi-User Interface jor Architectural Design. In: van Leeuwen J P and Timmermans H (Editors). Recent Advances in Design and Decision Support Systems in Architecture and Urban Planning. Dordrecht, Holland: Kluwer, 2004: 213–231.

[8] Shen Z J, Kawakami M, Kishimoto K. Study on Development of On-Line Cooperative Planning and Design System Using VRML And JAVA, A Case Study on Public Park Planning and Design. In: The 8th International Conference on Computers in Urban Planning and Urban Management (CUPUM), Sendai, Japan, 2003.

[9] Pettit C, Nelson A, Cartwright W. Using On-Line Geographical Visualisation Tools to Improve Land Use Decision-Making with a Bottom-Up Community Participatory Approach. In: van Leeuwen J P and Timmermans H (Editors). Recent Advances in Design and Decision Support Systems in Archi-

tecture and Urban Planning. Dordrecht, Holland: Kluwer, 2004: 53–68.
[10] Tanoue Y and Arima T. Development of VR System for Sharing the Image of Space in Community Design Workshops. In: Proceedings of the Twenty-Eighth Symposium on Computer Technology of Information Systems and Applications. 2005, 28: 49–54.
[11] Zamenopoulos T and Alexiou K. Structuring the Plan-Design Process as a Coordination Problem: The Paradigm of Distributed Learning Control Coordination. In: Longley P and Batty M (Editors). Advanced Spatial Analysis. Redlands, CA: ESRI Press, 2003: 407–426.
[12] Hamilton A , Trodd N, Zhang X, Fernando T, Watson K. Learning Through Visual Systems to Enhance the Urban Planning Process. Environment and Planning B: Planning and Design, 2001, 28: 833–845.
[13] Ieronutti L and Chittaro L. Employing Virtual Humans for Education and Training in X3D/VRML Worlds. Computers and Education, 2007, 49: 93–109.
[14] Shinobe H. Study on Community Design Learning at Primary Schools connected with Local Community-Practice and Evaluation of Community Design Learning on Park Design as a Project Based Learnin. Journal of the City Planning Institute of Japan, 2005, 40: 499–504.
[15] Shimura H and Satoh S. Process Design of the Simulation and Gaming for Community Planning. Journal of the City Planning Institute of Japan, 2001, 36: 691–696.
[16] Smith A, Dodge M, and Doyle S. Visual Communication in Urban Planning and Urban Design. Centre for Advanced Spatial Analysis (CASA), University College London, Working Paper Series, http://www.casa.ucl.ac.uk/working_papers/paper2.pdf [1998].
[17] Okabe A, Sato T, Okata J, Okunuki K. A Study on an Internet-Based Decision Support System for City Design: A Virtual Environment for Interactive Operations with Three Dimensional Objects. Center for Spatial Information Science, University of Tokyo, Tokyo, Japan. 1999.

Chapter 14 O' Jerusalem: A Decision Support System for Jerusalem (O JDSS)

Juval Portugali, Asaf Roz, Dan Gonen, Oz Shalom
and the ESLab team

"*... upon thy walls, O Jerusalem...*" (*Isaiah* 62:6)

The Political Issue

Since 2004, the Israeli government has been building a barrier between Israel and the "West Bank". Most of it is a fence as shown in Figure 1 and part of it, specifically in the urban areas, is a wall as shown in Figure 2. The most well known wall is the one that was (and is still being) built in Jerusalem. The construction of the fence/wall is associated with an on-going controversy in Israel and the world. The Palestinians object it for it disturbs the life of thousands in some parts of the annexed territories. The Israeli settlers object to it as it disconnects them from Israel. The majority of Israeli public supports it for it decreases suicide bombings but many have reservations about its location and course. The impression one gets from the international media, particularly in the West, is that the majority of the European public objects to it.

Figure 1 The fence

Figure 2 The wall

Systematic Distortions in Cognitive Maps

A *cognitive map* is the term first used by the psychologist Tolman[1] to information constructed in the mind/brain of animals, including humans, about large scale extended environments. Subsequent studies have applied the notion of cognitive maps to a variety of domains, including the way humans represent in their mind spatial objects such as buildings, neighborhoods, roads, boundaries, cities, landscapes and the like. "Systematic distortions in cognitive maps" is a phenomenon by which the mind/brain constructs distorted cognitive maps of the environment in question; not because of some malfunction in the mind/brain, but on the contrary—because it is working properly. The study of systematic distortions in cognitive maps is used, on the one hand, as means to expose the way the mind encodes, stores and decode geographical information. On the other hand, it is used to shed light on the differences and similarities between the cognitive map and the "real" structure of environments, and the implications thereof to humans' behavior in space[2-7].

In an empirical study, Israeli students were asked to draw the so-called "Green Line" which is the boundary between Israel and the West Bank[8]. The aim was to expose the students' cognitive maps of the Green Line. That is, what they have in mind when they hear, or talk about, notions such as the Green Line, the West Bank, Judea and Samaria; what images do they construct in their mind about the above geographical entities?

Figure 3A is an example of a 'correct' drawing (taking into consideration the variability in the drawing capabilities of the subjects). About one third of the subjects (33%) responded correctly. Another common response was the one presented in Figure 3B in which the West Bank is drawn as an area north of Jerusalem. About 23% of the students responded with such a map. The majority of the subjects responded in a variety of ways, all of which were distorted with some of the maps exaggerating the area of the West Bank, as in Figure 3C and some by drawing maps that minimized its size as in Figure 3D. The results indicate that the vast majority of the students' cognitive maps of the Green Line are distorted. This was found to be the case also with the vast majority of Israeli society and there is ground to assume that this is the case probably also with world society.

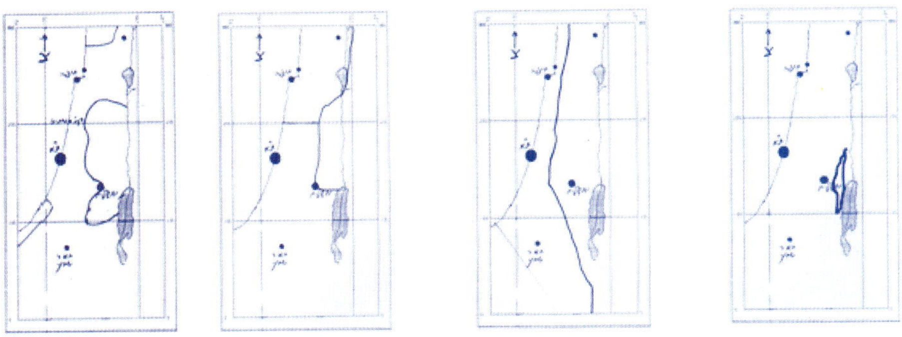

A B C D

Figure 3 Cognitive maps of the Green Line as drawn by students

There are three important implications from this study. First, with respect to the theory of cognitive maps: what we have here is a new form of systematic distortion—"Systematic distortions due to partial information", that is, the mind/brain constructs cognitive maps even when the information for their construction is insufficient. Second, with respect to the public discourse that takes place in Israel: one of the most controversial issues in Israel's social discourse is conducted on the basis of poor and distorted information. Third, with respect to the public discourse that takes place in the world: this controversial issue is conducted on the basis of poor and distorted information.

The above applies to the current controversy over the fence/wall that is currently under construction roughly along the Green Line. The building of the fence/wall is associated with demonstrations and heated public debates, in Israel and abroad, for and against the building of the fence/wall. The fact is, however, that the vast majority of the people involved in the controversy are not familiar with the basic facts: is it a fence or a wall? Where exactly it is a fence and where a wall? What is its course and where? What is its effect on the Palestinians? Israeli Arabs? The Israelis living east of it? west of it?

This situation provided the motivation for the development of O'Jerusalem: the Environmental Simulation Laboratory of Tel Aviv University (ESLab) in collaboration with the Center for Middle East Peace & Economic Cooperation (CPC)[1)] initiated the O'Jerusalem

1) CPC is a non-governmental, non-profit, non-partisan, private organization, based in Washington DC, and strives to promote a just resolution to the Israeli-Arab conflict. The project was supported also by the S. Daniel Abraham Center for International and Regional Studies, Tel Aviv University.

Decision Support System as a computerized tool that will serve governments, organizations, academic institutes, and the general public, in their debates regarding the fence-wall around Jerusalem.

The ESLab

On DSS and PSS

A typical DSS (decision support system) or PSS (planning support system) is built of three components as in Figure 4: a GIS, a family of environmental simulation models (SiMod) and a visualization component such as virtual reality (VR). In some cases, when the system is task specific, a monitoring unite is added to the system[9, 10].

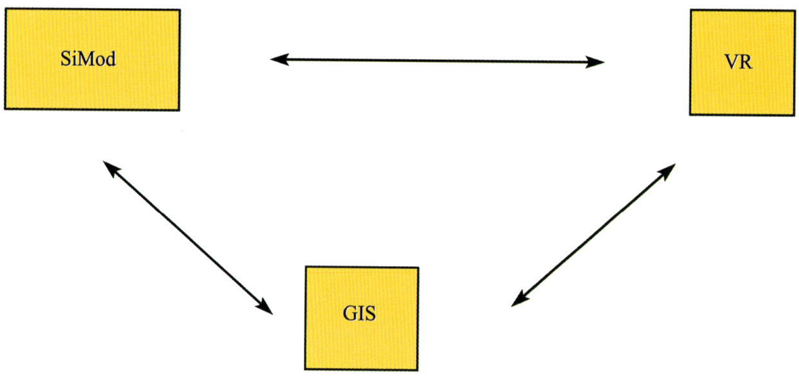

Figure 4 A typical PSS/DSS and the ESLab's GPSS

In line with the above, we have developed in our ESLab a General Purpose Support System (GPSS) composed of two virtual reality (VR) simulators—*MultiGen* and *Skyline*, a family of urban simulation models and geographic information systems (GIS). Figure 4 illustrates this system. Its description now follows.

The two VR simulators

MultiGen Paradigm (MPI) is an architecturally oriented VR simulator. It enables building large-scale virtual environments, such as cities and whole regions, with a high level of architectural realism. Its elementary parts are the *Creator*, which together with the creator terrain studio/generator (CTS), enables building real-time 3D content for use in visual simulation; the *SiteBuilder 3D* allows transformations of the GIS 2D map data into a 3D visual form and the *Vega Prime* enables simulation of movement in environments such as cities. Employing the *MultiGen*, we have already built V-Tel Aviv as we show in Figure 5 below and several other virtual environments which we present elsewhere (for specifications and applications, see Ref. [11].

Our second simulator is the *Skyline*. The main advantage of *Skyline* is that, unlike the

MultiGen, it can be viewed on the Internet. The three elementary parts of the *Skyline* are the *TerraBuilder* which allows users to create, edit and maintain 3D databases, the *TerraExplorer* which allows users to simulate movement through 3D Terra environments, and the *TerraGate*, a network data server technology designed to stream 3D geographic data in real time. Using the *Skyline*, we can simulate real time movement in or above entire countries, regions and cities (for applications, see ref. [11]).

The general process of building a virtual environment is basically the same in both *Skyline* and MPI. The starting step is a digital terrain model (DTM) layer of the area on top of which is superimposed, firstly, an air photograph (orthophoto) and secondly, GIS layers (dxf files) referring to heights of buildings, shapes, facades, symbols, labels and so on. In *Skyline*, a layer of links, labels and symbols can be conveniently added.

A major advantage of *Skyline* is, as noted, the possibility to post it on the Internet—making it an ideal tool for Internet-supported public participation (see below for examples). On the other hand, the architecturally elaborated *MultiGen Paradigm*, with its integrative link to VR helmet-mounted display (HMD) and similar devices, is an effective tool for producing genuine VR environments. The advantage of the ESLab system is its ability to integrate both VR simulators.

The family of urban simulation models

Our point of departure here is the family of urban simulation models as presented in the book *Self-Organization and the City* [12]. Within this project, we developed what we termed 'free agents in cellular space' (FACS) models and 'free agents in real space' (FARS) models. A typical FACS model is built of two layers: a cellular automata (CA) layer representing the infrastructure of the city–its buildings, road network *etc*., and an agent-based (AB) layer that is superimposed on the CA layer and mimics the behavior of the various agents operating in the city. In a FARS model, the abstract cellular space that represents the infrastructure of the city is replaced by the real space of the city as derived from GIS. A further description and elaboration is given in papers by one of the authors[6, 11, 12].

Building O'Jerusalem by Means of Skyline

O'Jerusalem was built by means of the ESLab's GPSS. So far, however, we have employed the GIS and the VR components (as we show in Figure 4 above). Movement in a virtual city (VCity) depends among other things on the realism of the VCity. More specifically, there is a trade-off between the level of details and the flexibility of movement in a VCity: the more details and realism you add, the slower and less flexible is the movement. The solution we developed in the past was to construct a generic building in the city concerned, apply it to the whole city, and then add a small number of detailed buildings. For example, in building virtual Tel Aviv (V-Tel Aviv) we first constructed a typical/generic "Tel-Avivian" building, built the vast majority of buildings in V-Tel Aviv as generic buildings and added a small number of detailed buildings. We have found that a few (less than 1%) detailed urban elements (buildings, squares, *etc.*) are sufficient for people to pattern recognize the city and specific locations in it. The result shown in Figure 5 is only partly satisfactory, however, since during movement at ground level (as a virtual walk or drive), all buildings look identical.

In O'Jerusalem, we developed a new method for producing generic buildings the effect of which is much more realistic and thus satisfactory. Figures 6, 7, and 8 illustrate the resulting VCities and their neighborhoods.

Figure 5　V-Tel Aviv, only a partly satisfactory rendition of building form

Figure 6　A snapshot from V-Ramalla

Figure 7 A snapshot from V-Jerusalem

Figure 8 A snapshot from an industrial zone Nave Yaakov

O'Jerusalem

O'Jerusalem, the software we have developed, allows the users to import existing GIS layers or create new ones as in Figure 9, add a line, present it in 3D Virtual Jerusalem as in Figure 10, transform the line into a fence or a wall as in Figure 11, and then move (“walk”, “drive” or “fly”) in real time along the fence/wall and see how it affects the city and its inhabitants. As can be seen, Figures 9 to 11 are snapshots from O’Jerusalem. Readers interested in experiencing O’Jersusalem are invited to do so by visiting the following website: http://www.eslab.tau.ac.il/. As noted above, so far we have employed GIS and VR components of the ESLab’s GPSS and in the future, we intend to add to O’Jerusalem the SiMod component (shown in Figure 4 above) to improve our ability to evaluate the impact of the fence/wall on all those affected.

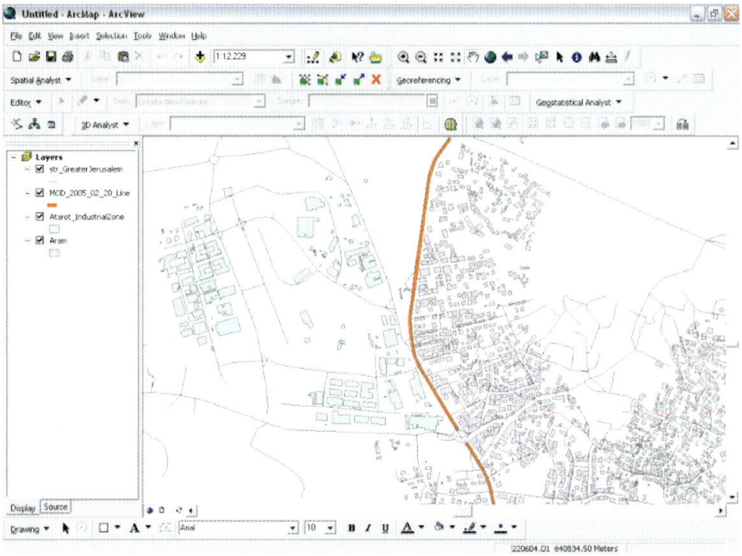

Figure 9　The line as a GIS Layer

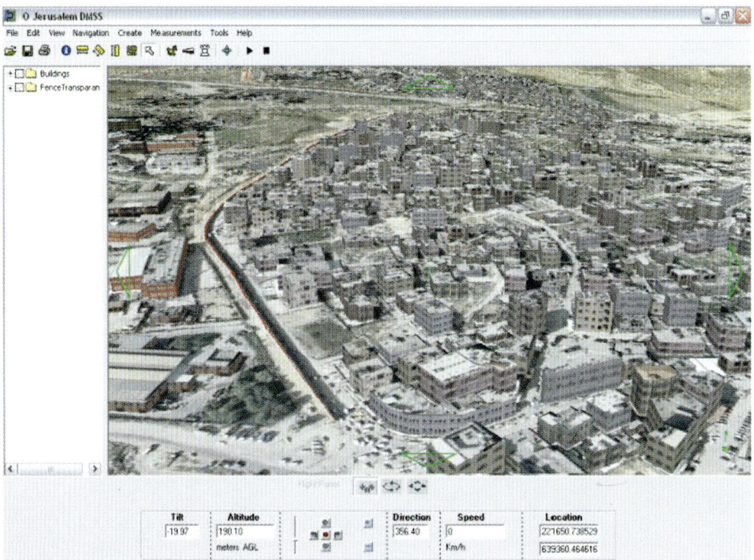

Figure 10　3D presentation of the line in V-Jerusalem

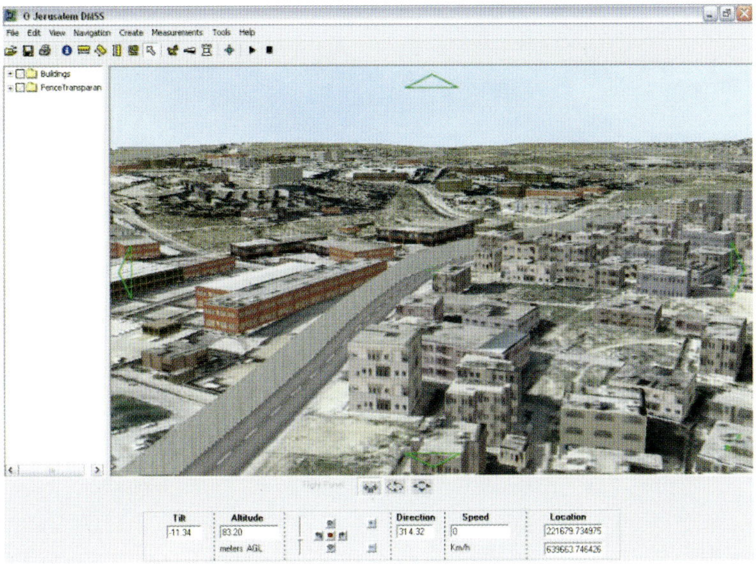

Figure 11 The transformation of the line into a Wall in V-Jerusalem

References

[1] Tolman E. Cognitive Maps in Rats and Men. Psychological Review, 1948, 56: 144–155.

[2] Tversky B. Distortions in Cognitive Maps. Geoforum, 1992, 23: 131–138.

[3] Kitchin R and Freundschuh S (Editors). Cognitive Mapping. London: Routledge, 2000.

[4] Kitchin R and Blades M. The Cognition of Geographic Space. London: I. B. Tauris, 2002.

[5] Portugali J (Editor). The Construction of Cognitive Maps. Dordrecht, Holland: Kluwer Academic Publishers, 1996.

[6] Portugali J. Toward A Cognitive Approach to Urban Modelling. Environment and Planning B: Planning and Design, 2004, 31: 589–613.

[7] Portugali J. Cognitive Maps Are Over 60. In: Lecture Notes in Computer Science. Berlin, DE: Springer-Verlag, 2005: 251–264.

[8] Portugali J. Implicate Relations: Society and Space in the Israeli-Palestinian Conflict. Tel Aviv, Israel: Hakibutz Hameuhad, 1996 (In Hebrew).

[9] Brail R K and Klosterman R E (Editors). Planning Support Systems. Redlands, CA: ESRI Press, 2001.

[10] Brail R K. Planning Support Systems Evolving: When the Rubber Hits the Road. In: Portugali J (Editor). Complex Artificial Environments. Berlin, DE: Springer, 2005: 307–318.

[11] Portugali J (Editor). Complex Artificial Environments. Berlin, DE: Springer, 2005.

[12] Portugali J. Self-Organization and the City. Berlin, DE: Springer, 2000.

Part 4

Constructing Mobile and Networked Virtual Geographic Environments

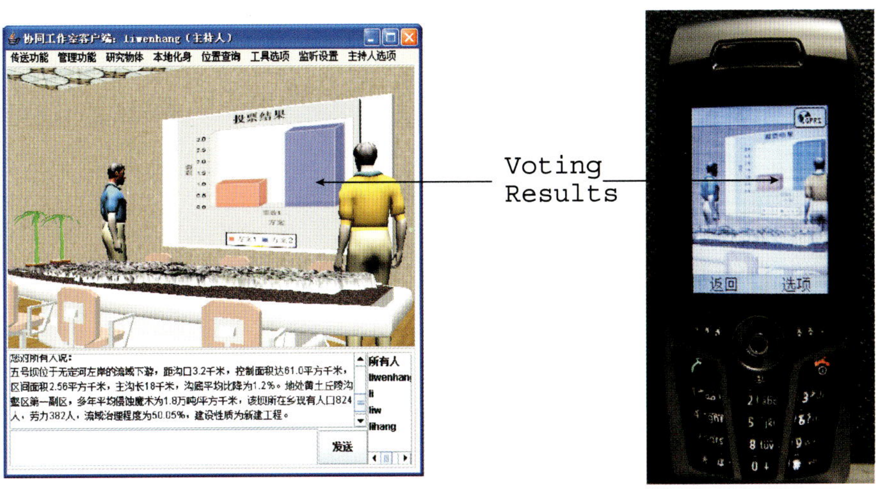

Just as the Internet and Web 2.0 technologies form one cutting edge of current research into VGEs, mobile devices form the other. In fact mobile devices represent not only the edge of hardware innovation at present but also the way in which communication and dissemination can be activated in wireless situations. Design issues are paramount and in this part of the book, the various contributions show how such VGEs are being fashioned where mobility and wireless communications are key to human-computer interaction. In one sense, mobility and wireless technologies form part of a more generic theme within GI Science which involves questions of process, dynamics and time across different scales and in different contexts. Users themselves engage in dynamic behaviors when accessing geographic information, particularly when they themselves are on-the-move. This information may be dynamic in itself and there are complicated interactions which involve mixing different dynamic regimes such as these. Add to this, methods for networking various technologies and for coping with uncertainty and it is clear that there are profound hardware, software and human computer interaction (HCI) construction problems in fashioning such VGEs.

The contributions in this part of the book are slightly more technical than those hitherto, but there are strong links to previous contributions particularly in terms of integrating different technologies, user interface issues and questions of searching and identifying different objects in terms of their information content. Once again mobile devices represent the cutting edge of these systems but also the use of mobile devices in data acquisition and the construction of more fixed systems poses new challenges. In this sense, remote sensing from satellite and related remote technologies are important in terms of the way these environments take data from the real world and the way they integrate it with new forms of networked technologies.

First, Li and his colleagues present two such examples of visualization which explore spatial uncertainty and which benefit from the use of VR technology. First, a real-time simulation of the Global Positioning Systems (GPS) satellite geometry is presented which has been carried out in the virtual environment of a university campus. The number of satellites visible to the receiver is modeled in real time as a user walks through the university campus. The Position Dilution of Precision (PDOP) from these satellites is displayed on the screen while the satellite geometry is visualized in both 3D and aerial views. Since the factors affecting PDOP are inherently three dimensional, communicating the spatial uncertainty of GPS coordinates within an immersive stereo environment has been viewed as a particularly powerful communication tool by those using the system, in this case, students. Secondly, elevations of raw LiDAR (*i.e.* Light Detection and Ranging) points are interpolated from irregularly spaced points at a 1m resolution and a continuous surface model is generated using a variety of interpolation techniques. The resulting surfaces are rendered in a real-time visual simulation environment in which navigation and user interaction are enhanced. Cross-validation results are visualized on top of the surface, while 25 centimeter aerial photos are draped over the terrain in order to provide a spatial context for examining interpolation cross-validation error. The rationale for this interpolation tool is to improve the communication of inherently spatial processes to students finding difficulties in mapping mathematical representations of surfaces and parameter options, in the setting of an interactive and immersive group discussion within a VR theater. In this sense, the user gains confidence that the environment is authentic and realistic.

Yanagisawa and Yoshikawa then present a method for quantifying information displayed on mobile phone screens. To focus on the load associated with searching for clues around destinations in city, they develop a method for information displayed on mobile phone screens through a case study which is applied to the spatial relationships between destination restaurants. The application is located around Tokyo's Shinjuku Station where clues for identifying their locations are sought out by the system. First, maps for areas that include the station are searched using PCs and recognizable buildings on large- and small-scale maps are identified. This system treats recognizable buildings as clues for identifying destinations, and the clues on each map are categorized. Maps for areas that include Shinjuku Station are searched using a mobile phone. Finally, focusing on the process of identifying the location of destination restaurants in Shinjuku by reading maps displayed on mobile phones, the load associated with searching for clues around the destination restaurants is identified and measured.

In the first part of the book, we identified the importance of the various digital earths that are being constructed as a means of focusing software efforts in GIS and VGEs. In this context, Li and his colleagues outline an interesting prospect for the public participation and interaction. Starting from the concept and technologies of digital earth, they analyze the architecture of geo-spatial information services according to developing requirements for geo-spatial information services. As a focus for digital earth, digital measurable images (DMI) from aero-space and land borne geo-referenced images are introduced and their working principles are provided in detail. By analyzing how DMI achieve the four functions of digital earth, which the authors define as being searchable, visible, measurable and being minable, the conclusion is that DMI can satisfy the needs of digital earth.

Finally, Jianhua and his team explore ways of building networked collaborative VGEs. They aim to explore the conceptual and technical framework for establishing a networked collaborative virtual geographic environment (CVGE) in terms of a workshop for metasynthetic engineering which grasps the essence of complex systems and their problems. The system they develop involves CSCW, mobile networks and Internet enabled virtual environments in the context of 3D geographic information representation and analysis. The authors argue that geocollaboration is involved in two major parts: collaborative modes and geo-problem representation. Interaction and dialogue among participants is highlighted in terms of a networked CVGE which is regarded as a heterogeneous system composed of the internet and mobile networks, which are heterogeneous clients of computers and mobile instruments. A prototype system is designed and developed with VRML, Java SE, Java ME, and Java 3D. Using the system, participants can connect the CVGE via the Internet and mobile phones, and communicate and interact with each other through 3D avatars, streamed media, text in collaborative space and geographic space, while in the field, they can implement dialogue and voting for different schemes involving collaborative decision making.

Chapter 15 Exploring Spatial Uncertainty of GPS Coordinates and DEM Interpolation in Virtual Environments

Jing Li, Claire H. Jarvis and Chris Brunsdon

Introduction

In virtual reality applications, images as display frames are usually rendered in real time as opposed to pre-rendered animations, which enables users to interact with the virtual objects in the scene. With the increased rendering capability of 3D graphics cards, VR technology has been widely used to explore spatial characteristics of geographical data [1]. In this chapter, we present two such examples which benefit from the use of VR technology. Firstly, a real-time simulation of Global Positioning Systems (GPS) satellite geometry is carried out in a virtual environment of a university campus. Secondly, we demonstrate the results from interpolating elevations of raw LiDAR (*i.e.* Light Detection and Ranging) data points to a continuous surface model using a variety of interpolation techniques. Both applications are used by faculty in their teaching of Masters level graduate students, presented through the media of an immersive VR theatre.

Visualization of GPS Satellites-Receiver Geometry

One of the factors that affects GPS accuracy is the strength of the GPS satellite-receiver geometry, Dilution of Precision (DOP) values are used to assess how strong a particular geometry is in relation to positioning accuracy. DOP values are a function of the diagonal elements of the covariance matrix (*i.e.* $(\boldsymbol{A}^{\mathrm{T}}\boldsymbol{A})^{-1}$) in a coordinate system where the matrix $\boldsymbol{A}$ contains the line of sight unit vectors for each satellite-receiver pair and unit vectors [2].

Among the DOPs, Position DOP (PDOP), Vertical DOP (VDOP) and Horizontal DOP (HDOP) are indicators of position, height and plane accuracy and the total positional error can therefore be estimated by $\sigma \times$ PDOP. A standard deviation of 2 m in pseudorange observations would give a standard deviation of $2 \times$ PDOP m. Thus, the smaller the PDOP, the more accurate the position would be. As a general rule, PDOP values larger than 5 are considered poor[3].

As shown in Figure 1 which is a typical screenshot in the real time simulation environment used here, a person holding a PDA equipped with a GPS receiver is standing in an urban canyon environment on the campus of the University of Leicester. Green rays indicate satellites visible to the receiver while the red rays point to satellites blocked by the buildings on either side of the road. The aerial view rendered in the upper right corner of the screen visualizes azimuth and elevation angles of the satellites. Note that the length of the rays indicate the satellite elevation angle, and thus longer rays represent low elevation satellites, although all the satellites are around 20,000 kilometers above the earth. The text at the bottom of the screen, which updates as the user moves across the campus environment, indicates that while there are currently eight satellites in the sky, only three of them are visible to the receiver. Therefore a 3D positional fix can not be obtained because a minimum of four satellites are required to get a 3D position fix.

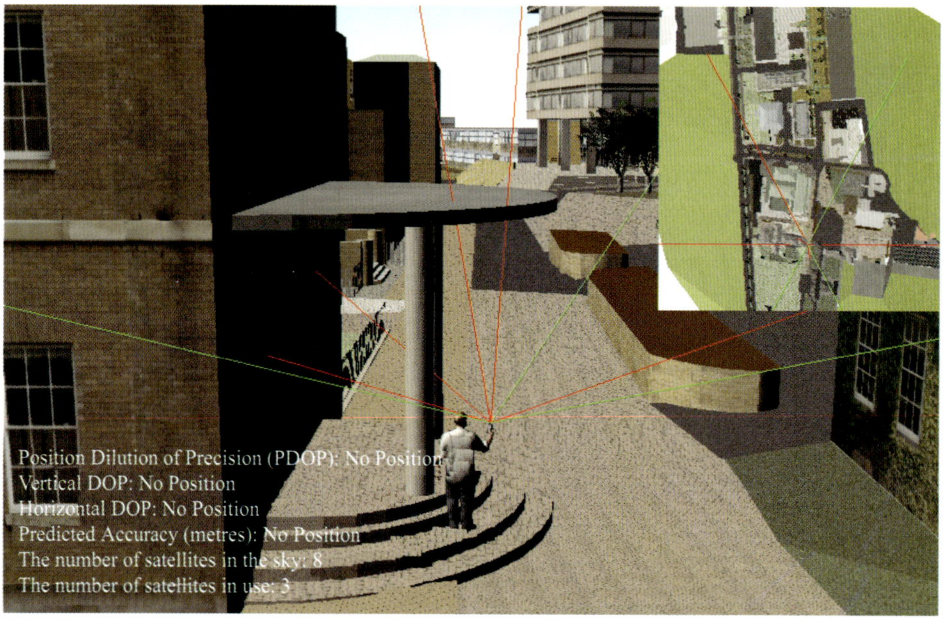

Figure 1　A screen shot of real-time simulation of satellites geometry

Figure 2 depicts a relatively strong satellite-receiver geometry with a small PDOP of 1.76 given six satellites visible. Assuming that the standard deviation in the pseudorange measurement error is 2 meters and this error is uncorrelated, the predicted position accuracy can be estimated at 2 multiplied by the PDOP, which is equal to 3.52 meters (as shown in Figure 2).

In general, the more satellites that are visible, the smaller the PDOP, and the more accurate the position should be. A smaller number of visible satellites are more likely to give relatively larger DOP values which may consequently produce unreliable coordinates. Figure 3 demonstrates scenarios at locations where only four satellites are visible. The very large PDOP of 12.86 shown in Figure 3 is due to both limited satellite visibility and poor satellite geometry. It should be noted that by simply looking at the green rays rendered both in 3D and the aerial view, this cannot provide a clear explanation as to why the large PDOP of 12.86 could occur in this particular urban environment. Therefore, there is a need to draw the tetrahedron formed by the four receiver-satellites unit vectors in order to give a clearer view of the occurrences of poor satellite-receiver geometries. It should be noted that the larger the tetrahedron's volume, the smaller the DOPs [4]. It is very clear from Figures 3 and 4 that the volume of the tetrahedron in Figure 3 is smaller than that in Figure 4. This explains why the satellite-receiver geometry drawn in Figure 4 is stronger and results in smaller DOP values (*i.e.* a PDOP of 2.54 in Figure 4 as opposed to a PDOP of 12.86 in Figure 3).

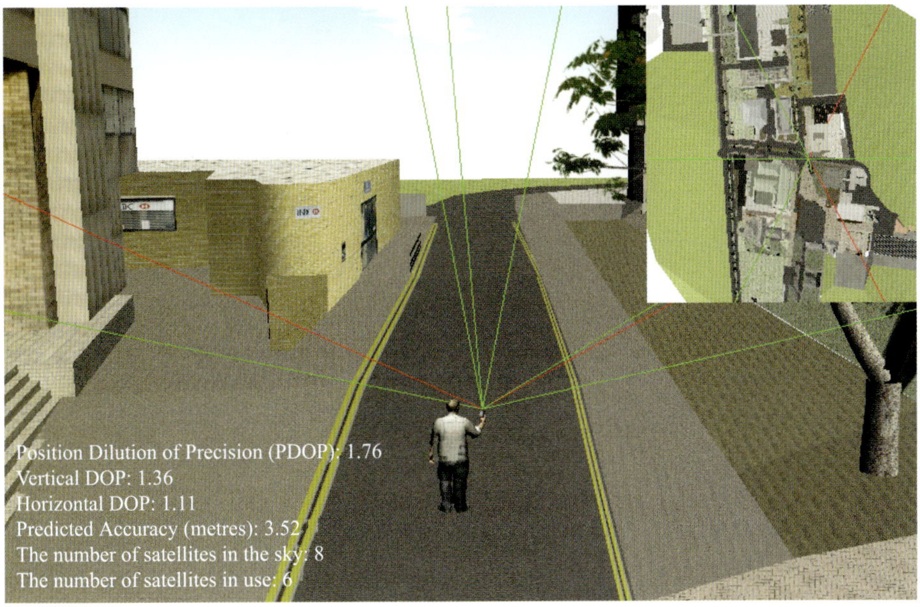

Figure 2 Six satellites visible, with a small PDOP of 1.76

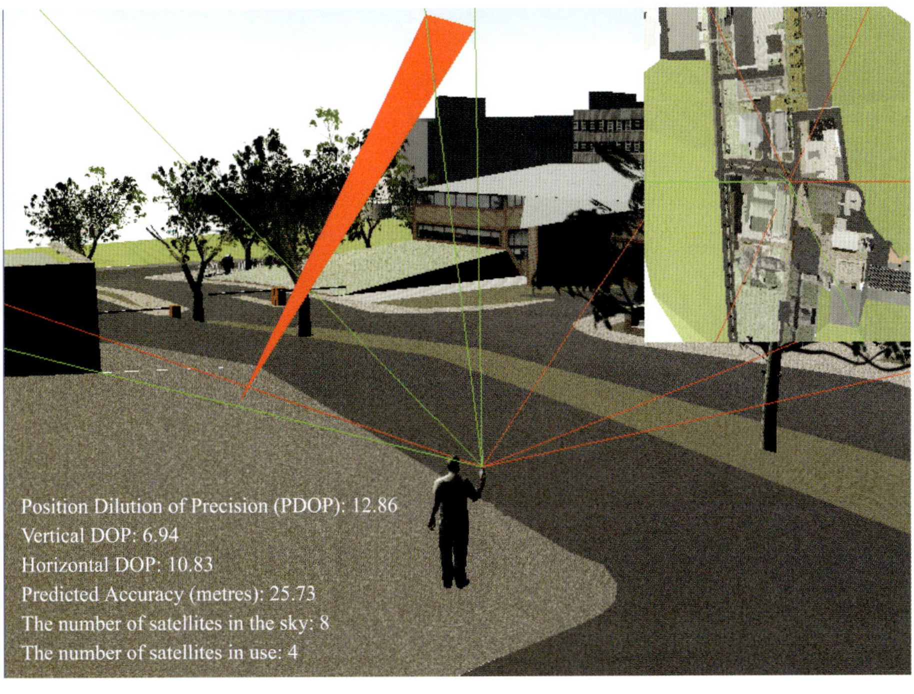

Figure 3 A large PDOP of 12.86, calculated from the signals from four visible satellites

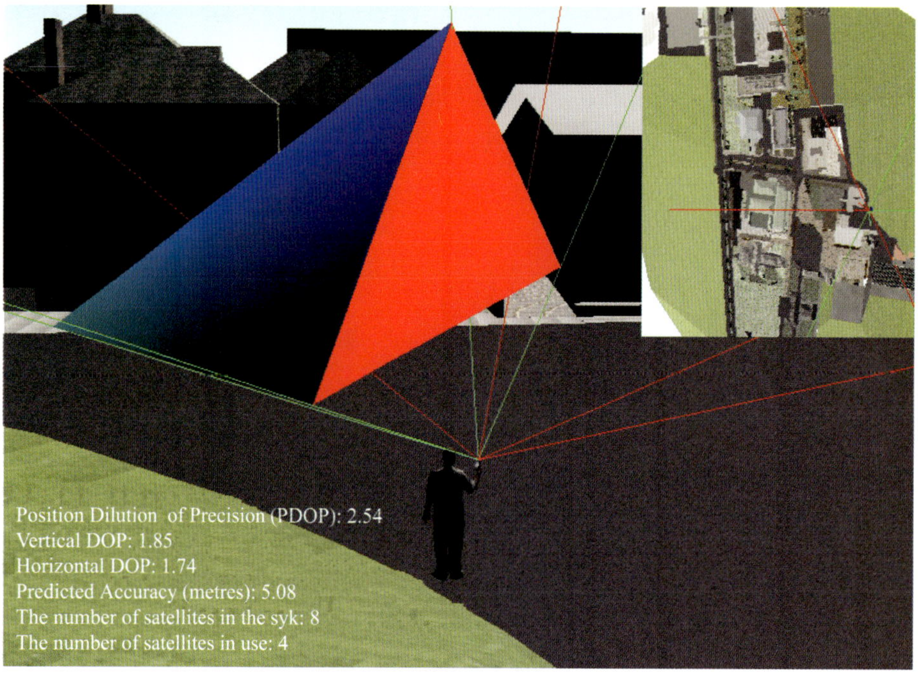

Figure 4 A very small PDOP of 2.54, resulting in a strong geometry with only four visible satellites

In summary, an improved understanding of GPS availability and accuracy in urban areas can be gained by viewing a real-time simulation in an immersive virtual environment. A variety of receiver-satellite geometries can be visualized in real-time in relation to how PDOP changes with respect to surrounding geographic environments, as the human model/avatar walks through the campus. The textured human and campus models add realism to the 3D simulation and therefore can provide enhanced learning and teaching experiences in the context of higher education.

Visualization of DEM Interpolation with LiDAR Data

Spatial interpolation of elevation data has long been the subject of debate in GI Science. Interpolation and surface fitting techniques such as kriging and splines have been thoroughly researched in the literature, and numerous experiments have been conducted to explore which method is most suitable in various spatial contexts[5] Nevertheless, exploring the visual (rather than mathematical) differences in LiDAR Digital Surface Models (DSMs) derived from various interpolation techniques in a virtual environment, would allow one to gain an insight or improved understanding in the characteristics of these interpolation methods. This is especially important in the context of postgraduate education, as students often have difficulties in understanding the fundamental principles of spatial interpolation. A typical student comments on this issue by saying: "... Spatial information science was challenging conceptually, but the practicals were not so difficult. It was very difficult conceptually, a lot of maths, a lot to get your head around, to imagine these interpolation techniques..."

The use of visualization here is pitched at bridging the gap between mathematical formulae and spatial form, filling the gap that previously had to be 'imagined'. With this interpolation tool, we "re-describe" academic knowledge in order to facilitate transfer at a level relating to ways of thinking (across space), particularly for those students with no geographical or mathematical background. To provide a context, we interpolate elevations of raw LiDAR points sampled at 1m resolution to generate a number of continuous surfaces. As shown in Figure 5, the raw LiDAR points cover an area of 300 m×455 m which comprises various field boundaries and vegetation in a rural area. The underlying terrain has a gently rising trend from east to west. Since there are very few man-made features in the data (*e.g.* buildings), the expected Root Mean Square Error (RMSE) would be smaller than that in highly built-up areas, as dramatic local variations would occur around building footprints in dense urban areas, which causes the RMSE to increase. In this experiment, the RMSE is obtained through cross validation (CV) which estimates the heights of sample points using a series of other remaining sample points in a 30 data point neighborhood, and then calculating the differences between the observed and estimated heights as cross-validation error.

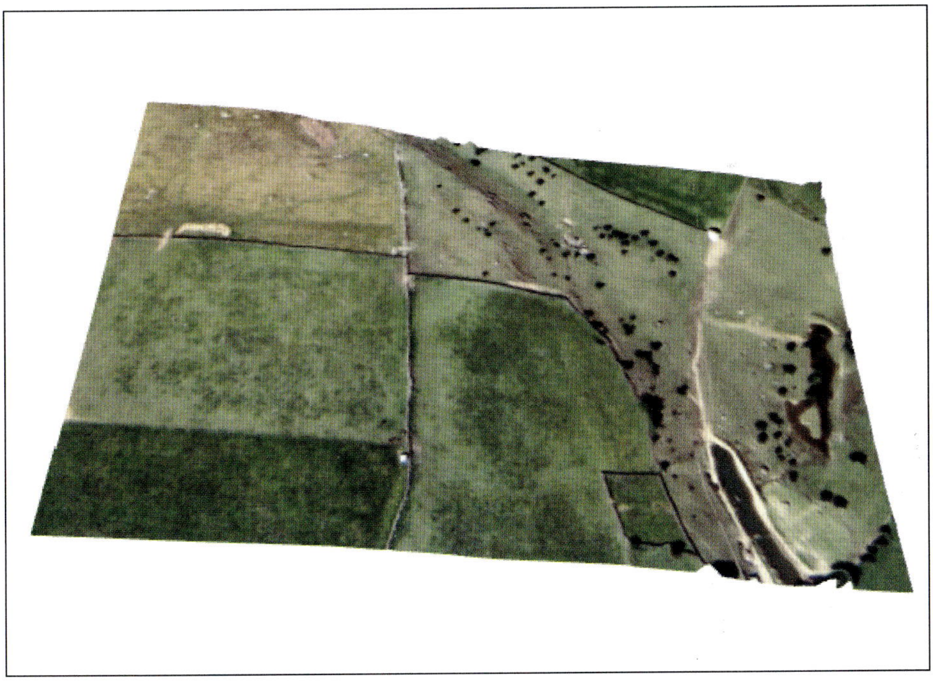

Figure 5 The study area, over which the underlying data set of raw LiDAR points is irregularly sampled at around 1 m resolution

Figures 6 to 8 visualize cross-validation results on top of the interpolated surfaces using the 'spline with tension' method with varying weights for the tension term. Note that only the sample points at which the cross-validation errors are larger than 1.5 standard deviation (*i.e.* around 0.44 m) are rendered using 3D spheres of various colors in the scene. The vertical lines starting from the centre of the spheres indicate the magnitude of the cross-validation error. For example, overshoot is represented by the vertical lines shooting upwards from the spheres which are usually located at the bottom of hilly areas and therefore blocked by the terrain; these can also be viewed in wireframe mode. Undershoot occurs when the lines shoot downwards from the spheres. It is evident from Figures 6 to 8 that the spline surfaces become progressively smoother as the weights increase. The smallest RMSE of 0.292 m is obtained with a weight of 10 as shown in Figure 7. The other two surfaces shown in Figure 6 and Figure 8 have similar RMSE values, but the appearances of the two spline surfaces are dramatically different as shown in these two figures. This is because adopting the smoother spline surface with a weight of 10,000 in Figure 8 caused large cross-validation errors to occur in more "bumpy" areas such as hills and deep slopes where local variations in height are significant. This further suggests that overshoot occurs at the bottom of more hilly areas and undershoot happens at the top of the hills across the study area. However, only undershoots are more likely to occur in less bumpy areas as on the gentle slopes of Figure 8. This visual observation agrees with theory in that a spline surface with a weight of 0.0018 in Figure 6 is more "stiff " than that in Figure 8 and therefore tends to produce both overshoot and undershoot even in less bumpy areas where only very gentle slopes are present. Overall, the RMSE can only be used as a global measure of interpolation error as it aggregates local variations in error across the study area. In Figure 6 and Figure 8, the large cross-validation errors in the two spline surfaces tend to even out owing to the fact that the two spline techniques perform differently according to the amount of local variation in heights.

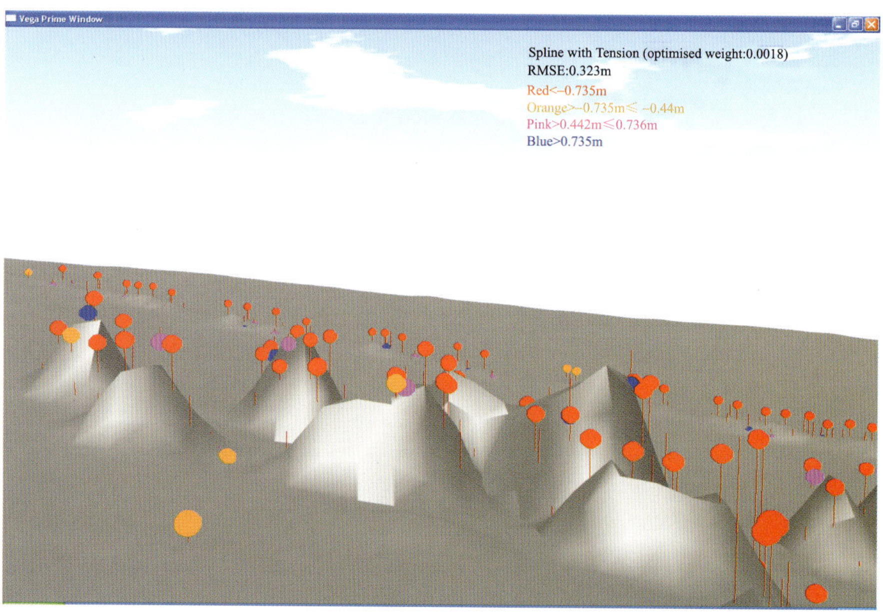

Figure 6 Cross-validation error calculated using spline with tension (RMSE: 0.323 m, weight: 0.0018)

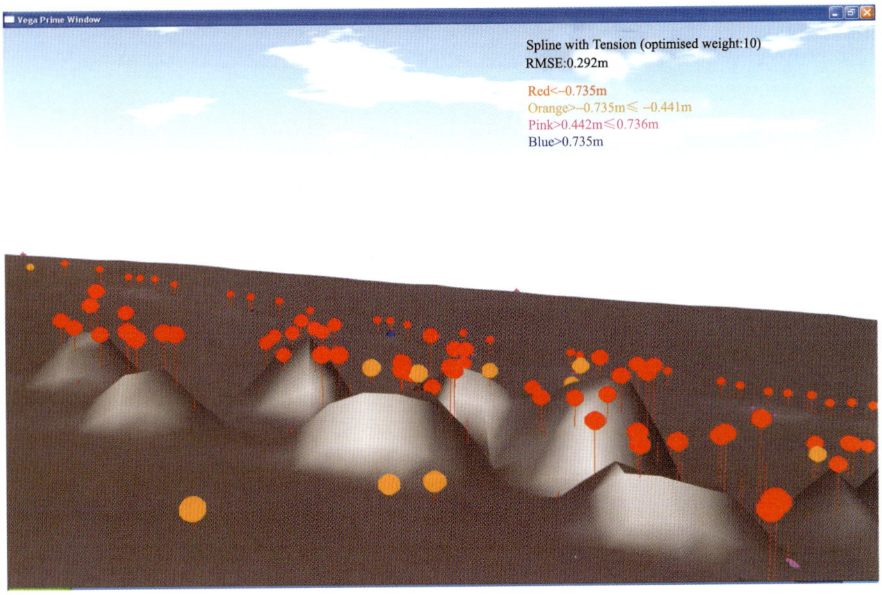

Figure 7 Cross-validation error calculated using spline with tension (RMSE: 0.292 m, weight:10)

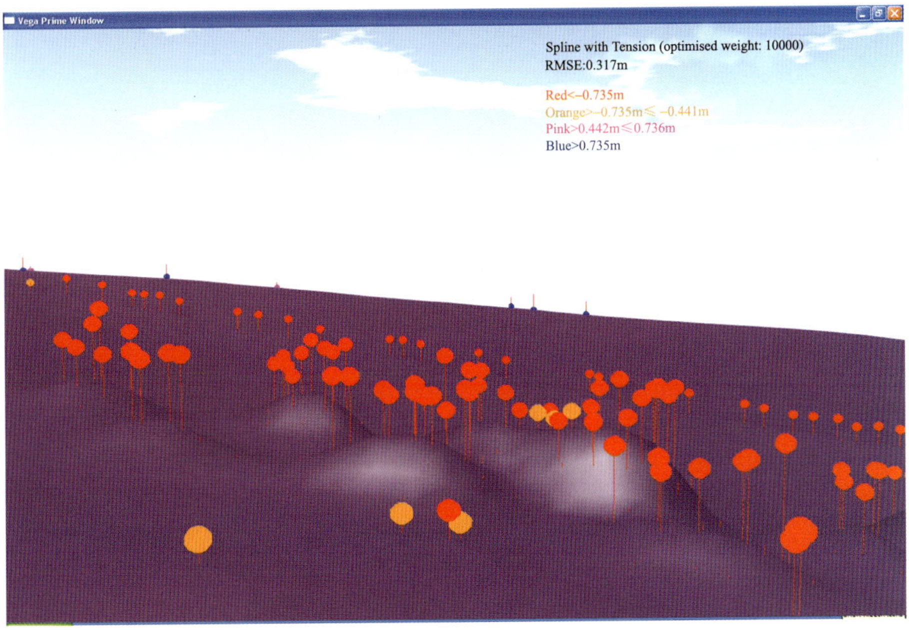

Figure 8 Cross-validation error calculated using spline with tension (RMSE: 0.317 m, weight: 10000)

A second example of spatial interpolation is concerned with the sampling process after interpolation has taken place. As shown in Figures 9 and 10, the two universal kriging surfaces derived from the same geostatistical model without nugget, are sampled at 1 m and

0.5 m resolution respectively. It is very clear from Figures 10 and 11 that there is considerably more detail in the 0.5 m grid than that in 1m grid, despite the fact that both are derived from the same model and the 0.5 grid resolution is even finer than the original 1m raw LiDAR points. The 0.5 m grid is also closer to the sample points. Although in theory both surfaces should go through the sample points, the final surface used for visualization is based on the regular grid, which may not coincide with the original sample points. It is thus easy to see that 3D visualization can raise the awareness of the scale issues in DEM interpolation which is often lacking among graduates studying GI Science.

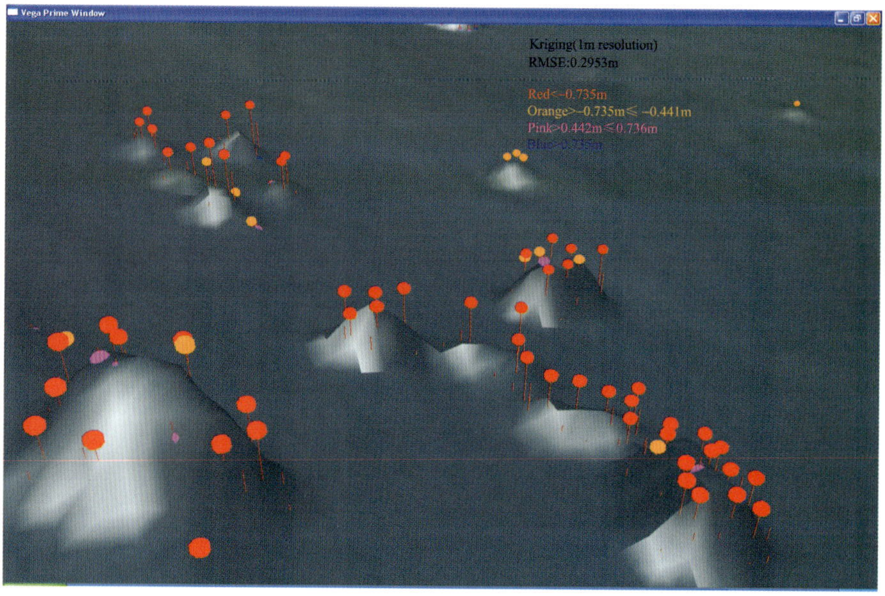

Figure 9 Cross-validation error calculated by kriging
(RMSE: 0.2953 m, 1 m resolution)

To sum up, an improved understanding of the three different spline methods and scale issues in spatial interpolation has been gained through enhanced interactive real-time visualization as apposed to the conventional desktop 3D GIS. No window GUIs are required to be switched on or off for a particular surface and this improves interactivity with the surfaces in real-time full screen mode. A simple keystroke allows users to flick through all the surfaces accessing cross-validation results currently present in the memory in a timely manner, thus enhancing memory retention as well as facilitating comparisons between different surfaces. The use of lighting, materials, colors, various types of geometries (*e.g.* lines and spheres), 2D texts (on the upper right hand corner of the screen) and a 'game motion' model enables the user to inspect the surfaces from any viewing angle and further enhance the learning experience. The approach here differs from past work visualizing interpolation results in its use of a theatre setting to tap a community approach to sense-making and data exploration and it encourages the explicit linkage by staff between mathematical descriptions and surface form in a blended manner. However, the degree to which this tool does indeed assist a wider range of students with specific spatial literacy related learning needs, transferring knowledge from theory and formula to spatially explicit applications, requires further testing.

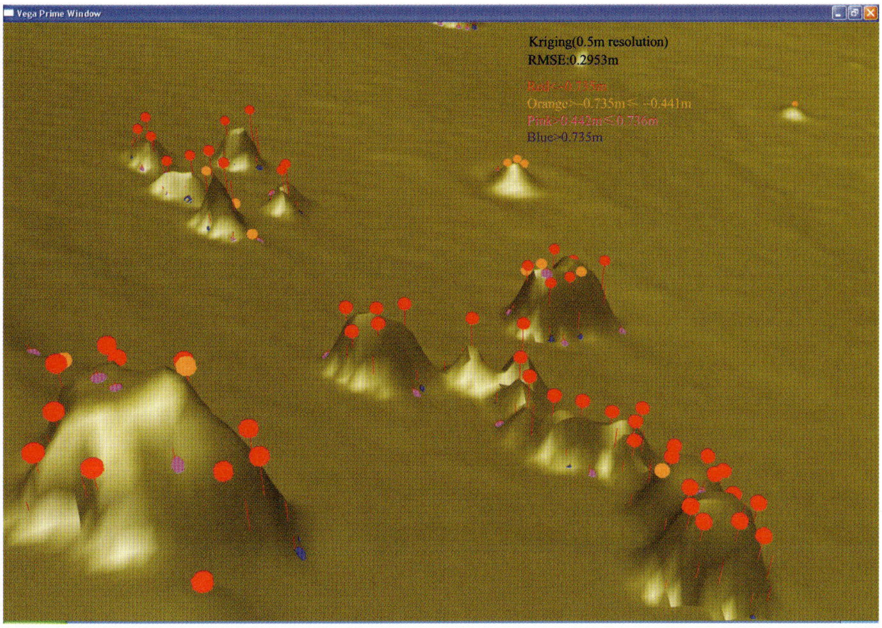

Figure 10 Cross-validation error calculated by kriging
(RMSE: 0.2953 m, 0.5 m resolution)

3D Immersive Visualization in VR Theatre

An added advantage of 3D visualization is to create virtual environments in which the user's feeling of immersion is enhanced through improved learning and teaching experiences. As illustrated in Figure 11, our VR setup is a 6-image generator/projector system utilizing a curved semi-immersive screen which has a 160° field of view angle (FOV) equivalent to the

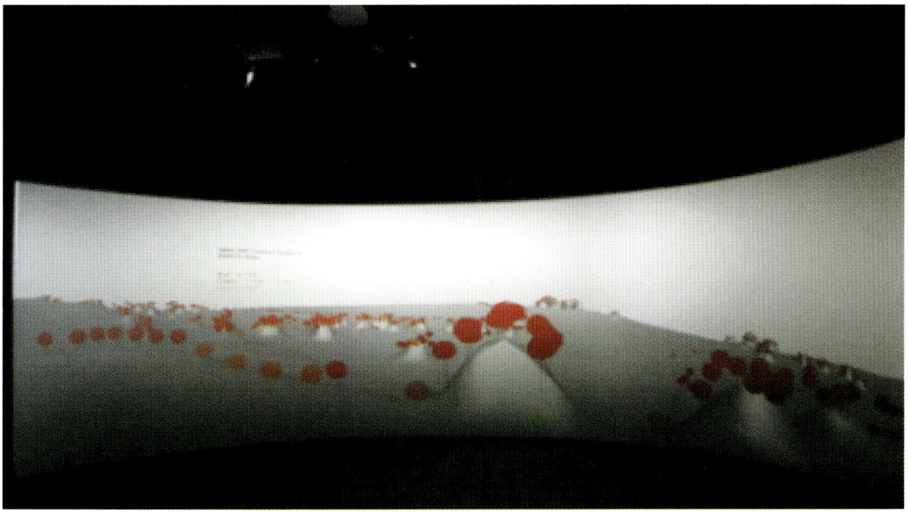

Figure 11 Interpolation tool in VR theatre

FOV angle of human eyes. This distributed rendering system significantly reduces the computational overhead on the image generator/projector system since each PC only renders one image for either the right or left eye. This is in contrast to a single machine stereo system in which one PC has to generate two separate images for the right and left eye respectively. A wider FOV angle has significant advantages over desktop 3D GIS in that users can explore a larger area of the terrain in 3D stereo and interact with virtual environments with minimum efforts.

Conclusions

This chapter has described two examples of using 3D immersive VR technology to improve teaching and learning experiences in GIS education. It is evident that the power of 3D visualization enables us to discover some hidden characteristics of geographic data and to improve our understanding of some more abstract concepts, therefore bridging the gap between theory and human understanding. Two such examples have been presented. Firstly, the strength of arbitrary receiver-satellite geometry in urban areas is visualized both in 3D and aerial views with the continual updating of the PDOP values. Secondly, the cross-validation errors are overlaid onto the interpolated surfaces, which provide visual cues to explore the behavior of the different interpolation techniques across the study area. Future work will focus on further evaluation of the two teaching tools reported here and as well as developing more novel applications of 3D immersive VR in the context of the geosciences.

References

[1] Unwin D and Fisher P (Editors). Virtual Reality in Geography. London: Taylor & Francis, 2001.

[2] Langley R B. The Mathematics of GPS. GPS World, 1991, 2: 45–50.

[3] Taylor G and Blewitt G. Intelligent Positioning: GIS-GPS Unification. London: John Wiley and Sons, 2006.

[4] Langley R B. Dilution of Precision. GPS World, 1999, 10: 52–59.

[5] Fisher P. Valediction. International Journal of Geographical Information Science, 2007, 21: 1165–1170.

Chapter 16 Quantifying Information Displayed on Mobile Phone Screens

Kazuki Yanagisawa and Tohru Yoshikawa

Introduction

Recent technical innovations in mobile phones have been brought about by the rapid development of information and communications technology. It is now possible to connect to and access various sites on the Internet using mobile phones, regardless of user location, to obtain desired information. Maps and other local information are examples of useful data that can be accessed from mobile phones and reading maps on a mobile phone is a very convenient method to obtain such locality information when away from the home or workplace.

Due to the small size of mobile phone displays, however, reading maps on mobile phones can be inconvenient and frustrating compared to reading maps on PC displays, which have much larger screens. This is partly because digital maps displayed on mobile phones were originally developed for PC displays but it is also because the display range of mobile phones is often too small to view complete map images. Therefore, some images may have to be split into parts. If the scale is relatively large, only partial images of large buildings are displayed. Even when large-scale maps can be displayed, it is often not possible to identify locations because the image of the building containing the destination is not displayed in their entirety. However, when the scale size is smaller, complete images of small buildings are displayed along with neighboring buildings. The spatial relationship between the destination building and neighboring buildings might help in the identification of the destination on a mobile phone map, despite the smallness of the map image.

In the present study, by considering the relationship among the displayed image, the display range, and the process of identifying a destination, we develop a method for quantifying information displayed on mobile phone screens. The proposed method can offer new ways to improve the usability of maps on mobile phones with functions similar to PCs.

Research Method

Reaching a destination requires the identification of its location. In addition, it is necessary to obtain the spatial relationship between the destination and its nearby geographical features in order to identify its exact location. For example, pedestrian navigation by mobile phone provides sequential directions starting from an actual location and ending at the destination. We can thus identify the destination correctly and reach the destination directly. In considering small displays of mobile phones, maps for pedestrian navigation are often deformed. In a previous study related to route guides and deformed maps, Fujii *et al.*[1] described a system for generating route guide maps for mobile terminals. Ninomiya *et al.*[2] proposed effective deformed maps for mobile devices with small displays. Kakuchi *et al.*[3] developed a method for evaluating the advantage of position based on route information. Furthermore, in connection with deformed maps, the relationship between display scale and display object has been studied extensively[4–7].

Without sequential support by a route guide system, we can reach a destination on the basis of geographical information about the surroundings, such as landmarks, and can identify destinations on maps. For example, Nihon Computer Graphic1) developed a system for generating deformed maps that chooses important geographical information within the region containing the destination and present location. However, this system does not consider the effect of information-seeking behavior on mobile phones. We can search for nearby geographical features while reading a map on a mobile phone by scrolling the viewing area of the map. In a previous study related to information-seeking behavior, Yanagisawa and Yoshikawa[8] formulated routes and calculated the load associated with a search for shops based on visual information in an urban space. In a related context, the intervening opportunities model of spatial interaction[9–11] considers the relationship between the information-seeking behavior and the choice of destination.

However, whether information-seeking behavior can be formulated when we identify the destination by reading maps on mobile phones remains unclear. In addition, it is unclear how the process of identifying the destination will be affected by information about surrounding geographical features. These questions are important in considering the readability of maps on mobile phones. Therefore, to focus on the load associated with searching for clues around a destination, the present chapter develops a method for quantifying information displayed on mobile phone screens. As a case study, this method is applied to the spatial relationship between destination restaurants around Shinjuku Station and clues for identifying their locations. First, maps around Tokyo's Shinjuku Station are searched using PCs. In addition, recognizable buildings on a large- or small-scale map are obtained. As an example, this chapter treats recognizable buildings as clues for identifying destinations, and clues on each map are categorized. Second, maps around Shinjuku Station are searched using mobile phones. Finally, with a focus on the process of identifying the location of destination restaurants in Shinjuku by reading maps on mobile phones, the load associated with searching for clues around the restaurants is formulated.

Recognizable Buildings on Maps with Scales of 1/8,000 or 1/1,500

Types of clues for identifying the location of a destination depend on the personal perspec-

1) Nihon Computer Graphic Co., Ltd. "Generation system of deformed map for direction" Patent No. 2001-215871, Industrial Property Digital Library, Japan Patent Office.

tive of the user. Also, understanding the degree of importance and priority for each type of clue is also important for quantification of displayed information. Keeping these points in mind, the present chapter treats recognizable buildings on maps served by Yahoo! Japan[1], as clues, as a case study for quantifying information displayed on mobile phone screens. Consideration of the effect of other types of clue such as rail stations, bus stops, and intersections, is an issue for future research.

The maps used in the present study were supplied by the map service of Yahoo! Japan, one of the largest Japanese portal sites. The maps are read using a PC having a display resolution of 1,280×1,024 pixels in order to categorize recognizable buildings. The objective region is the area around Shinjuku Station that can be read on a scale of 1/1,500. Even though the size of the displayed images varies with display resolution, the maps read by a PC can be read using mobile phones, despite the decrease in the visible area. Figure 1 shows recognizable buildings extracted from the map display supplied by the map service of Yahoo! Japan. In the figure, buildings on these maps are considered recognizable if they meet the following two conditions: 1) They are brightly colored and highly visible, as compared to other buildings, and 2) They have a clearly visible name. Recognizable buildings are important clues for identifying destinations on maps. Figure 1 was originally constructed from a cartographic database using GIS[2]. The shaded buildings in Figure 1 indicate buildings recognizable at a scale of 1/8,000 and 1/1,500, respectively.

In the present chapter, polygon-like features that are recognizable buildings and destination buildings are represented by their centroids. In Figures 1A and 1B, centroids of recognizable buildings are shown. In addition, Figures 1A and 1B contain the centroid of the destination restaurant, for which information is retrievable by the map service of Yahoo! Japan.

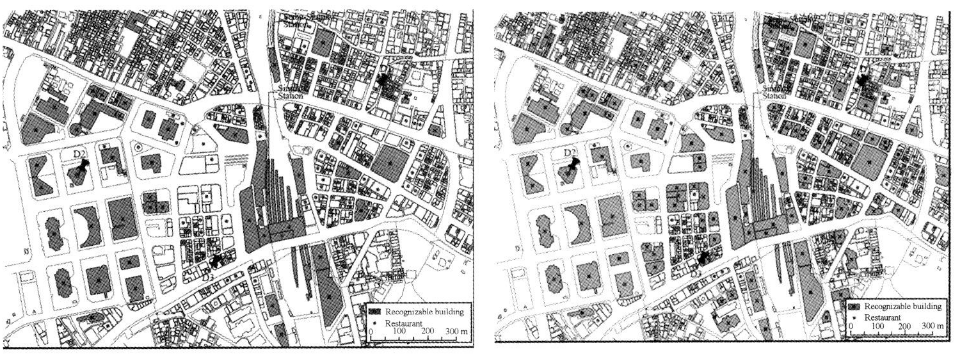

A. 1/8,000 scale B. 1/1,500 scale

Figure 1 Recognizable buildings and restaurants

The total number of recognizable buildings at scales of 1/8,000 and 1/1,500 are 43 and 121, respectively, and the number of restaurants is 205. Figure 2 shows the frequency distribution of the rooftop area of these three building groups. Figures 1 and 2 show that the

1) Yahoo! Japan http://www.yahoo.co.jp.

2) Tokyo Metropolitan Government City Planning GIS.

number of recognizable buildings is greater on a large-scale map than on a small-scale map. In addition, smaller buildings are recognizable on a large-scale map. Since the keyword "Shinjuku Station" indicates a huge building complex that includes a rail station and a commercial complex that encompasses the rail station, part of the rail station of Shinjuku Station is observable at a scale of 1/8,000. However, at a scale of 1/1,500, individual components of the commercial complex are recognizable.

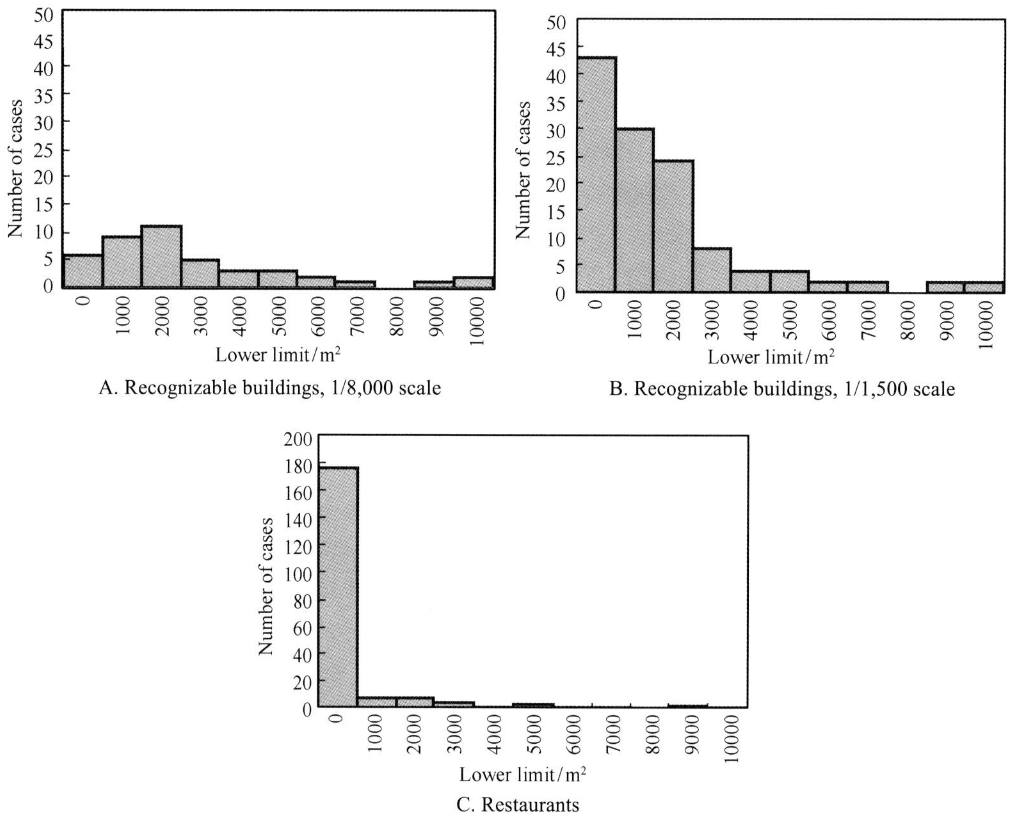

Figure 2 Frequency distribution chart of rooftop area of recognizable buildings and restaurants

Map Display on Mobile Phones

As a case study, the present study considers maps displayed on the SoftBank 705P[1)] (Matsushita Electric Industrial Co.) mobile phone. Maps for mobile phones supplied by Yahoo! Japan are based on the same cartographic database as maps for PCs. However, the viewable area of maps for mobile phones is very small. Figure 3 shows the results of a search by mobile phone using the keyword "Shinjuku Station." The mobile phone has a 2.2-inch LCD display, the resolution of which is 320×240 pixels. The viewable area at a scale of 1/8,000 is 613 m×613 m as in Figure 3B, and the viewable area at a scale of 1/1,500 is 115 m × 115 m as in Figure 3C.

---

1) http://panasonic.jp/mobile/705p/.

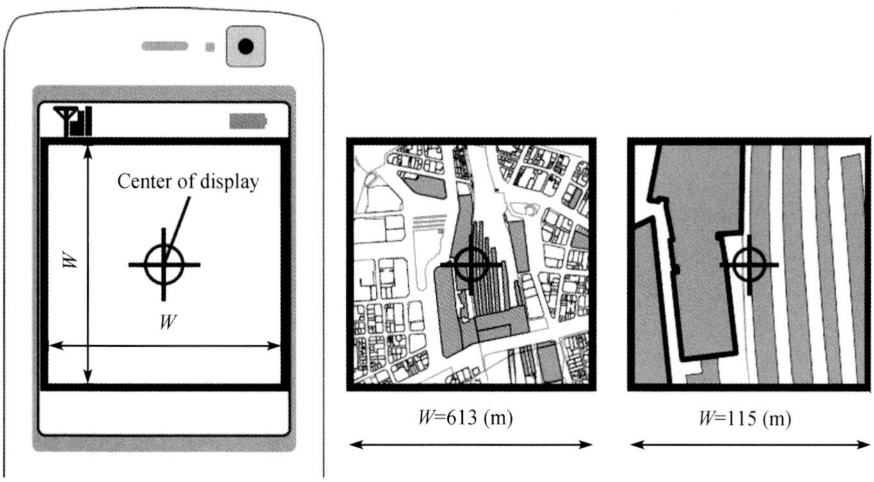

A. Exact size of the mobile phone used in the present study
B. Displayed area at 1/8,000 scale
C. Displayed area at 1/1,500 scale

Figure 3 Result of the search 'Shinjuku Station' using the SoftBank 705P mobile phone

Formulation of Load Associated with Searching for Clues by Reading Maps on Mobile Phones

In order to formulate the load associated with searching for clues, let us analyze the process of identifying the locations of destination restaurants in Shinjuku by reading maps on a mobile phone. As an example, let us assume that the target restaurants are D1, D2, and D3 in Figure 1, and that their locations are identified on the basis of clues in the form of nearby recognizable buildings. The process of identifying the location of a destination restaurant is considered as follows. First, we retrieve the destination restaurant using a mobile Internet map service. As a result, the restaurant's location is shown on a map on the mobile phone. Second, by scrolling the viewing area of the map, we start to search for clues. Clues that indicate locations closer to the destination are more helpful in identifying its precise location than clues that indicate locations farther from the destination. With respect to reading maps on mobile phones, if a clue is located near the destination, we can search for the clue in a small range by scrolling the viewing area of the map. However, if there are no clues near the destination, we have to continue to scroll the viewing area of the map until a clue appears. In addition, the present study assumes that if one clue is obtained, more clues will eventually be obtained. This is because an increase in the number of clues is helpful for accurately identifying the destination's location. Finally, we identify the destination based on the spatial relationship between the clues and the destination. Therefore, as a process for identifying the destination's location, this chapter assumes that we search for clues continuously from the clue nearest the destination to the clue farthest from the destination as in Figure 4. The assumption of this counting method is similar to that of cross K-function analysis [12].

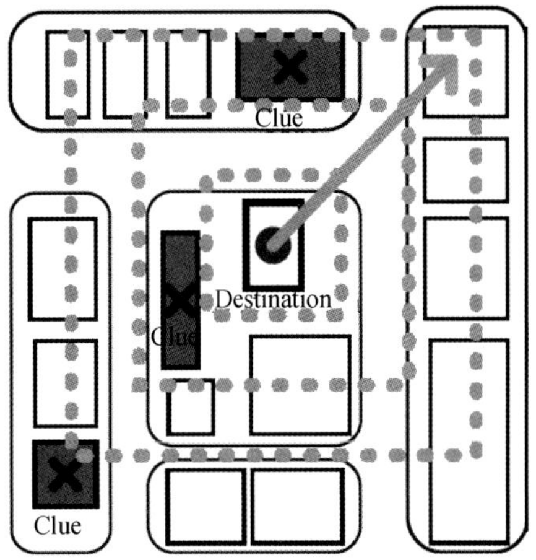

Figure 4 Process of searching for clues near the destination

Based on the assumptions mentioned above, we formulate the load associated with searching for a clue near the destination as in Figure 5. In the figure, D_i denotes the destination, and C_j denotes the clue. Let us calculate the distance between D_i and C_j. The absolute value of the difference in x coordinates between D_i and C_j is termed mx. Similarly, we calculate the absolute value, termed my, of the difference in y coordinates. In addition, R denotes the maximum value of either mx or my. To search for C_j, we must read in its entirety a map that includes D_i and C_j, the area of which is a large square area having sides of $2R+W$ m. However, the map range that can be read using a mobile phone is limited to W m × W m square area as in Figure 3, where W denotes one side of the square viewable area of a mobile phone. Consequently, a large square area having sides of $2R + W$ m must be filled with a limited square area W meters on a side in order to search for C_j. Based on this assumption, the load associated with searching for C_j is calculated as $(2R+W)^2/W^2$. However, points of destination and clues and the centroids of destination buildings are not always displayed in the center of the screen. To be exact, we incorporate a margin of error related to the display situation in this model. The assumption "$2R+W$" adopted is somewhat arbitrary; however, the present chapter develops a basic quantitative method according to this assumption.

The calculated load may be considered as the number of times that the viewing area of map must be scrolled. This load is based on the worst case, when the maximum effort is expended to obtain a clue. If we can obtain a clue with minimum effort, the difference between the maximum and minimum efforts is calculated as shown in Figure 6.

This calculation method is applied to the spatial relationships among destinations D1, D2, and D3, and clues that comprise recognizable buildings in Figure 1. Figures 7, 8 and 9 show the calculation results. In Figures 7B, 8B, and 9B, clues are separated into those recognizable at a scale of 1/8,000 and those recognizable at a scale of 1/1,500. In addition, in order to calculate the inconvenience of reading maps on mobile phones compared with reading maps on PCs, the formulated method is applied to the process of searching for clues using PCs only at a scale of 1/1,500. This is because the viewable area on a PC at a scale of 1/8,000 is far greater than the objective region shown in Figure 1. On a PC with a display resolution of 1,280 × 1,024 pixels, the viewable area of a map of scale 1/1,500 covers 373 m × 373 m. Figures 7C, 8C, and 9C are extracted from Figures 7B, 8B, and 9B for comparison with Figures 7C, 8C, and 9C at the same scale. The 1/1,500 scale results (Figures 7C, 8C, and 9C) obtained using a mobile phone suggest that there are more clues at this scale than at 1/8,000 (Figures 7A, 8A, and 9A). However, the viewing area of a map of scale 1/1,500 is smaller than that of scale 1/8,000 as in Figure 3. Therefore, if we search for the same clue, *e.g.*, Shinjuku Station, on maps of both scales, the load associated with searching on a large-scale map is larger than that associated with searching on a small-scale map. This is due to the amount of information available on a large-scale map, which may contain as many as three times the number of clues on a small-scale map as in Figure 2. Note that this increase in the number of clues is much smaller than the increase in the total number of points, which is 28, that is the square of the increase in the rate of the scale. This indicates that as the scale is enlarged from 1/8,000 to 1/1,500, the increase in the number of recognizable buildings is not as great as the increase in the total number of buildings. Thus, using a small-scale map, we obtain fewer clues in the process of finding Shinjuku Station. The load curves for Figures 9A and 9C are more moderate than the other curves because of the interaction between location D3, the distribution pattern of clues near D3, and the calculation method.

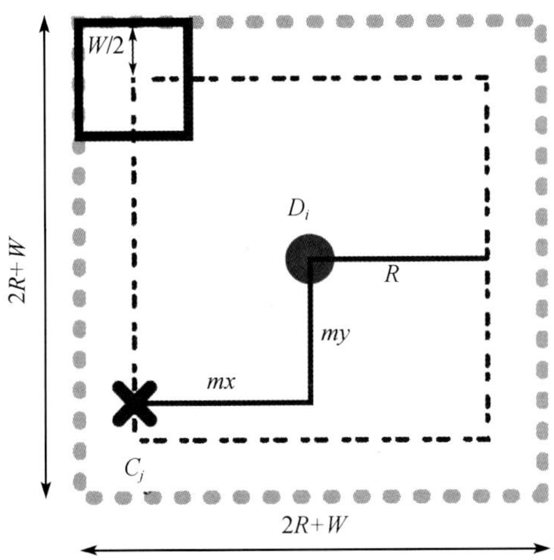

Figure 5 Formulation method of load associated with searching for a clue

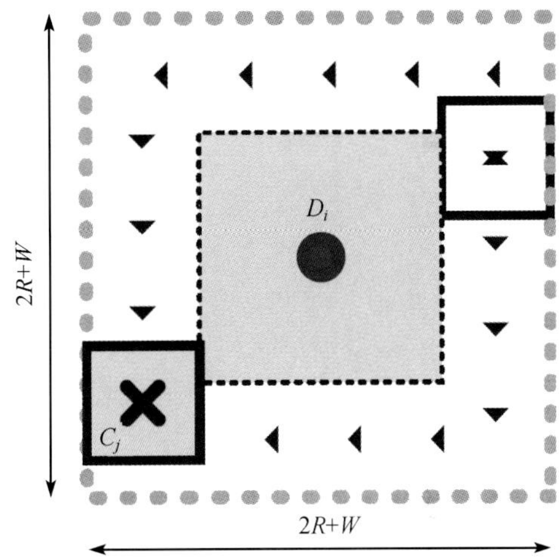

Figure 6 Difference between maximum and minimum efforts (shaded area)

Next, we focus on the differences between the load for a mobile phone and the load for a PC while viewing a map of scale 1/1,500. The load curves for a PC in Figures 7B, 8B, and 9B are more moderate than those for a mobile phone because the viewable area of the PC is approximately 10 times larger than that of the mobile phone. Note that the load for a PC at a scale of 1/1,500 is similar to that for a mobile phone at a scale of 1/8,000, while the number of obtained clues is relatively small, perhaps around 10 or 20. This can be explained as follows. The visible area of a PC is about 10 times larger than that of a mobile phone. Furthermore, the number of clues included in the window of a fixed device when viewing a 1/1,500 scale map is 1/10 of that included when viewing a 1/8,000 scale map because, the visible area shrinks to 1/28 whereas the number of clues per unit area increases about three-fold. This implies that the total load for a PC at a scale of 1/1,500 may be slightly larger than that for a mobile phone at a scale of 1/8,000.

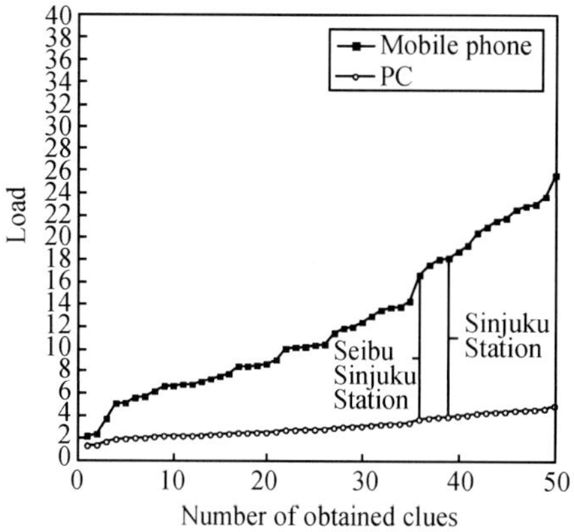

C. D1 (1/1,500 scale, mobile phone and PC)

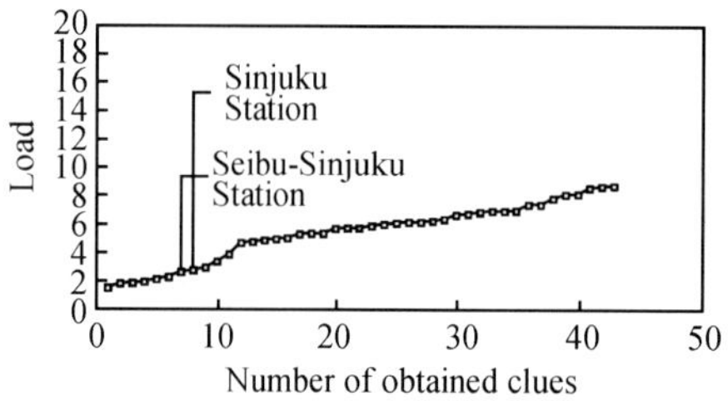

A. D1 (1/8,000 scale, mobile phone)

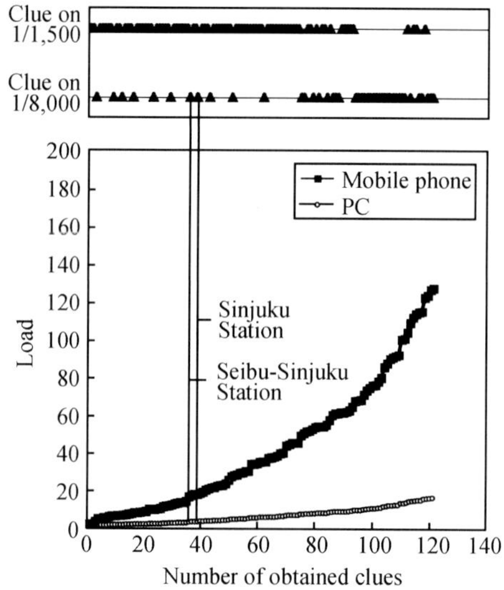

B. D1 (1/1,500 scale, mobile phone and PC)

Figure 7　Load associated with searching for clues

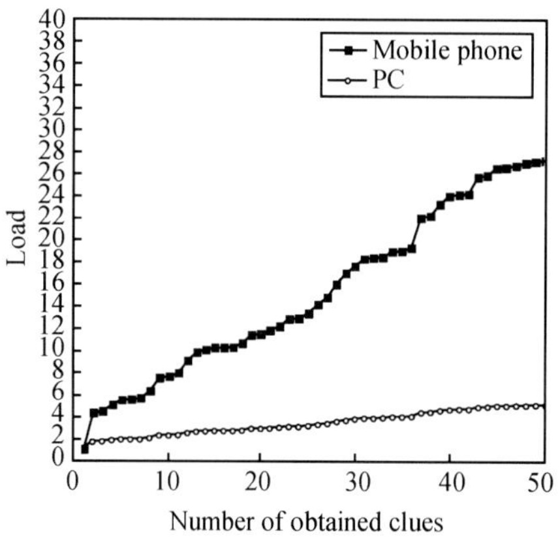

C. D2 (1/1,500 scale, mobile phone and PC)

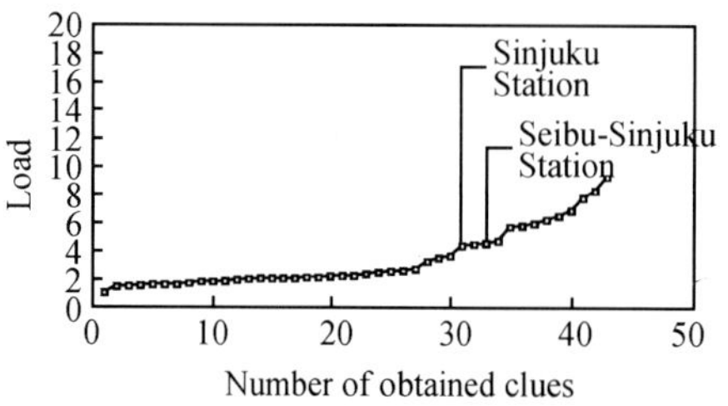

A. D2 (1/8,000 scale, mobile phone)

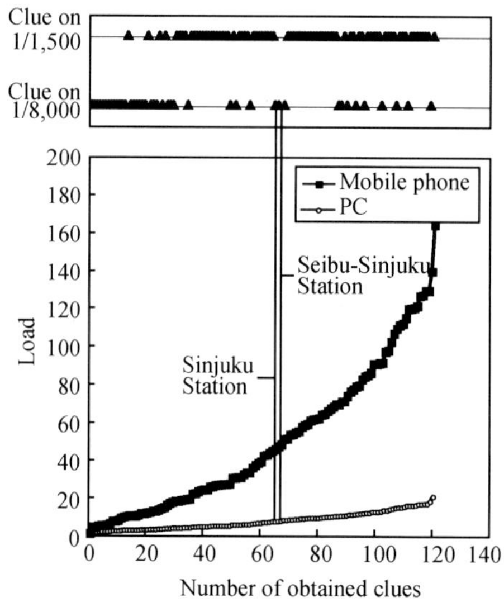

B. D2 (1/1,500 scale, mobile phone and PC)

Figure 8　Load associated with searching for clues (D2)

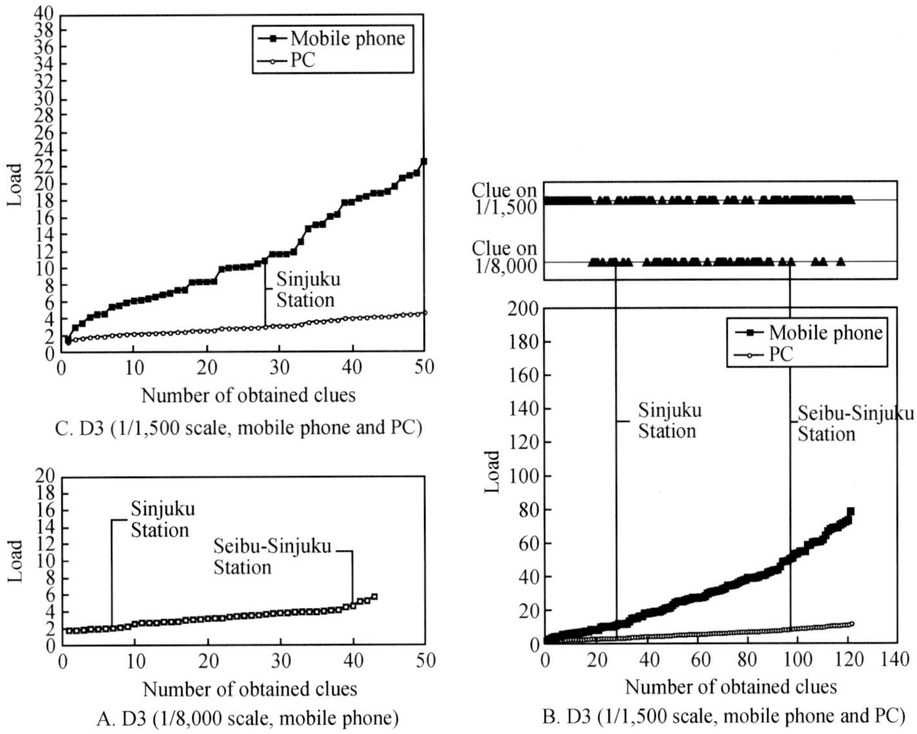

Figure 9 Load associated with searching for clues (D3)

Concluding Remarks

The present study developed a quantification method for information displayed on mobile phone screens. This method was constructed by combining existing frameworks, including cross K-function analysis, which is used to analyze the spatial relationship between two different distributions of point-like features, and information-seeking behavior, which is often discussed as the basis for the intervening opportunities model. Furthermore, this method provides a simple technique by which the essential problem of reading maps of limited size can be dealt with.

The results are based on the arbitrary basic calculation model. Only one type of clue, recognizable buildings selected by the map service of Yahoo! Japan, is adopted; however, the effect of clue combination and the interaction of various clues on identifying a destination needs to be considered. For example, the number of recognizable intersections would be a quadratically enhanced factor for identifying the destination. Furthermore, nearby rail stations and bus stops also enhance the identification of the destination as well as the nearby intersections that act as the secondary clues for identifying the primary clues, *i.e.*, recognizable buildings. In conjunction with further consideration for scrolling behavior on using mobile maps, the proposed method can provide new ways to improve the readability of maps on mobile phones with functions similar to PCs.

The following issues remain to be addressed and will be the subject of further research.

First we intend to apply the proposed calculation method to other areas, this generalizing the method, we then intend to calculate the optimum relationship between the quantity of displayed clues and their variety on maps for mobile phones, and finally investigate information-seeking behavior in a relatively large viewable area.

Acknowledgements This project was supported by a Tokyo Metropolitan University International Research Grant. This project was also supported by Research Fellowships for Young Scientists provided by the Japan Society for the Promotion of Science. We would like to thank the reviewers for providing valuable comments.

References

[1] Fujii K and Sugiyama K. Route Guide Map Generation System for Mobile Communication. Transactions of Information Processing Society of Japan, 2000, 41: 2394–2403 (In Japanese).

[2] Ninomiya N, Togawa N, Yanagisawa M, Ohtsuki T. A Deformed Map Generation Algorithm Considering Visibility in a Small Display And Easiness of Route Understanding for Pedestrian Navigation Information. Processing Society of Japan (IPSJ) Technical Reports, 2006, 103: 111–116 (In Japanese).

[3] Kakuchi and Yoshikawa T. Amount of Route Information Description as a Criterion of the Advantage of Position. Journal of Architecture and Planning (Transactions of Architectural Institute of Japan), 2002, 562: 217–223 (In Japanese).

[4] Teraki. Display of Building at Middle Scale on GIS. City Planning Review, Special Issue, Papers on City Planning, 2000, 35: 1123–1128 (In Japanese).

[5] Kano T, Ota K, Gota M. Study on the Informed Density in the Image using Scale-Texture—Part 2 Urban Space Notation by Informed Density. Summaries of Technical Papers of Annual Meeting (Architectural Institute of Japan), 2003, F-1: 967–968 (In Japanese).

[6] Kano, T, Ota K, Gota M. A Space Notation with Urban and Architectural Scale—Part 1 A Basic Study with Tokyo GIS Data. Summaries of Technical Papers of Annual Meeting (Architectural Institute of Japan), 2004, F-1: 485–486 (In Japanese).

[7] Ota K and Gota M. Basic Study on Urban Space Composition using High-Resolution Satellite Images—Part 1 Hierarchical Notation of Physical Elements in Urban Space. Summaries of Technical Papers of Annual Meeting (Architectural Institute of Japan), 2005, F-1: 1193–1194 (In Japanese).

[8] Yanagisawa K and Yoshikawa T. Formulation of Routes and Load of Quest for Objects in Urban Space with Excessive Visual Information: A Case Study of Searching Goods in Akihabara. Journal of Architecture and Planning (Transactions of Architectural Institute of Japan), 2005, 597: 127–133(In Japanese).

[9] Stouffer A. Intervening Opportunities: A Theory Relating Mobility and Distance.American Sociological Review, 1940, 5: 845–867.

[10] Stouffer S A. Intervening Opportunities and Competing Migrants. Journal of Regional Science, 1960, 2: 1–26.

[11] Okabe. A Theoretical Comparison of the Opportunity and Gravity Models. Regional Science and Urban Economics, 1976, 6: 381–397.

[12] Okabe and Yoshikawa T. SAINF: A Toolbox for Analyzing the Effect of Point-Like, Line-Like and Polygon-Like Infrastructural Features on the Distribution of Point-Like Non-Infrastructural Features. Journal of Geographical Systems, 2003, 5: 407–413.

Chapter 17 Digital Earth with Digital Measurable Images*

Deren Li, Zhenfeng Shao, Qingwu Hu and Mi Wang

Introduction of Digital Earth and its Four Functions

What is Digital Earth? This interesting scientific topic is a visionary concept popularized by the former US Vice President Al Gore, for the virtual and 3D representation of the Earth. Such a digital earth is spatially referenced and interconnected with digital knowledge archives from around the planet with vast amounts of scientific, natural, and cultural information used to describe and understand the Earth, its systems, and human activities. Digital Earth thus represents a rich convergence of technological advances, active visionaries and recognition of the paramount need for humans to better understand their planet.

The industry pioneers who lead the 3D Earth visualization software are beginning to present their corporate philosophies and investments for building a true Digital Earth. Agency leaders from NOAA, NASA, and the United Nations have defined the governance aspects of programs that are helping evolve the Digital Earth Vision. This relates to such supporting technologies as scientific calculating, mass storage, remote sensing, wide-band network, inter-operability and metadata. Virtual reality, geographic first, Digital Earth is a virtual reality system that makes people feel that they are personally on the scene. In the next place, Digital Earth organizes vast amounts of geo-spatial information through GIS. Second, a global information network is established to realize resource sharing.

Three layers are necessary for the construction of Digital Earth, as shown in Figure 1. The first layer is information infrastructure composed of a space-based information network and the internet. The next layer is spatial information infrastructure. The third layer is thematic data including data from humanities, geography and economy and their corresponding data mining tools.

What can Digital Earth do for us? We will examine the functions of the internet at first. The internet is something more than information for it provides a platform of information sharing, searching and releasing. Currently, almost all mainstream business on the internet is involved with information.

As a media which carries more than 80% of information related to human activities, geo-spatial information plays a more and more important role in the internet services. On the

* Supported by the National Key Basic Research and Development Program of China (Grant No. 2004CB318206), and the Chinese national 863 Project (Grant No. 2006AA12A115).

other hand, the rapid development of network (which is now in its 3rd generation), 3G—the 3rd generation of communication systems—and grid technology, and the coming of Google Earth makes geographic information and 3S (GPS, GIS and RS) technologies available to many which originally could only be within the reach of professional users. All internet users can access various kinds of web services, a majority of which are free, on a uniform web platform. Public geo-spatial services will improve applications and reduce funds devoted to such applications, thereby promoting the explosive increase of the whole business. From a macro point of view, global geo-spatial information sharing plays a central role in economic growth.

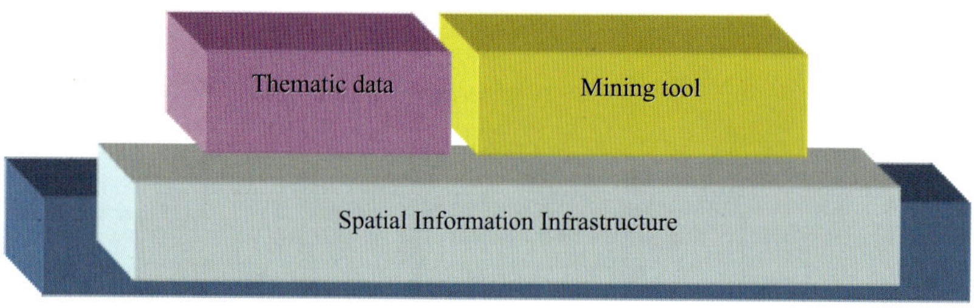

Figure 1 Three layers for the construction of Digital Earth

At present, the realization of a geo-spatial information services platform mainly relies on 4D products such as digital elevation models, digital raster graphics, digital orthophoto quadrangles, digital line graphics and so on. With the improvement of social information, requirements for geo-spatial information are increasing. The integrity, exactness and up-to-date nature of geo-spatial data play a critical role in ensuring the quality and effectiveness of geo-spatial information services. With an eye to the developing trend of requirements for geo-spatial information services, data requirements need to be massive, high-resolution, visible and immediately data minable.

With the third internet tidal wave, the concept of Digital Earth and corresponding technologies including the Grid brings new working mechanisms for all kinds of business applications. According to the concept of Digital Earth, services provided for users need to satisfy the following six characteristics: certainty, experience, communication, variation, creativity and relationships to applications. With regard to geo-spatial services, being searchable is the basic attribute; visibility is the basis of experience (*e.g.* Google Earth and Microsoft Virtual Earth); measurability is the guarantee of variation and creativity; and being minable is the guarantee of relating to real applications. So we conclude that being searchable, visible, measurable, and minable are the four functions of Digital Earth.

A service platform based on the geo-spatial grid can integrate Web 2.0 technologies such as Ajax and provide interactive intercommunion services for users. The architecture of geo-spatial information services of Digital Earth can be illustrated in Figure 2 below.

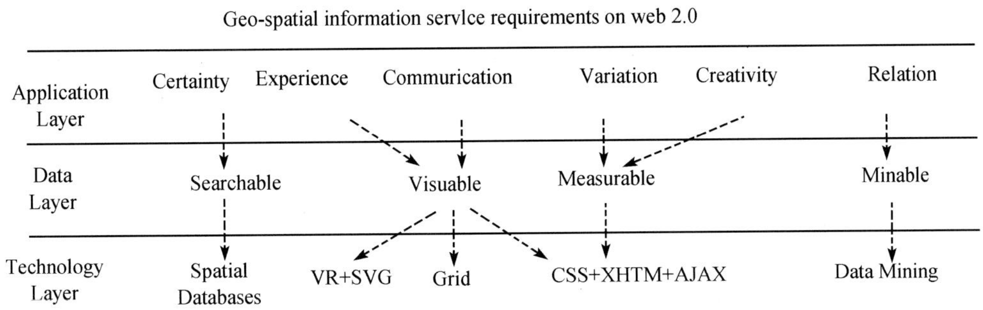

Figure 2 Geo-spatial information service requirements of Digital Earth in Web 2.0

Digital Measurable Images (DMI) Integrated with Seamless Stereo Images Can Satisfy the Needs of Digital Earth

The principle of "measurable seamless stereo image"

In the field of photogrammetry and remote sensing, digital differential rectification is a very mature technology which is mainly used for the production of ortho-images. The digital ortho-image not only has correct planar position, but also holds abundant image information. However, it only contains two-dimensional information, not three-dimensional. Though overlaying contour lines on the ortho-image can partly make up for this omission, it is impossible to substitute for the stereo sense by means of stereoscopic observation. Therefore, we can artificially produce a so-called Digital Stereo-orthophoto Partner (DSP) vs. Digital Orthophoto Quadrangles (DOQ). DOQ and DSP are together called the Stereo Orthophoto Pair[1, 2]. The basic principle of a measurable seamless stereo image is based on the stereo orthophoto pair. Figure 3 summarily shows the basic principle of measurable seamless stereo image[3].

As shown in Figure 3, the coordinate Z of the grid XY points forms a regular digital elevation model (DEM). Figure 3A shows the procedure for making an ortho-image, directly transforming the elevation of a grid point to the original photo according to the collinearity equation to obtain image data for the ortho-image.

In order to obtain a stereo landscape model, the x-parallaxes are artificially introduced according to the DEM. The simplest method for parallax introduction is to use parallel rays with a certain angle between Z-axis as in Figure 3B. The artificial parallaxes reflect the feature of the terrain and form an auxiliary image, which can construct a stereo model with ortho-images. Figure 4 is a vertical plane diagram of projection. Take the ground point P as an example, the height difference relative to the plane of projection is Z, p_0 is the ortho projection point of P and then p_1 is the slope parallel projection point of P. An orthophoto is obtained by ortho projection and the Stereo Orthophoto Partner is obtained by the slope parallel projection. The parallax P is obtained by stereo measurement, so it is obvious that

$$P = \tan\alpha \cdot Z = k \cdot Z \tag{1}$$

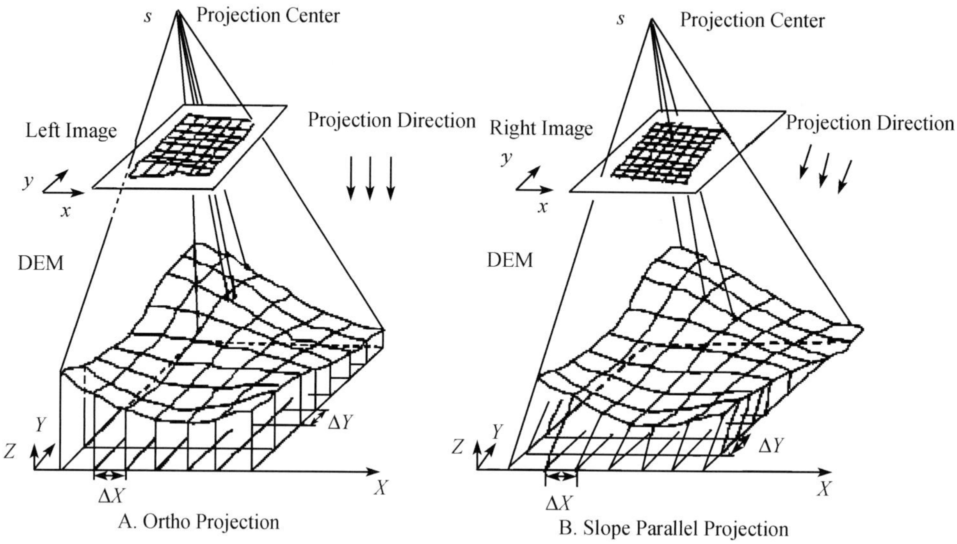

Figure3 The basic principle of measurable seamless stereo image (From Kraus, 1984)

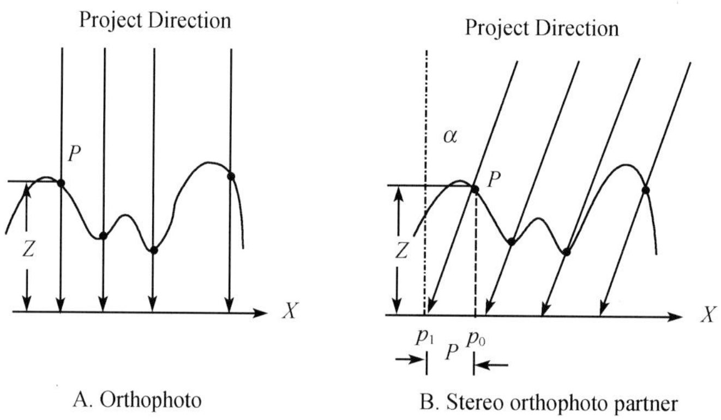

Figure 4 The parallel projection method

As the direction of the slope parallel projection parallels with the *XZ* plane, there are only *x*-parallaxes between the orthophoto and the Stereo Orthophoto Partner. Because they are generated from left and right photos of a stereo pair separately, it meets the prerequisite condition of the stereo measurement and can provide not only the planimetric correct position but also the height of the ground point.

In order to make the measurement, the angle α must be accordant with the parallaxes of the original stereo model. Generally, it is taken as $\tan\alpha = \frac{B}{H}$ where *B* is the base line of stereo model, *H* is the flight height. The height of *P* can be obtained according to formula (2)

$$Z_i = \frac{P_i}{k} M + Z_0 \tag{2}$$

Where *M* is photo scale, Z_0 is the height of reference plane, *k* equals tan α, *and* P_i is the

x-parallaxes measured from the stereo orthophoto pair.

For a photographic block, many stereo models can be formed by stereo pairs. Figure 5 illustrates a block consisting of three strips, with six stereo models in each strip.

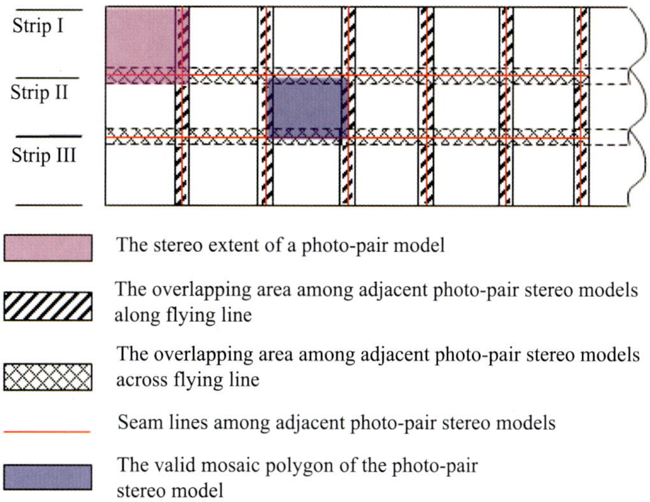

Figure 5 The arrangements of photo-pair models

In this figure, the overlapping area between two adjacent images is called the stereo extent, while a line in the overlapping areas between two adjacent stereo models is called a seam line. The polygon area formed by the seam lines of every stereo model is called valid mosaic polygon of the model. The mosaic of ortho-images of all the valid mosaic polygons in effect covers the whole block seamlessly. Two such seamless ortho-images could be produced. One is produced from all the left images, leading to a left mosaic, and the other produced from all the right images, leading to the right mosaic.

To produce a stereo view, parallax needs to be introduced to one of the mosaics (*e.g.* the right mosaic) to form a so-called mosaic stereomate. The value of the parallax for each pixel is a function of the height at its position. Then the left mosaic and the mosaic stereomate form a seamless stereo model. Such a seamless stereo model is good enough for visualization purposes. To make the accuracy as a high a measurement as possible, the following strategy is adopted in this study.

1. The six orientation parameters of each image are recorded in a file (*i.e.* orientation file).
2. The lineage of each pixel on both mosaic ortho-image and mosaic stereomate is recorded. That is, for each point (pixel) on the seamless stereo model, the system knows 1) the (original) left and right images where it is derived; 2) the orientation parameters of both images; and 3) its image coordinates on both images.
3. With such information when measurement is performed, the ground coordinates of each point on the seamless stereo model are directly computed from original stereo-pair instead of the image coordinates on the mosaic ortho-images and mosaic stereomate.

Such a seamless stereo model is called the measurable seamless stereo model as we show in Figure 6.

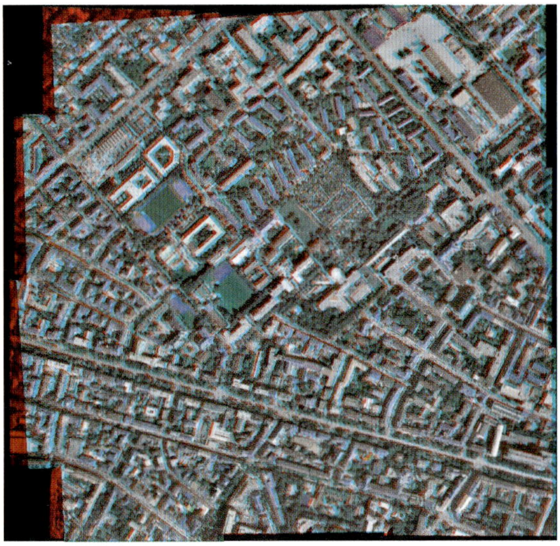

Figure 6 A block of measurable seamless stereo models with three strips

How to generate digital measurable images?

In this chapter, a system called VisionCruiser™ is designed and manufactured to generate seamless digital measurable images. The system integrates geodetic quality GPS, inertial navigation systems (INS) and multiple digital stereo cameras that are mounted on a land vehicle. It proves to be the most rapid, convenient, accurate and economic tool for collecting and updating geomatic data. VisionCruiser™ has been used on over 220,000 km rights of way in China.

VisionCruiser™ contains spatial semantic information which cannot be described by traditional maps. VisionCruiser™ represents the physical situation of the true Earth and catches knowledge-based GIS. Thus, a plentiful information base consisting of geography, economy and humanities data contained in VisionCruiser™ is the data source for many geo-spatial services. The working principle of VisionCruiser™ is illustrated in Figure 7.

In Figure 7, data is collected from a VisionCruiser™ vehicle travelling at standard road speed. The vehicle captures high-accuracy GPS data and best-in-its-class consecutive images while travelling along the defined routes. As a new surveying technology, VisionCruiser™ provides a Mobile Mapping System (MMS) collecting Digital Measurable Images (DMI) of roads and sideward objects. It is a vehicle which is furnished with such sensors and equipment as GPS, INS/DR and CCDs running at high speed. Users can implement measurement-on-demand between various elements, especially the elements beside the roadside, according to concrete applications based on these referenced DMI.

Figure 7 The working principle of VisionCruiser™

The integration of DMI and seamless stereo images

Firstly a seamless stereo image database is built and through its navigation, functions similar to Google Earth and Microsoft Virtual Earth are realized. When a region of interest (ROI) is selected, a concept called VisionCruiser™ is put forward, which is related to the integration of DMI and seamless stereo images for the sake of browsing, query, measurement, and information mining as shown in Figure 8.

Figure 8 The integration of DMI and seamless stereo image

VisionCruiser™ Can Provide the Four Functions of Digital Earth

Being searchable

The function of being searchable makes convenient, easy, and quick query of interesting information a reality. DMI is integrated with spaceborne/airborne image resources or users' existing data resources seamlessly in VisionCruiser™, as shown in Figure 9, which results in Point of Interest (POI) services on Digital Earth as easy as on the general digital map. That is to say, when users roam in image space built on the basis of digital images acquired by the Mobile Mapping System (MMS), corresponding properties of certain interesting objects (*i.e.* a bank or shopping mall) can by obtained just by clicking on the screen. The operation is as easy as on a digital map. The difference is that query by images is more social acceptable and more accordant with the requirements of Digital Earth. As illustrated in Figure 10, a query result of the position of POI, a hotel here, is returned.

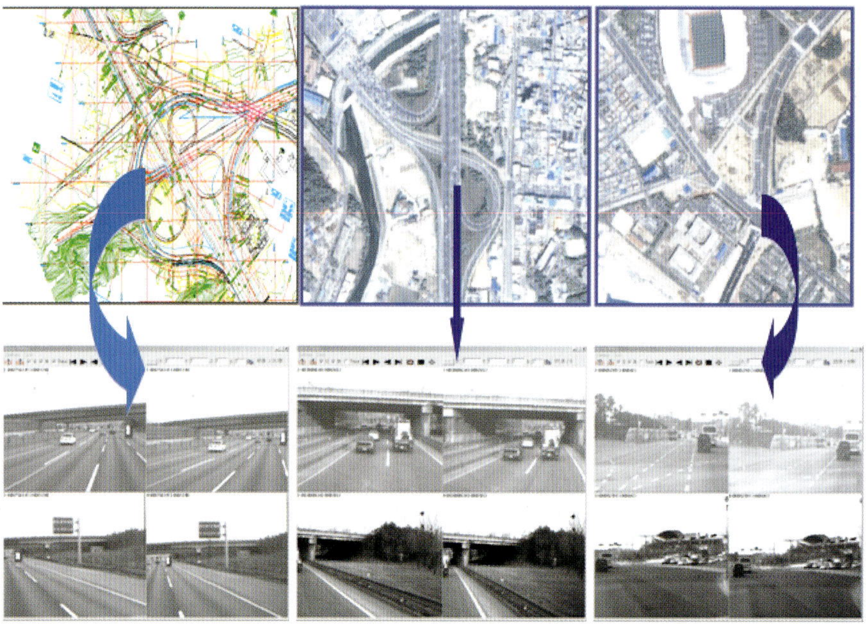

Figure 9 Seamless integration of different data resources

Also VisionCruiser™ can turn us from data users to data creators by providing the most rapid, convenient, accurate and economic tool for collecting and updating geomatic data. For example, Figure 11 shows an interface for annotating a shop which is possible for an end user.

Measurability

The contents provided by VisionCruiser™ provide a general definition for a 4D product and Digital Measurable Images (the fifth dimension product other than the existing 4 digital products DEM/DOQ/DLG/DRG). Users can customize concrete services according to their own requirements and implement directly and convenient browsing, relative surveying, absolute

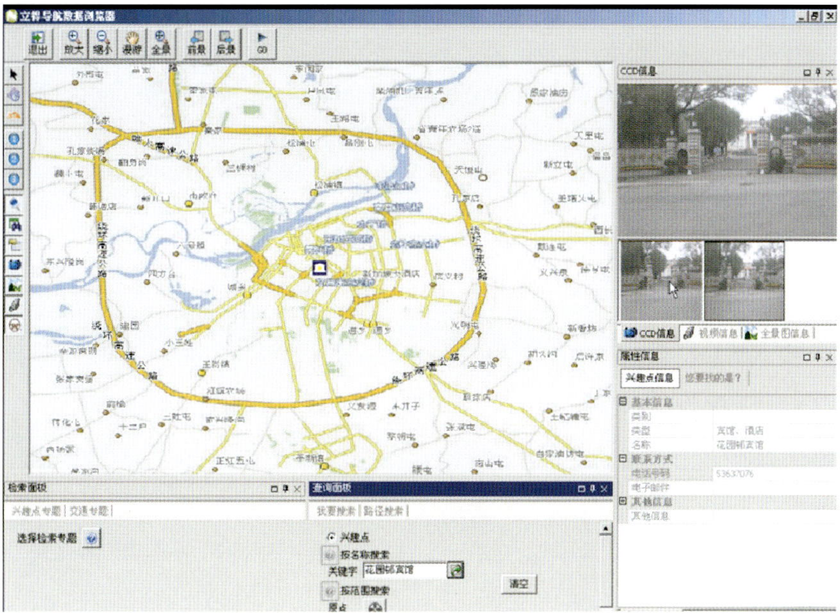

Figure 10 Searching POI images (a hotel)

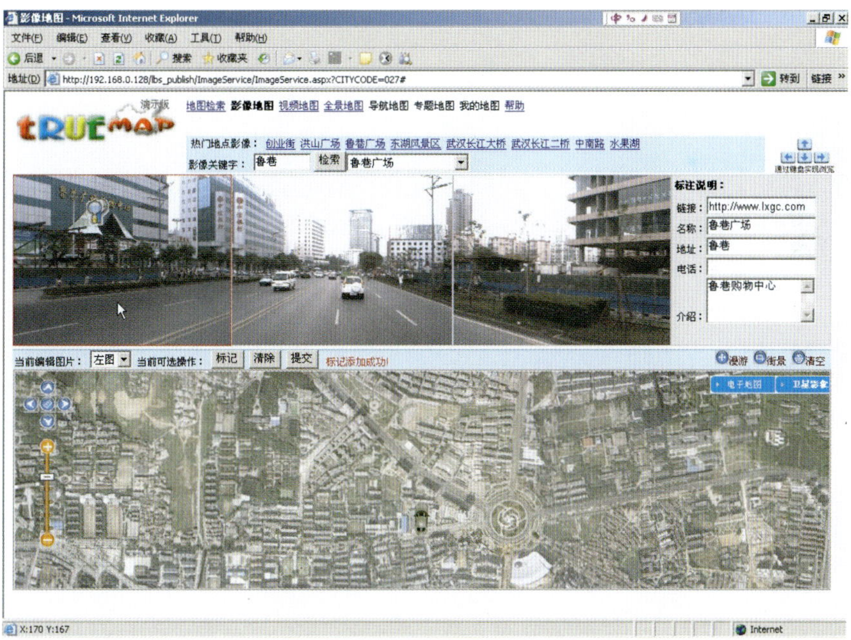

Figure 11 The annotation of a shop

posing, analytical surveying and so on. DMI gathered by MMS carries such data as external orientation elements and attitude elements. Cooperating with accurate time parameters, these data constitute the basis for surveying on multiple levels and seamless merging of the database.

Generally, the height and area information of a building is difficult to measure and the usual method is to access it through querying an attribute database which saves these information using traditional surveying in advance. That is to say, if the information we are interested in is not stored in database beforehand, it cannot be reached through query operation. However in VisionCruiser™, all information about an object including its length, area, perimeter and so on can be obtained in real time. Figure 12 shows a result of a real-time measurement by an end user on the Web. As shown in Figure 13, the distance from the appointed site to the telecom building is measured.

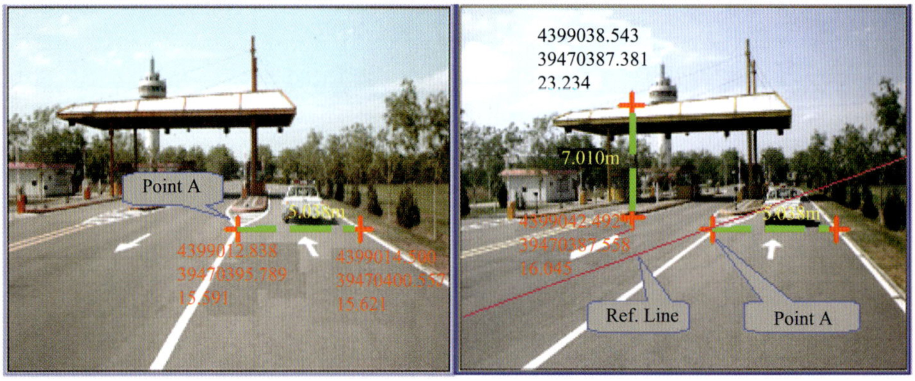

Figure 12 Measurement of road width and object height by end users on the Web

Figure 13 The measured distance from appointed site to telecom building

Visibility

Here measurable 3D images are contained in the content of the visualization other than vector data and raster image data. The goal is to make the representation of objects more vivid and comprehensive. DMI collected by MMS are known as a new product introduced to the spatial database while mass referenced DMIs accord with the customary human activities which cover data about the geography, culture and economy. These have become the preferred product of our images series with visibility.

Visualization is the most intuitive characteristic of VisionCruiser™. A series of video images, which are obtained by double CCD cameras, are matched through image correlation technology. Therefore stereoscopic images are generated to meet the mandate of "What You See is What You Get". Figure 14 shows a visualization from the result of using this technology on the Qinghai-Tibet Railway.

Figure 14 The visualization result of Qinghai-Tibet Railway, China

The total length of Qinghai-Tibet Railway is 1,142 km, data collection took 7 days, data processing took 10 days, and the resources used include 2 operators, 1 MMS vehicle and 1 railcar.

Figures 15 and 16 illustrate two other results of visualization. While the visualization of Figure 15 is the integration result of DMI and aerial image, the visualization of Figure 16 is the associated result of DMI and the digital map in vector format. Obviously, the integration results accord with human vision more than single images or vector graphics do.

Being minable

While the scene exhibited by the 4D product is a projection of the real world on a 2D surface, the scene exhibited by VisionCruiser™ is three-dimensional. This provides physical and social information which can be mined. Users can achieve attribute information mining through the corresponding application software, plug-ins and APIs according to their particular applications. VisionCruiser™ offers extensive data support for further applications including visibility analysis, transportation analysis, and business position selection.

The information that many users needs is related to professional applications and personal functions such as equipment involving electric power departments, municipal facilities of

Figure 15 The integration result of DMI and an aerial image

Figure 16 The integration result of DMI and the digital map in vector format

administrative departments, safeguards of the police department, traffic information of the transportation department, location of restaurant services and so on. All the information cannot be recovered by the traditional 4D product perfectly and effectively but it can be by VisionCruiser™. For example, certain rules can be discovered from texture images according to content-based retrieval. Figure 17 is the mined result of attribution information through POI retrieval.

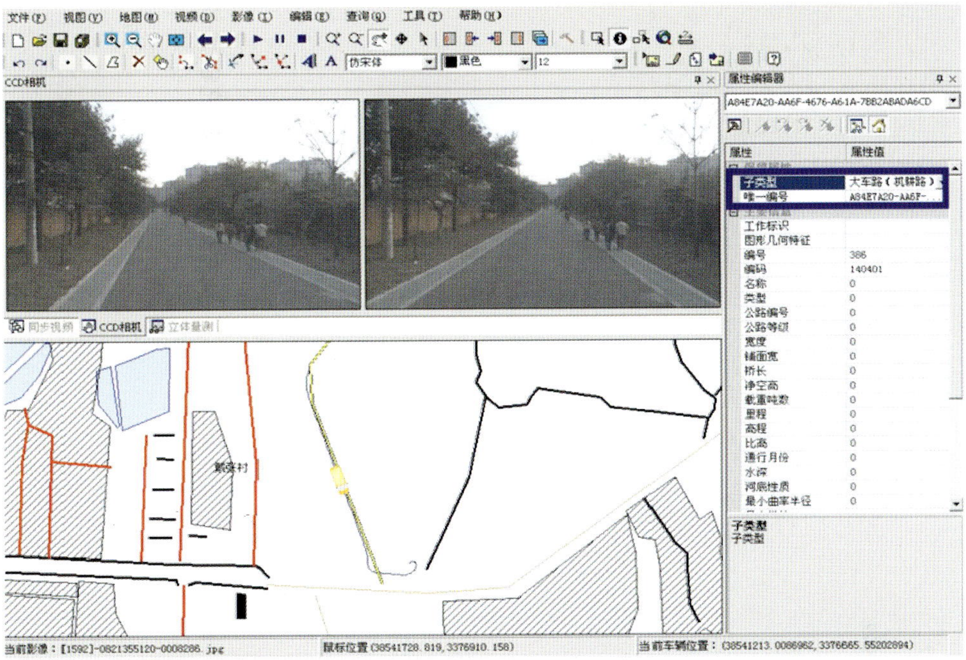

Figure 17 The mined result of attribution information through POI retrieval

Conclusions

Based on developing trends in requirements for geo-spatial information services, this chapter has illustrated the concept and technologies of Digital Earth, through its information architecture. The concept of a measurable seamless stereo model has been proposed and the experimental results illustrate that such a model can be achieved. An integrated system, the VisionCruiser™ which integrates a measurable seamless stereo model is introduced to provide a solution to the four functions define for Digital Earth. The DMIs collected by MMS provide promising prospects for GIS applications and geo-services over existing 4D products.

In conclusion, with the ability to meet the requirements of measurement-on-demand and geo-spatial information services for end users, DMI is not only a promising technology for Digital Earth but also an essential component in virtual geographic environments (VGE).

References

[1] Blachut T J. Further Extension of the Orthophoto Technique. The Canadian Surveyor, 1968, 22: 206–220.

[2] Blachut T J. The Stereo-Orthophoto Technique in Cadastral and General Mapping. Photogrammetric Engineering and Remote Sensing, 1976, 42: 1511–1519.

[3] Kraus K. Photogrammetrie: Band2. Bonn, DE: Duemmlers Verlag, 1984.

Chapter 18 Networked Collaborative Virtual Geographic Environments: Design and Implementation*

Jianhua Gong, Wenhang Li, Yabin Li and Mingxiang Huang

Introduction

3D visualization and virtual reality are increasingly important in geographic information science and play a wide role in applications of geographic information technology to the public including managers, policy analysts, geography teachers, children and many other niche groups[1–6]. Up to now, Google Earth has been downloaded more than 200 million times, and has widely been used in daily life by hundreds of thousands of people [7]. On the other hand, in view of the requirements of geo-collaboration by group of users dealing with complicated geo-problems such as geographic planning of regional sustainable development, impacts of global environment change, and emergency response, more and more effort is being placed on collaborative visualization and collaborative GIS[8–10].

By considering the integration of network-based 3D geo-visualization like Google Earth in the context of collaborative visualization, what are the forms of such new systems? In order to deal with open, large, and very complex interacting computer and human systems, Qian and Dai [11] put forward the concept of a workshop hall for metasynthetic engineering which would integrate human group intelligence and networked computing power. Based on the concept of virtual geographic environments (VGE)[3,12] and the workshop hall metasynthetic engineering, and incorporating geo-collaboration and distributed virtual environment technologies[13], this chapter addresses the basic conceptual framework for a networked collaborative VGE (CVGE), as well as developing a prototype which illustrates the form and basic functions of the CVGE.

Virtual Geographic Environments and Geo-collaboration

Features of virtual geographic environments

A virtual geographic environment is defined as a shared virtual space into which digital geographic objects and environments, as well as 3D avatars, representing users or people in

* Supported by the National High-Tech R&D Program (863 Program) (Grant No. 2006AA12Z204 and No. 2008AA121600), and the National Natural Science Foundation Project (Grant No. 40871181).

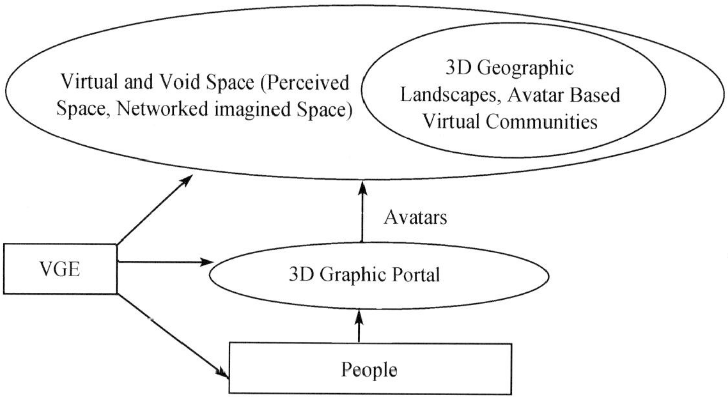

Figure 1 Fundamental features of a VGE

the physical environments, are transplanted from the real worlds. Thus, a VGE has the following characteristics which we illustrate in Figure 1.

1. A 3D virtual but void space first exists. The virtual space is of two kinds: one is the perceived or imagined space; the other is the mathematical world which may be infinitely large containing varied 3D digital objects. Therefore, in a virtual mathematical world, spaces are occupied by digital geo-referenced objects and environments simply represent different kinds of geospatial objects. Moreover there may be any number of such objects as well as their counterparts which are non geo-spatial referenced spaces.
2. Avatars are focused. People are centered in virtual environments, and avatars depict the identities of the online users who are observed people in real worlds. Virtual humans in the virtual space are an obvious property of VGE.
3. A 3D portal is necessary for entering the virtual space and environments. A 3D portal is also the human-computer 3D graphical interface which gives us basic and constant memory of the virtual worlds.

Geo-collaboration

Geo-collaboration can be defined as a group of people working together in both the same or different geographical locations and times to accomplish geo-tasks or to solve complex geo-problems. Geo-collaboration has four distinct components: geographic environments, a series of geo-tasks, task-related geo-problems, and multi-participants. Multi-participants mediated by collaborative tools act in unison to accomplish geo-tasks and solve geo-problems contained within such geographic environments. The major differences between geo-collaboration and general collaboration are that the former is involved in the context of complicated, multi-scale geographic spaces and environments whereas the latter are involved with non-spatial environments. Combined with the VGE system, geocollaboration could be fulfilled through avatars in a networked, three-dimensional, virtual space based on the representation of geographic environments and their associated geo-problems.

The Framework for a Networked CVGE

Current 3D virtual worlds

Google Earth and Second Life are currently well-known 3D virtual worlds on the Internet[7, 14] as we show in Figure 2. Google Earth is the representation and visualization of the real physical Earth mainly via multi-resolution remotely sensed images. Second Life is a 3D virtual community within which thousands of virtual humans live and work. The major differences between them are that in Google's digital earth, no people populate the virtual landscape which is in essence a simulation of real geographic environments, while in the Second Life world, many avatar-based individuals populate in the virtual scenes which may have no relation with the physical world but which are rooted in locations within the real world. Considering the properties of both simulations of physical and human worlds, a typical VGE should be designed through the integration of two or more virtual worlds such Google Earth and Second Life which represent the current prototypes.

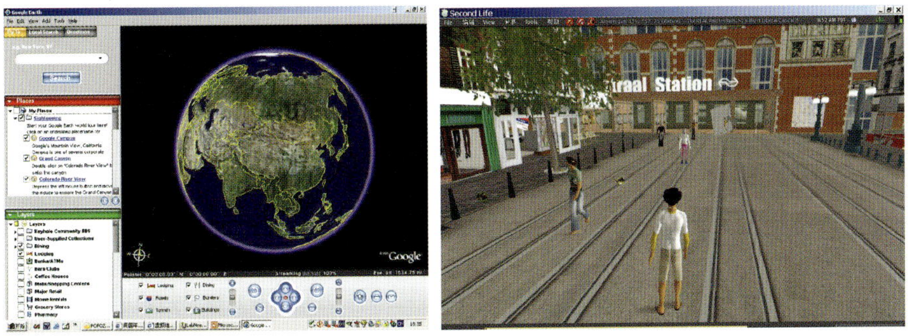

Figure 2 Google Earth and Second Life

Mobile communication and mobile CSCW

In recent years, mobile communication technology has been growing rapidly and mobile instruments play an increasingly more important role in human daily life. Mobile CSCW is a new research area involving Computer Supported Cooperative Work (CSCW) which explore working modes on mobile devices. Moreover, with the development of 3D rendering on mobile instruments, there is considerable interest in 3D visualization. The Web3D 2006 Symposium for example showed four 3D browsers running on mobile PDA devices[15] and it is out of contention that supporting the motto of "Working Anytime and Anywhere (WATAW)", networked CVGE should include geo-collaboration among users on Internet and mobile networks.

Conceptual framework

There is the ideal mode called "4A" for pursuing life-time distance learning, which is that *anyone* can learn *anything* in an *anytime/anywhere* environment [16]. In the chapter, the "5A" model is proposed for geographical research from point of view of group collaboration. "5A" is *Anyone*, *Anytime*, *Anywhere*, *Anything* and *Association*. The flexible "5A" model has few restrictions with participants, space (including movement status), time (in terms of discussion periods), and the content of geographic issues. All these are played out with a tight association between one another.

Based on the VGE concept and "5A" mode, and considering virtual worlds like Google Earth and Second Life, as well as mobile CSCW, the framework of a networked CVGE is established as we show in Figure 3. There are three components: virtual geographic landscapes, a virtual consulting studio(s), and a field studio(s).

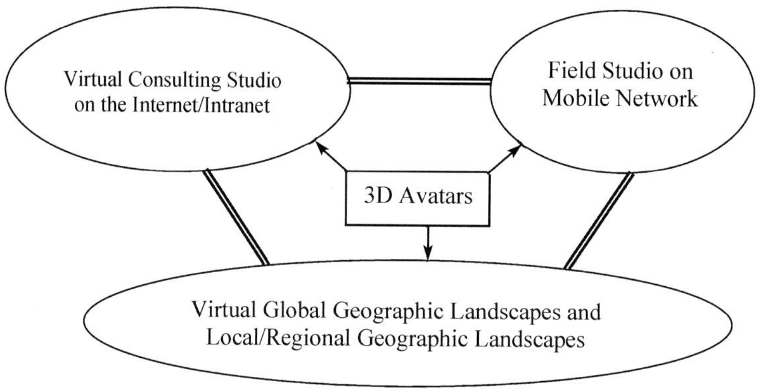

Figure 3　A framework for a networked VGE

Virtual geographic landscapes are the representation and visual simulation of global and local geographic environments on the Earth. Spaces in VGE are geospatial and geo-referenced ones that have geographic coordinate systems. Users can navigate in these virtual geospatial spaces and implement communications and interactions between each other. Multi-users can utilize either the 3D avatar or other mediums such as IDs and names for collaboration. If a 3D avatar is used for a user's ID in virtual space, the avatar size and location should be consistent with the geo-referenced space. In a VGE, the first things considered are geographic landscapes, geospatial relationships and distributions, geo-processes, *etc.*, while communications and collaboration among users are in second place.

The virtual consulting studio is a 3D virtual space on the Internet/Intranet which may bear no relation to geospatial space, and mainly provides a human-scale room for easily engaging interaction and collaboration. 3D avatars combined with video images of users are required for more effective and intensive communication. Specifically, a scaled geospatial space could be nested in a virtual consulting studio, which may allow multi-users to conveniently observe geographic objects, processes and events.

A field studio on a mobile network enables users to participate in collaborative consultation, and discussing discussions and meetings at any time and any where. Due to the limit capacity of a mobile network and hand held devices at present, field collaboration can be carried out in many ways, for instance, though 2D graphics, 3D graphics, and stream media images. Data size and transferring, system functions, and visualization modes are needed for specific designs for effective communications across the network.

In a networked CVGE, avatars are important. They can appear in virtual geographic environments, virtual consulting studios and even in 3D mobile virtual spaces. Avatars can be depicted by text name, 2D symbols, 3D graphics, video-images, and even audio-images. They can represent online users or human in physical worlds. They are the spirits of the networked CVGE.

Key Issues of CVGE

To date, there are no implementation systems related to networked CVGE. Many issues need to be dealt with in realizing an Internet/Intranet and mobile network based on CVGE. From technological standpoints, the key issues are as follows.

1. Handling of heterogeneous and asymmetric environments: a CVGE is positioned as a heterogeneous system, composed of heterogeneous networks of Internet/Intranet and mobile network and of heterogeneous clients of computers and mobile instruments. Typical asymmetries of communication capability, computation power, storage capacity, display capability, and so on need to be dealt with in implementing a networked CVGE.
2. Representation of large, multi-scale geographic environments and processes on the network: it is still a challenge for a CVGE system to be capable of coping with network-based 3D browsing and editing of geographic environments ranging from the global earth to local and small city environments, as well as intensive geospatial model computation, model process steering, and interactive visualization at varied levels.
3. Technical challenges in collaborative processes among online users: as to the group communication and interaction, research efforts should be placed on multi-perceptive group interactions, data organization and transfer, cooperative process sharing, scene consistency maintenance, and robust connection maintenance mechanisms for mobile instruments.

A Prototype System

Architecture of networked CVGE

A client/server architecture for a networked CVGE system has been designed along the lines shown in Figure 4. On the server side, the PC and mobile server services are built in terms of heterogeneous collaborative clients and communication networks. The PC server service is a multi-thread Java ServerSocket executed in a Tomcat Container serving computer terminals on the Internet or Intranet, while the Mobile Server Service is a multi-thread Java Servlet using HTTP as protocols serving wireless terminals such as mobile phones and PDAs (Personal Digital Assistants). There are message bridges between the PC and mobile servers for collaboration message exchange, thus collaborative messages can be dispatched amongst computers and wireless terminals.

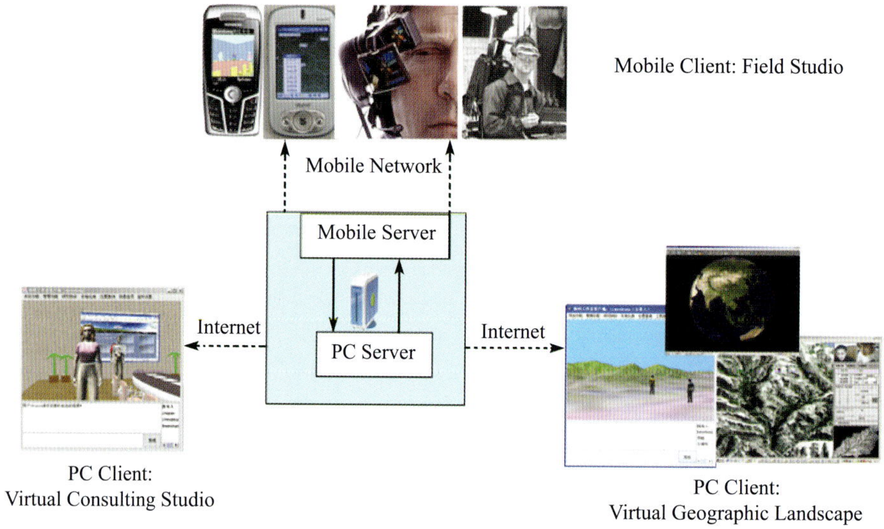

Figure 4 Architecture of the networked CVGE

Operations at the computer or wireless terminal (termed as the "sender") will be captured and packaged descriptions of actions with their parameters, which are the collaborative messages. These collaborative messages will be sent to the server first and then transferred to all other terminals. When a terminal (termed as a "receiver") gets the message, the packaged message will be recognized and decoded as an actual operation from the sender according to its operation and parameter protocols. This is the way communication is implemented and such descriptions with actions and their parameters are called "Collaboration Protocols". Sampled protocols are listed in Table 1.

Table 1 Sampled protocols in CVGE

	Action	Parameters			
Intenet/Intranet	Join	User Name	Avatar Figure	Avatar Location	
	Addavatar	User ID	User Name	Avatar Figure	Avatar Location
	Moveavatar	User ID	User Name	New Avatar Location	
	Delete avatar	User ID			
	Talk	User ID	User Name	Message Content	
Wireless Nework	Readvideo	nLength	Byte	VideoContent	
	Readpic	nLength	Byte	PicContent	
	Readtext	String	TextContent		
	Sendpic	nLength	Byte	PicContent	
	Sendtext	String	TextContent		

According to the conceptual framework, the client architecture of a networked CVGE system is composed of three major components: global/local virtual geographic landscapes, virtual consulting studios, and field studios.

A prototype system of networked CVGE

Java SE, Java ME, Java 3D, and VRML are used for developing the prototype system of the networked CVGE. A virtual consulting studio and field studio are also included in the environment. Figure 5 shows the client login interface for PC and mobile participants of the CVGE system. Figure 6 demonstrates the 3D virtual consulting studio with a 3D avatar combined with a real-time video image and scene collaboration on a mobile phone. In Figure 6, major objects in the virtual consulting studio are avatars, a 3D digital sand-table for placing studied object in the virtual geographic environment, a whiteboard for displaying and editing, and a text box for online user interaction through audio. Considering the limit capacity of the mobile devices, a stream media-based approach is employed to implement geo-collaboration on the mobile client.

Figure 5 Login interface for the PC and mobile participants

① User's login name; ② Avatar options; ③ Director's option; ④ Target server IP and Port

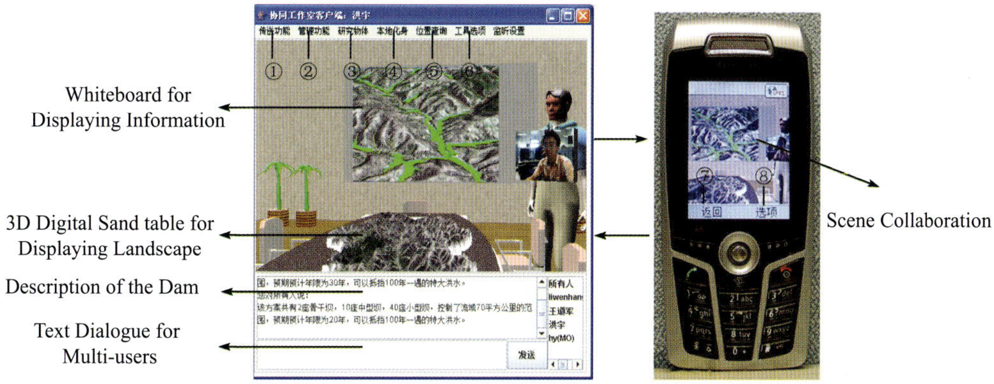

Figure 6 Interfaces for PC and mobile clients

① Data transfer; ② Scene management; ③ Study objects functions; ④ Avatar functions; ⑤ Location query functions; ⑥ System options; ⑦ Return button; ⑧ Options of mobile functions

Functions of the prototype system

A digital elevation model (DEM) of a small watershed on the digital sand-table is modeled in VRML, and users can either observe the 3D landforms of the watershed or, as participants, navigate in the 3D watershed space embodied by the small avatars in an immersive fashion.

This function is involved in the change of the users' spatial reference system. The prototype also can implement state-based and process-based record and playback of contents added by users or produced by the consulting process. Meantime, the system can help organize choices of several schemes via anonymous voting. Figure 7 describes the voting process. Firstly, the chairman of the meeting announces the start of voting and the number of schemes to be selected as shown in Figure 7A. The online client users on the PC and mobile networks all receive messages from the chairman, and are asked to make choices (Figure 7B, Figure 7C). Figure 8 demonstrates the statistical results of voting displayed in a virtual scene on the online user terminals at the same time.

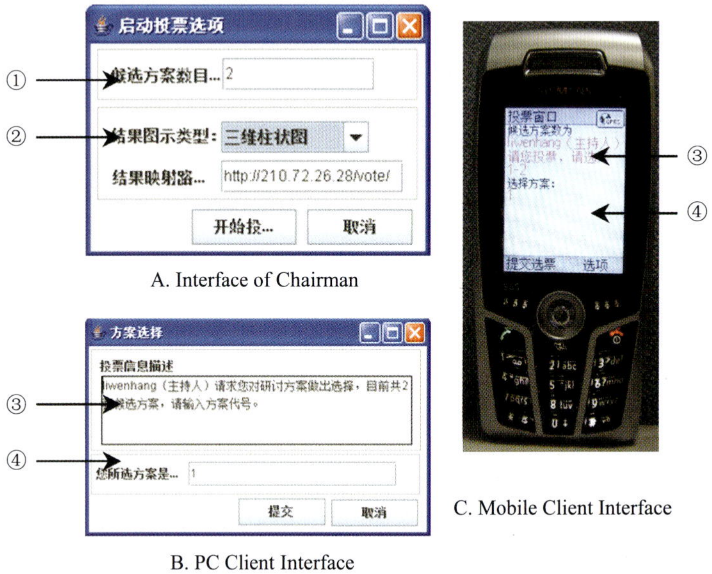

A. Interface of Chairman

B. PC Client Interface

C. Mobile Client Interface

Figure 7 PC and mobile client interface

① Alternative number; ② Result statistic graph type; ③ Scheme choice; ④ The selected number

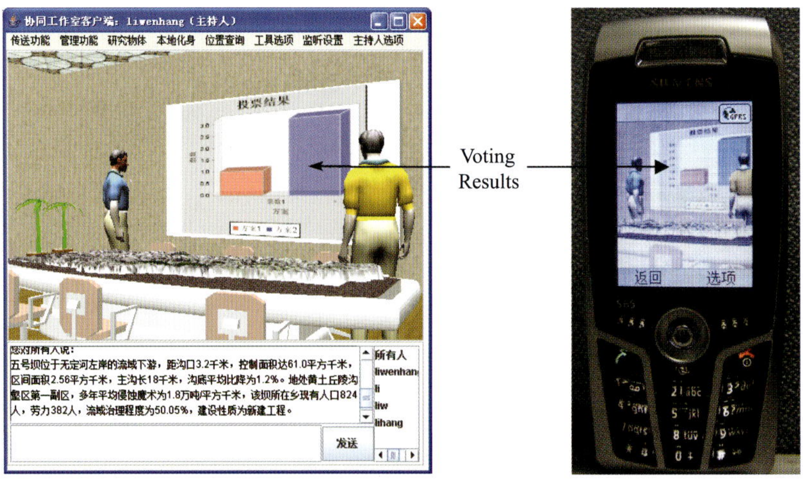

Figure 8 Voting result after statistical computation

Conclusions

Distributed virtual environments and CSCW are the main driving forces for the development of collaborative VGEs. Digital earth products like Google Earth and Virtual Earth also have an enormous impact on 3D Internet-based visualization of the global and local geographic environments. Goodchild[17] argues that the digital earth paradigm could be taken as one of the grand challenge of geographic information science. From the characteristics of virtual reality, geo-collaboration, and mobile CSCW, this chapter has aimed to illustrate a future architecture and 3D portal forms for collaborative geospatial visualization and virtual reality. The CVGE framework, including virtual geographic landscapes, virtual consulting studios, and the field studio on Internet and mobile networks has been presented. The chapter has addressed key issues and fundamental architecture of networked CVGE. A prototype system has been successfully developed for testing the feasibility of major techniques regarding network communication, network-based data organization and transfer, 3D visualization of virtual consulting studios and DEM, real-time video image-based avatars, interaction between online PC users and mobile users, collaboration processes and so on. As a rudimentary foray into this field, further research efforts should be directed towards the building of global and local virtual geographic landscapes based on the digital earth technologies, while the integration of virtual consulting and field studios in such CVGEs should ensure that such systems become increasingly efficient and stable as more and more users come to regard them as routine applications of GI Science.

References

[1] Benko H, Ishak E, Feiner S. Collaborative Visualization of an Archaeological Excavation. In: NSF Lake Tahoe Workshop on Collaborative Virtual Reality and Visualization (CVRV 2003), October 26–28, http://www1.cs.columbia.edu/graphics/projects/ArcheoVis/ [2003].

[2] Brodlie K W, *et al.* Distributed and Collaborative Visualization. Computer Graphics Forum, 2004, 23: 223–251.

[3] Gong J H and Lin H. Virtual Geographic Environments: A Geographic Perspective on Online Virtual Reality. Beijing: High Education Press, 2001 (In Chinese).

[4] Gore A. The Digital Earth: Understanding our Planet in the 21st Century. California Science Center, Los Angeles,California, http://www.digitalearth.gov/VP19980131.html [1998–1–13].

[5] Lin H and Gong J H. Distributed Virtual Environments for Managing Country Parks in Hong Kong —A Case Study of the Shing Mun Country Park. Photogrammetric Engineering & Remote Sensing, 2006, 68: 369–377.

[6] MacEachren A M, Edsall R, Haug D, Baxter R, Otto G S, Masters R, Fuhrmann S, and Qian L. Virtual Environments for Geographic Visualization: Potential and Challenges. In: Proceedings of the ACM Workshop on New Paradigms in Information Visualization and Manipulation. Kansas City, KS, 1999: 35–40.

[7] Google Earth. http://earth.google.com/ [2008].

[8] MacEachren A M and Brewer I. Developing a Conceptual Framework for Visually-Enabled Geo-collaboration. International Journal of Geographical Information Science, 2004, 34: 1–34.

[9] Gong J H and Lin H. A Collaborative VGE: Design and Development. In: Shivanand Balram and Suzana Dragicevic (Editors). Collaborative Geographic Information Systems. Canada: Simon Fraser University, 2006:186–206.

[10] Wood J,Wright H, Brodlie K. Collaborative Visualization. IEEE Information Visualization '97, Phoenix, AZ, October 19–24, 1997: 253–259.

[11] Qian S S and Dai L W. On Great Wisdom of Information Space-Fusion of Thinking Science: Literature Art, and Information Network. Shanghai: Press of Shanghai Jiaotong University, 2007 (In Chinese).

[12] Gong J H and Lin H. Virtual Geographical Environments and Virtual Geography. In: Forer P, Yeh A G O, and He J (Editors). Proceedings 9th International Symposium on Spatial Data Handling. Beijing, China, August 10–12, 2000: 28–39.

[13] Singhal S and Zyda M. Networked Virtual Environments: Design and Implementation. SIGGRAPH Series. New York: ACM Press, 1999: 1–17.

[14] Second Life. http://secondlife.com/ [2007]

[15] Nadalutti D, Chittaro L, Buttussi F. Rendering of X3D Content on Mobile Devices with OpenGL ES. In: Proceedings of Web3D 2006, 11th International Conference on 3D Web Technology. New York: ACM Press, 2006: 19–26.

[16] Bouras C, Philopoulos A, Tsiatsos T. E-Learning through Distributed Virtual Environments. Journal of Network and Computer Applications, 2001, 24: 175–199.

[17] Goodchild M F. Geographic Information Science: The Grand Challenges. In: Wilson J P and Fotheringham A S (Editors). The Handbook of Geographic Information Science. Malden, MA: Blackwell, 2008: 596–608.

Part 5

Mobility and Dynamics in Visualization

Mobility and dynamics are a recurrent theme in VGEs and in the last part of this book, we grapple directly with movements in the form of virtual environments which seek to show how people move in their local environments. This is an area that relates largely to traffic flow but in terms of personal movement, the methods are rather different from those in transportation planning, involving factors that pertain to cognition and geometry as well as economy. Moreover the technologies used to develop these kinds of model and their computer environment are rich in data and require rather special data systems for their implementation. Detailed infrastructure at the local scale involve unusual mixes of geometry and heterogeneous objects that have to be accessed and treated as if they were sometimes fixed, sometimes mobile while the actors who inhabit these environments have their own motion. Again as we saw in Parts 3 and 4, there are several types of dynamics that need to be modeled in VGEs which involve overlapping and intersecting motion and different time scales. One feature that must now be clear to the reader is that the complexity of these virtual environments is such that what is urgently needed are frameworks that enable us to look more reflectively at their structure and form. The state of the art as presented in this snapshot suggests that we urgently need to move to the construction of VGEs that are more integrated and where some of these thorny issues about motion, dynamics, temporality and scale are resolved in more satisfactory and more all encompassing ways.

Once again when mobility is measure and modeled at fine scale, remote devices such as GPS are all important in capturing the data that is needed to define and realize these kinds of simulation. Moreover linking such dynamics to 3D and visualizing patterns in this data are important issues that are only broached for the first time when all these components are put together. Video games technology has much to recommend itself here for motion as well as realism—if not functional realism—have always been central to such gaming. These kinds of virtual environment can teach us an awful lot about rendering, speed, storage and user interfaces. Moreover that fact that such games are being opened for more serious uses simply adds to the arsenal of virtual realities that are now available for fashioning VGEs.

Arikawa and Fujita first present a new framework based on advanced geocoding methods. These enable local environments which are relevant for spatial movement to be encoded as higher dimensional spatial objects from natural spatial descriptions that include route descriptions and scene descriptions in natural language. The current dissemination of pedestrian navigation services on mobile phones in Japan is based on rich sidewalk network databases which include a large number of detailed POIs (points of interest) such as the entrances to buildings, crosswalks and foot bridges. They propose the formal spatial association statement (FSAS) which represents the list data structures composed of nouns concerning places and other parts which may be descriptions of relationships between the places. FSAS is simple and flexible so that FSAS instances can be generated in both rough and precise terms depending on the quality of parsing natural spatial descriptions and of matching them with sidewalk networks. They demonstrate the basic concept of data structures and algorithms for high-dimensional geocoding for natural spatial description through FSAS.

Moreover the prototype systems of high-dimensional geocoding are introduced to prove whether or not these proposed methods are useful and feasible.

Wu and his colleagues then present a multi-agent-based simulation of pedestrian crowds at outdoor events using the example of fireworks displays in Hong Kong. The multi-agent-based system simulates pedestrian movement during outdoor events based on GIS data and data on the fireworks display over Hong Kong harbor at Tsim Sha Tsui. The event is decomposed into different scenarios, including arrival and dispersal and since individual motion is significant to the simulation, different sets of rules are designed for the behavior of each agent, including the pedestrian and crowd manager involved in different scenarios. The agents act intellectually and communicate with other agents, based upon the rules extracted from the analysis of observed information. The entire region was rasterized into grid cells at a minimum area suitable for one pedestrian with each pedestrian agent generated from randomly chosen subway station, and one fireworks display selected as the favorite attractor agent. During each time step, the agents searched available routes to the attractors and took steps according to the selected routes. Pedestrians do not stop acting until they arrive at their destination cells. While in dispersal, the start points and the attractors are exchanged when pedestrians start moving from the sites where they watched the display towards the subway stations. Meanwhile, several crowd manager agents were set to move within the region sending instructions to the pedestrians to keep them away from confirmed overcrowded districts. On the other hand, the simulation process was designed to be tuned through the interface to different parameters such as the maximum number of pedestrian agents. The system is clearly helpful for users in preparing for emergency situations and is considerably more effective than the conventional sand table game. Throughout the development of the model, technical issues including path finding and time step slice problems were discussed and future improvements to behavior in the model are presented.

Gong and Mackett focus on how behavior can be visualized from data extracted from children walking to school using GPS technologies. Previous behavioral research involving walking has focused mainly on the personal and social factors that can affect physical activity. More recently, it has been increasingly argued that the built environment has also strong impacts on personal physical activity level, although there is poor understanding of the relationship between urban form and physical activity in children. Nowadays, children enjoy less freedom in the urban setting than previously, and it is becoming clear that less independence and mobility constrain children's development. A decline in children's physical activity levels is also apparent, which parallels the increase in the number of obese children and the growth in sedentary activities. Thus, there is an urgent need to understand better children's movement patterns in the built environment. In their chapter, 3D visualization is shown to be a very effective tool to display the local travel patterns that children make on foot, the places children go to, and the places where children are most physically active. Children's daily movement pathways are then characterized using a variety of measures, such as gender, age, energy expenditure, land use location, turning angles, traveling speed and so on. The results indicate how boys and girls interact differently with their local environment, and how children behave differently in different types of environment.

In the last chapter, Shepherd and Shepherd speculate on the contributions that games might make to the construction of VGEs. They go further; they argue that videogames might indeed be the new paradigm for GIS, largely because motion and dynamics are central to such endeavors and because games designers have grappled with how to represent such issues rather successfully for many years. User immersion is central as is the question of interface, analytics less so. Videogames and GIS have more in common than might be expected. Indeed, the authors advance the proposition that videogame technology may not only be considered as a kind of GIS, but that in several important respects its real world modeling capabilities out-perform those of most GIS. They examine some of the key differences between videogames and GIS, explore a number of perhaps surprising similarities between their technologies, and consider which ideas might profitably be borrowed from videogames to improve GIS functionality and usability. In many senses, their contribution is a fitting conclusion to all these chapters which provide a comprehensive snapshot of the state of the art in progressing GIS to VGEs.

Chapter 19 Simple High-dimensional Geocoding by Matching Natural Spatial Descriptions with Sidewalk Network Databases

Masatoshi Arikawa and Hideyuki Fujita

Introduction

Geographic search engines such as Google Maps are widely popular. They enable us to retrieve web pages by spatial keys such as geographic extents of map displays and place names. Most web pages do not include *direct geographic descriptions* such as geographic coordinates, but include *indirect geographic descriptions* such as addresses, place names, postal codes and telephone numbers. To realize geographic search engines, it is necessary to generate databases of points of interest (POIs) which represent locations of the contents of web pages[1]. The POI databases can be used as representative points of web pages for indirectly retrieving web pages by spatial keys. The POI databases of web pages are generated by two main information processing devices: 1) geoparsing and 2) geocoding. Geoparsing is to extract indirect geographic descriptions from web pages. Geocoding is to convert indirect geographic descriptions into direct geographic descriptions. Thus, web pages and digital documents on the Internet including both direct and indirect geographic descriptions can be retrieved through web browsers with map interfaces. The geographic search engines enhance our spatial capability of finding relevant digital content on the Internet.

In this chapter, we present a next framework for extending the geoparsing and geocoding methods of generating geographic points to the ones of generating *higher dimensional spatial objects* such as geographic lines, arrows, areas and so on. For example, if a digital document includes a route guide description composed of multiple POIs, the current geoparsing and geocoding methods can generate only a collection of POIs, but the series of POIs are meaningful and imply a route showing a polyline or a line of interest with important passing points. Another example is a scene description to indicate landmarks and geographic objects in a user's view. A scene description like "AAA building in front of the user, BBB building on the left, and CCC building on the right" can be used for detecting the position and direction of the user. As a result, an arrow representing the user's position and direction can be generated as the geocoding of the scene description. Figure 1 shows an overview of high-dimensional geocoding of natural spatial descriptions through FSAS (formal spatial association statements) which is our proposed intermediate representation of natural spatial description to be explained in detail later, using POI and sidewalk network databases and rule bases for FSAS. If the framework becomes general, the capabilities of spatial search

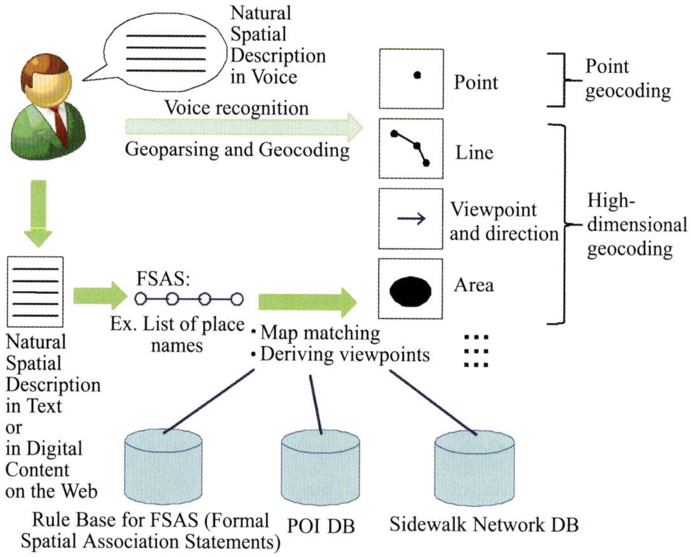

Figure 1 Overview of geocoding natural spatial descriptions into high-dimensional spatial objects

engines will be increased and we will be able to make our daily communications more efficient with spatial information technology.

Machine understanding of natural language is still an unsolved problem. The technology of realizing a search engine allows us to retrieve web pages by key words using only simple and heuristic technology of the partial understanding of natural language. Our proposed geoparsing and geocoding methods are also the same as using simple and heuristic analyzing methods for partially processing natural language, but not using complicated and rigid methods for fully analyzing natural language. Our algorithm used for generating FSAS extracts only nouns from documents, and judged as if the nouns are geographic names using geographic dictionaries for covering pedestrian's behavior, that is, these are sidewalk network databases developed for location based services.

Section 2 proposes FSAS to represent and process natural spatial descriptions in natural language by a computer. Section 3 explains the sidewalk network databases. In Section 4, we demonstrate a method for geocoding natural route descriptions and a prototype system. In Section 5, we apply FSAS to geocoding for scene descriptions. The result of geocoding a scene description is to detect the position and direction of a user's view. Finally, we draw the conclusion and sketch future work.

Natural and Formal Spatial Descriptions

This section explains two kinds of spatial description. One is *natural route description* to represent a route from place to place by natural language. The other is the *formal spatial association statement* (FSAS) as a list-structured graph representation for spatial descriptions.

Structure of natural route descriptions

The Japanese language has no space between the words. It is not simple to separate a sentence into words by a computer, but there is an effective method for the separation of words in Japanese using morphological analysis tool such that illustrated by ChaSen[2]. A natural route description can be divided into noun and verbal phrases. For example, the Japanese description "渋谷駅ハチ公口を出て道玄坂を上り交差点を右へ曲がる(You exit from the Shibuya Hachiko Exit and go up Dogen-zaka Slope Street, and then turn right at the intersection)" is divided as follows:

渋谷駅ハチ公口を出て道玄坂を上り交差点を右へ曲がる
↓
[渋谷駅ハチ公口][を出て][道玄坂][を上り][交差点][を右へ曲がる]
↓
noun phrases: [渋谷駅ハチ公口], [道玄坂], [交差点]
verbal phrases: [を出て], [を上り], [を右へ曲がる]

All of the above noun phrases have a place name. A place name refers to a unique place in the real world and works as a reference point for the next verbal phrase. The above verbal phrase has a preposition such as "を(at, from or through)", a verb such as "曲がる(turn)" or "出る(exit)" and an adjunct such as "右へ(right)", "100 メートル(100 meter)" or "5 分(5 minutes)". It is usual for Japanese people that an agent noun for the verbal phrase is often omitted in a description. A person who is referred to by an omitted agent noun is a walker who is a writer of a description or third person.

We call a noun phrase in a natural spatial description a *spatial anchor description*. The spatial anchor description includes unique nouns and general nouns. A general noun such as "the intersection" is deduced from the context of a description. In Table 1, we collect typical

Table 1 Examples of spatial anchor descriptions in Shibuya

Spatial Anchor	
Building	
	タワーレコード(Tower Records), エイチエムブイ(HMV), QFRONT (QFRONT), 道玄坂上交番 (Dogen-zaka-ue Police Box)...
Station Exit	
	JR 渋谷駅東口/ハチ公口/宮益口 (JR Shibuya Station East Exit / Hachiko Exit / Miyamasu-guchi Exit), 京王井の頭線渋谷駅(Keio Inokashira Line Shibuya Station) ...
Street	
	明治通り (Meiji Street), 公園通り (Park Street), 井の頭通り (Inokashira Street), 文化村通り(Bunkamura Street)...
Slope Street	
	宮益坂 (Miyamasu-zaka), スペイン坂 (Spain-zaka), 道玄坂 (Dogen-zaka) ... ("zaka" or "saka" means a slope street in Japanese)
Intersection	
	渋谷スクランブル交差点 (Shibuya Scramble Intersection), 道玄坂上交番前交差点 (Dogenzaka-ue Koban-mae Intersection)...

places which are referred to by spatial anchor descriptions in Shibuya, one of the big cities in Tokyo, Japan. Spatial anchor descriptions referring to those places can be used as start points, passage points, and end points of routes in natural route descriptions. Table 2 gives instances of verbal phrases. These verbal phrases are the expressions of a walker's action on the route. Walker's actions on the route denote the topology between places. On the other hand, we call a verbal phrase in a natural route description a *spatial relationship description.*

Table 2 Examples of spatial relationship descriptions

Spatial Relationship	Examples
Direction	まっすぐ進み，直進し(go forward, go ahead, advance) 右へ曲がり，右手に進み(turn to the right, on the right, on one's right) 左へ曲がり，左手に進み(turn to the left, on the left, on one's left)
Distance	200 メートル(in 200 meter), 5 分(in 5 minutes)

Formal Spatial Association Statements (FSAS)

A simple model of natural route descriptions

We simplify route descriptions as a list structure shown in Figure 2. A description of spatial anchor corresponds to an ellipse, and a spatial relationship description corresponds to a rectangle in Figure 2. Figure 3 illustrates the image of geocoding a natural route description with spatial databases. Constructing a natural route description can be compared to sticking of pins and stretching of a thread between pins on a map. Using this simple model, we formalize natural route descriptions.

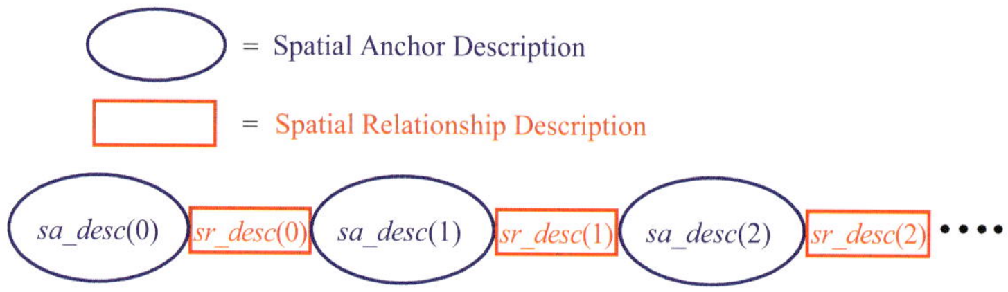

Figure 2 Image of model for articulating a natural description as a formal spatial association statement

Grammar of formal spatial association statement

Formal statements are necessary for computers to indirectly deal with natural route descriptions. On the assumption that all of natural route descriptions can be expressed with nodes and links, we propose formal spatial association statement (FSAS) to represent and process natural route descriptions. FSAS also work as a query language for sidewalk network databases. Using FSAS, a natural route description is represented as a sub-graph of a directed graph of sidewalk networks as we show in Figure 4. Table 3 shows the grammar of FSAS. Generalization tables are indispensable for converting various casual descriptions into regular ones, one of which is FSAS as in Table 4. A use case of the generalization table is to make an instance of the spatial relationship description as a value of the attribute

Figure 3 Image of geocoding a natural route description using a sidewalk network database through the intermediary of simple model

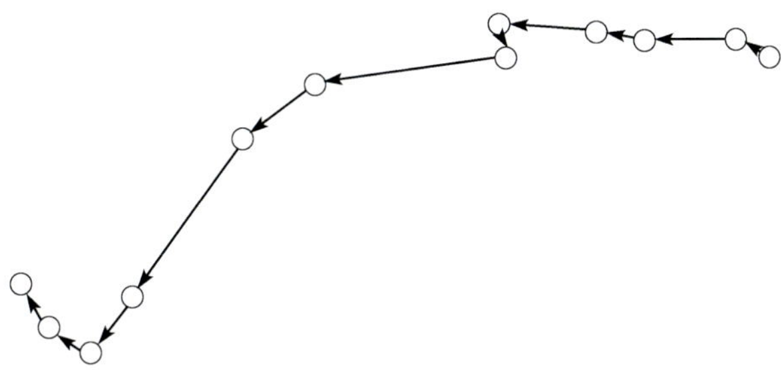

Figure 4 Sub-graph *G* which is the result of *FSAS*(*route_desc*), where *route_desc* is a text string of a natural route description

Table 3 Grammar of a formal spatial association statement

FSAS ::=	*sa_desc*(0)(:*sr_desc*(*i*):*sa_desc*(*i*+1))* [*i*={0,⋯,*n*}]
sa_desc(*i*) ::=	*node*(*i*).*node_attribute_list*
node_attribute_list ::=	*none* | *node_attribute_value* (&*node_attribute_value*)*
node_attibute_value ::=	*node_attribute* = *value*
node_attribute ::=	*id* | *name* | *coordinate* | *class* | *status*
value ::=	*numerical_value* | *string_value* | *url* | *status_values* | *connect_values*
status_values ::=	*start* | *end* | *via*
connect_values::=	*straight* | *right* | *left* | *right(angle)* | *left*(*angle*)
sr_desc(*i*) ::=	*link*(*i*).*link_attribute_list*
link_attribute_list ::=	*none* | *link_attribute_value* (&*link_attribute_value*)*
link_attibute_value ::=	*link_attribute* = *value*
link_attribute ::=	*id* | *start_node*(*id*) | *end_node*(*id*) | *direction* | *connect* | *distance*

Table 4 Example of a generalization table for descriptions of spatial relationship and values of the attribute "*link.connect*"

Specialized descriptions for spatial relationships	Generalized descriptions for spatial relationship (values of *link.connect*)
go forward, go ahead, advance	Straight
Turn to the right, on the right, on one's right	Right
turn to the left, on the left, on one's left	Left

"*link.connect*" as in Table 3. The attribute "*link.connect*" plays an important role in finding a spatial object when a name referring to the next place is omitted.

Sidewalk Network Databases

Walking is fundamental behavior for a human being and the sidewalk (or pavement) is a space which we walk on a daily basis. Sidewalk network databases are large-scale databases, which are regarded as digital copies of our walking environment. In this section, we explain the schema of sidewalk network databases and a prototype system for managing and editing them.

What are sidewalk network databases?

Sidewalk network databases store spatial data concerning sidewalks, underground walks, footbridges and cross walks for pedestrians. Sidewalk network databases are simply structured as nodes and links. They are provided as commercial products by Shobunsha Publications Inc. and Zenrin Co. Ltd.[3, 4]. The commercial sidewalk network databases presently cover major cities in Japan. We redesign the schema of sidewalk network databases in order to increase flexibility in advanced applications.

Before the emergence of sidewalk network databases, there have been popular geographic network databases such as road networks for car navigations as in Figure 5, railroads, facilities networks and so on. However, the traditional geographic network databases were all designed for small scale uses, not for large scale uses such as human navigation. On the other hand, sidewalk network databases for walkers as we show in Figure 6 are becoming popular for human navigation systems in Japan. Some services and products using sidewalk network databases are already been on the market. "EZnaviwalk" provided by KDDI au is one of the most popular human navigation services using a mobile phone, GPS and/or electric compass[5]. Using sidewalk network databases, train timetables and airline timetables, EZnaviwalk software will find the most direct, time-saving, money-saving, or CO_2-saving routes. This service began in October 2003. In October 2008, there were about two million dues-paying members of EZnaviwalk.

Schema of sidewalk network database

A sidewalk network database is a directed graph representing sidewalks in urban environment. It consists of the following two components:

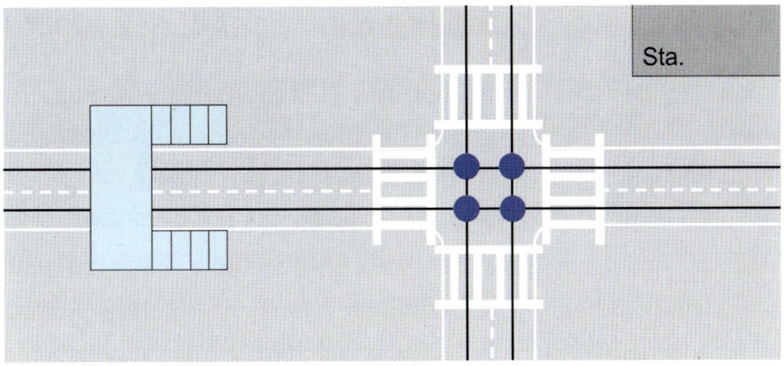

Figure 5 Schematic diagram of a road network database

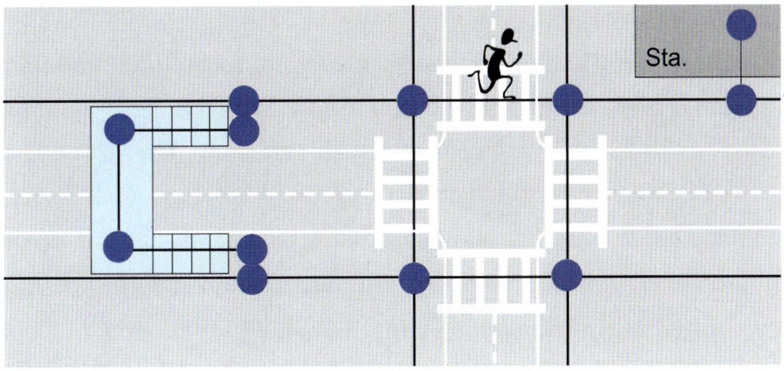

Figure 6 Schematic diagram of a sidewalk network database

1. a set N of nodes.
2. a set L of links.

A link is a binary relation on N and it is also a pair of two nodes. A tuple (N, L) comprises directed graphs.

We define spatial objects as data units corresponding to entities in the real world (*e.g.*, sidewalks, footbridges, crosswalks and entrances of buildings). An example of a sidewalk network database is shown in Figure 7. Each node is represented by a circle. In Figure 7, a set N is the set of node data n_i.

$$N = \{n_1, n_2, n_3, n_4, n_5, n_6, \ldots\}$$

A node data n_i has the following structure.

$$n_i = (id_i, name_i, class_i, m_i, uri_i, in\_link_i, out\_link_i)$$

The elements of n_i are the followings:

id_i the identifier of the node

$name_i$ a name of an instance of spatial objects

$class_i$ a class of spatial objects

$m_i = (x_{m_i}, y_{m_i})$ geographic coordinates

uri_i information for the point of interest like a picture image

in_link_i the identifiers of the incoming links to n_i

out_link_i the identifiers of the outgoing links from n_i

A $name_i$ is text data referring to the instance of spatial objects. Nodes composing the specific instance of spatial objects have $name_i$ value only if the place has a name. Spatial objects in the database can be classified into four main classes according to their role in the city or geographic characters[6]. Each class has its own characteristics and restrictions. Spatial objects of the same type are categorized as classes such as "exit", "intersection", "street" and "slope street". For example, the node n_2 has "α Station Exit" as $name_2$ and "station exit" as $class_2$. Composing β Intersection on the map, the nodes n_3, n_4, n_5 and n_6 have "β Intersection" as $name_i$ and "intersection" as $class_i$.

Next, in Figure 7, a set L is the set of link data l_i.

$$L = \{l_1, l_2, l_3, \ldots\}$$

A link data l_i has the following structure:

$$l_i = (start\_node_i, end\_node_i)$$

The elements of l_i are the following:

$start\_node_i$ the identifier of the node which l_i starts from

end_node_i the identifier of the node which l_i arrives at

Each link in L is represented by an arrow from $start\_node_i$ to end_node_i. For example, $n_3 \to n_6$ is the link l_7 and $n_6 \to n_3$ is the link l_{110} in Figure 7. It is a notable fact that there is no link $n_2 \to n_1$ because sidewalk network databases do not include sidewalks leading to the inside of buildings.

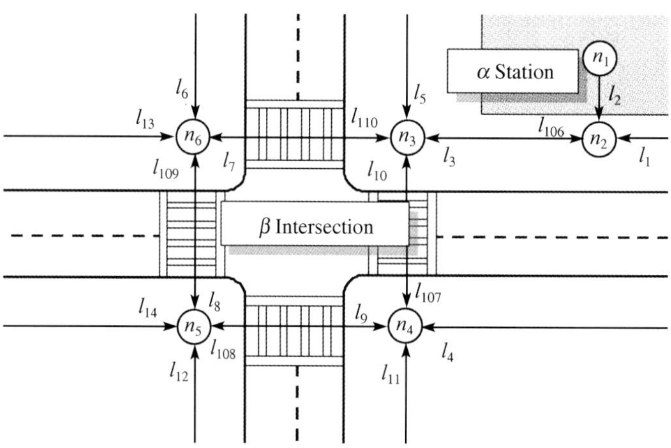

Figure 7 Example of a sidewalk network database

Classes of spatial objects

Spatial objects in the sidewalk network database can be classified into four main classes: exit, intersection, street and slope street. An attribute value of class is available as one of the parameters for deducing valid spatial relationship descriptions. As simple examples, an intersection class is a clue to deduce "turn right at", "turn left at" and "cross at". The slope street class is also a clue to deduce "go up" and "go down".

Prototype of sidewalk network database management system

We have developed a naive management and editing system for sidewalk network databases. We explain each component in the user interface in Figures 8 and 9 below as follows.

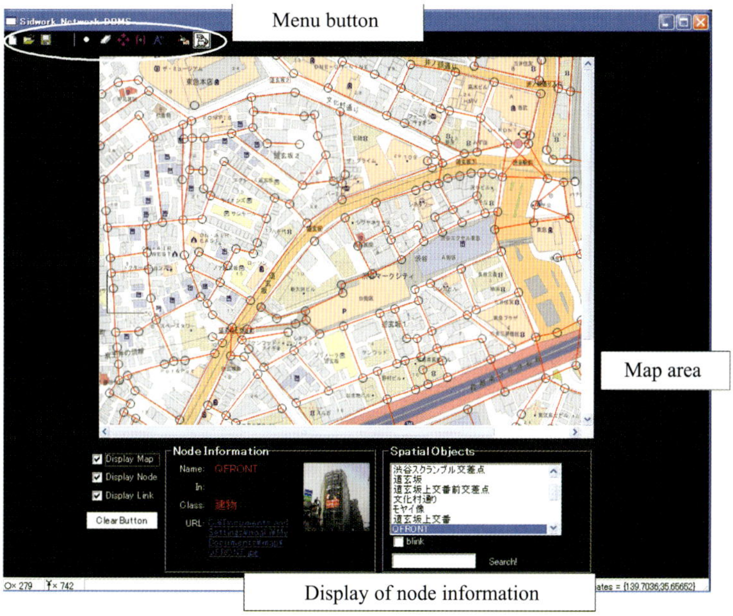

Figure 8 Graphical user interface of the prototype system for managing sidewalk network databases

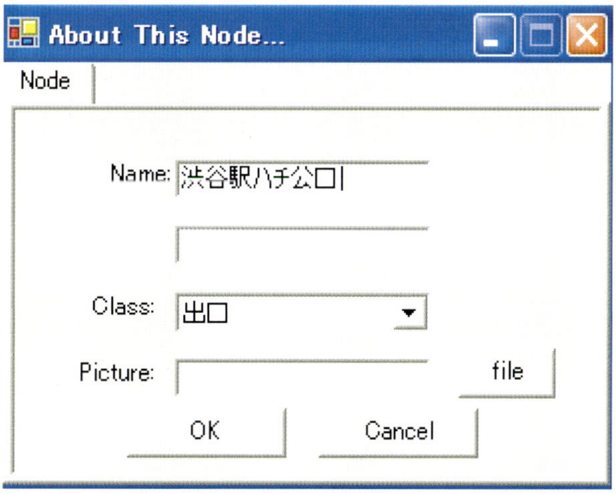

Figure 9 Dialog box of entry form for node information

1. Menu button: Users can change the operation mode by the menu button. Main functions are 1) loading and saving sidewalk network data which are XML formatted and 2) creating, deleting, modifying and moving both nodes and links. Furthermore, we

select functions of referring to node information (*e.g.*, a name of a place) and filling it using the entry form.

2. Map area: Sidewalk network databases and a map image in the selected area are overlapped and visualized. Figure 9 shows a dialog box for an entry form of node information, which allows users to add values in name and class attributes for spatial objects to nodes. This window appears when users push the entry button in menu buttons and click a node on the map area.
3. Display of node information: This area allows users to see values of attributes, a name, a class and a picture image for an instance of a spatial object in the map area.

Geocoding Method for Natural Route Descriptions

Geocoding is a process of querying a region in the real world which is referred to by natural spatial descriptions. We define the geocoding process as follows:

Input data: a text string g of natural spatial description

Output data: a region r (or a point p) corresponding to g in the real world

A geocoding process G is defined as a function which meets the following decision rule.

$$r = G(g)$$

This section provides a solution method for the geocoding process using sidewalk network databases. The first subsection explains the outline of geocoding process for natural route descriptions. The second and third subsections go into details about geocoding process. The fourth subsection describes the prototype system and experimental demonstration.

Outline of geocoding process

An address description, which is a kind of natural spatial description, includes both some numbers for representing the locations of buildings, and some names for representing administrative areas where someone lives or works. An address description corresponds to a closed region on a large-scale map or a point on a small-scale map. On the other hand, a natural route description includes both some noun phrases referring to places and verbal phrases for indirectly representing spatial relationships between places. A natural route description usually corresponds to a polyline on a map. The geocoding process for natural route descriptions is not only converting place names into geographic coordinates, but also verifying spatial validity between places.

The geocoding for natural route descriptions consists of the following two processes. We go into details of them in the following subsections:

(1) Converting a natural route description into a formal spatial association statement.

(2) Validating a formal spatial association statement.

A FSAS is a geometric model for representing natural route descriptions. Computers can deal with natural route descriptions indirectly through FSAS as an intermediate description.

Converting natural route description into formal spatial association statement

The procedure to convert natural route descriptions into FSAS can be classified into two sub-processes. At this stage, FSAS need not have geographic coordinates.

Separating spatial anchor descriptions and spatial relationship descriptions in a natural route description

In the case that the input text is "ハチ公口を出て道玄坂を上り交差点を右へ", the text is separated as follows:

sa_desc(0) = "ハチ公口": "Hachiko Exit"

sr_desc(0) = "を出て": "exit from"

sa_desc(1) = "道玄坂": "Dogenzaka (slope street)"

sr_desc(1) = "を上り": "go up"

sa_desc(2) = "交差点": "Intersection"

sr_desc(2) = "を右へ": "to right"

sa_desc(3) = *none*

Generalizing spatial anchors and spatial relationship descriptions

In the above example, *sa_desc*(0), *sa_desc*(2) and *sr_desc*(2) are invalid descriptions for matching with sidewalk network databases. They are generalized as follow:

sa_desc(0) = "JR 渋谷駅ハチ公口": "JR Shibuya Station Hachiko Exit"

sa_desc(2) = "道玄坂上交番前交差点": "Dogenz-aka-ue Koban-mae Intersection"

sr_desc(2) = "を右へ曲がる": "turn right"

For the process of parsing a natural route statement, there are already significant achievements in the research field of natural language processing[7]. We use a Japanese morphological analysis system devised by ChaSen[2] for the sub-process. For the sub-process of dealing with incomplete or ambiguous information in this section, we plan to realize the functions to complement the incomplete place names and to correct inadequate natural route descriptions using sophisticated geographic thesaurus and generalization rules for place names. Thus, the further investigation of completing inadequate natural spatial descriptions in detail is a subject for future work in our research.

Validating Formal Spatial Association Statement

This stage verifies the spatial validity between places which are referred by a set of *sa_desc* of FSAS. The procedure to validate FSAS can be classified into two sub-procedures for *sa_desc* and *sr_desc*.

A *sa_desc*(*k*) is needed to match with the set of valid spatial anchors. The set of valid spatial anchors is the result from the following function.

$$valid\_sa\ (passedlink,\ sr\_desc(k-1)) \quad k=1,\cdots,n$$

In the function *valid_sa*, *passedelink* is $node(p-1) \rightarrow node(p)$. The *node*(*p*) is the last node which is matched at the time. *node*(*p*) is derived by the following geocoding function *G*.

$$node(p) = G(sa\_desc(k))$$

A text *sa_desc*(*k*) as a noun corresponds to a spatial object *node*(*p*) stored in a sidewalk network database. Thus, *k* and *p* are indices which indicate the same spatial entity in the real world in text space and database space, respectively. Figure 10 shows an example of deducing matched nodes with both spatial anchor descriptions and spatial relationship descriptions. For example in Figure 10, the plane is divided in quarters using the angle of *passedlink*. Depending on the meaning of *sr_desc*(*k*–1), the target of matching with *sa_desc*(*k*) is changed. Figure 10 shows that three target areas for matching a set of nodes with *sa_desc*(*k*) as follows:

(1) within the plane of I when *sr_desc*(*k*–1) means "right".

(2) within the plane of IV when *sr_desc*(*k*–1) means "left".

(3) within the plane of II+III when *sr_desc*(*k*–1) means "straight".

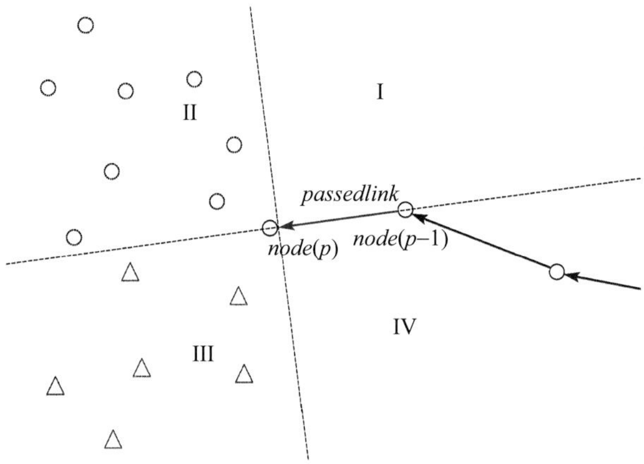

Figure 10 Quarter's division of the plane for matching with valid spatial anchors

The *sa_desc*(0), which is the first element in a set of *sa_desc*, is matched with the set of all spatial anchors in a sidewalk network database because there is no *passedlink*. Furthermore, there are the following two types of matchings with *sa_desc*. Both matchings are executed simultaneously.

(1) Spatial anchor point matching: the results of the queries are a set of nodes corresponding to the spatial anchors.

(2) Path matching: this step decides the sequence order of places in the route using the rule which minimizes each of the moving costs between neighbor places.

A *sr_desc*(*k*) is needed to match with the set of valid spatial relationship descriptions. The set of valid spatial relationship descriptions is the result of the following function *valid_sr*.

$$valid\_sr(passedlink, node(p).out\_link)$$

Depending on an angle formed by *passedlink* and *node*(*p*).*out_link*, the target of matching with *link*(*p*).*connect* of *sr_desc*(*k*) is changed. Figure 11 shows that three target areas for matching a set of links with *sr_desc*(*k*) as follows:

(1) "right" is the target when one or more link data of *node*(*p*).*out_link* are within the region I.

(2) "straight" is the target when one or more link data of *node*(*p*).*out_link* are within the region II.

(3) "left" is the target when one or more link data of *node*(*p*).*out_link* are within the region III.

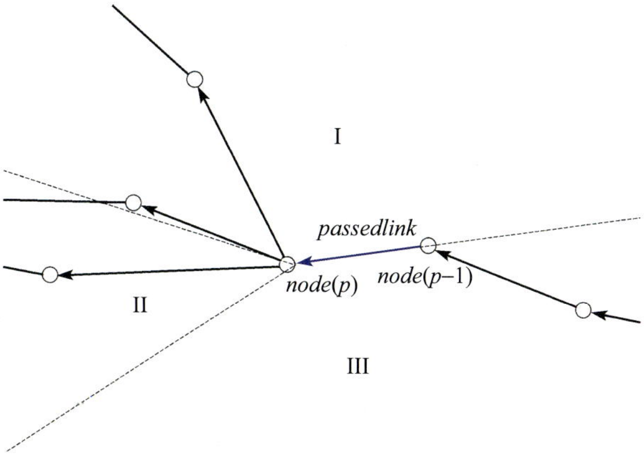

Figure 11 Division of the plane for matching with link.connect

Prototype system

Our proposed framework has been verified through developing a prototype system. Figure 12 shows the user interface of the prototype system.

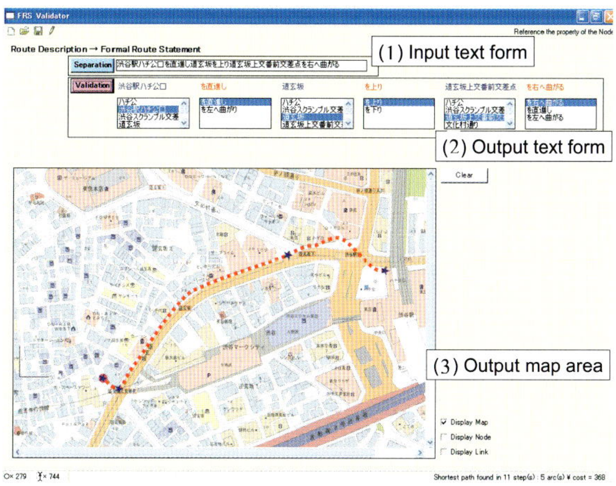

Figure 12 Graphical user interface of the prototype system for geocoding natural route descriptions

Overview of prototype system

We have developed a prototype system, which processes a natural route description in Japanese, and then it can be visualized as a polyline on the map using a sidewalk network database. Each component in the user interface in Figure 12 is as follows:

(1) Input text form: Users can input a natural route description using this text form.
(2) Output text form: A result of separating input text and validating separated elements is displayed in this form.
(3) Output map area: A route is visualized on the sidewalk network database as a result of geocoding a natural route description.

Experimental demonstration

Using a natural route description used in the section *Converting natural route description into formal spatial association statement* as an example of an input text, Figures 13A, B and C show the behavior of the geocoding process for an example of a description. The captions of the figures explain the details of the behavior. The following are explanations about interactive operations on the prototype system to geocode a Japanese natural route description into a polyline on a map.

1. Input a Japanese natural route description "渋谷駅ハチ公口を直進し道玄坂を上がり道玄坂交番前交差点を右へ曲がる" in the field next to the button "Separation" in the first line of the application window as in Figure 13A.
2. Press the button "Separation" in the first line of the application window as in Figure 13A to divide the description into nouns and others using the morphological analysis system by ChaSen[2]. Separated Japanese words with spaces between words are displayed in the second line of the application window as the result of the dividing the description. The first, third and fifth blue-colored word segments are noun phrases, and the second, fourth and sixth red-colored word segments are considered as others between noun phrases. The others are considered as verbal phrases.
3. Press the button "Validation" in the second line of the application window in Figure 13A to match the first noun, that is, the first spatial anchor, with the sidewalk network database and to find the corresponding POI. The leftmost text box in the third lines shows a set of candidates for the first noun in the description. At the same time, the corresponding star shaped icon in the right side appears on the map in Figure 13A. The highlighted text in the leftmost text box means the most relevant candidate for the first noun, and other texts in the same text box are other candidates. The highlighted text can be changed by clicking another candidate text in the same box. The bottom part below a text box of highlighting a noun is used for showing the class of the noun. For example, "class=出口" means "class=Exit" in Japanese.
4. The second leftmost text box shows candidates of spatial relationships which are deduced from the sidewalk network database after deciding passing points. The dashed line corresponding to the highlighted spatial relationship in the second leftmost text box appears on the map in Figure 13A. The highlighted text in the second leftmost text box means the most relevant spatial relationships for the second word in the description. The highlighted text can be changed by clicking another candidate text in the same box.

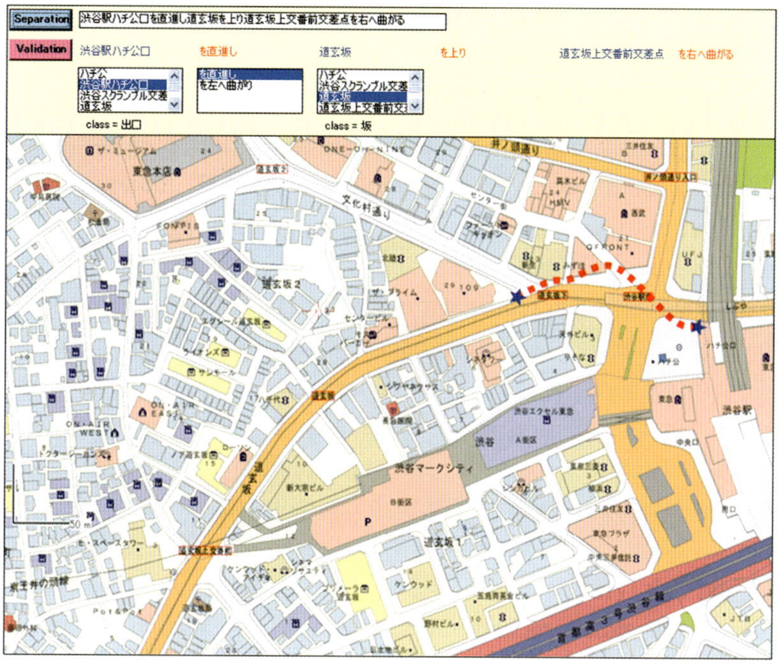

Figure 13A Example of geocoding a natural route description by a prototype system (Step 1)

The prototype system searches the shortest path (the dashed line) from the node matched *sa_desc*(0) to the nearest one of the nodes making up *sa_desc*(1). *sr_desc*(0) is also matched with the sidewalk network database. This shows the result of processing *sa_desc*(0) + *sr_desc*(0) + *sa_desc*(1).

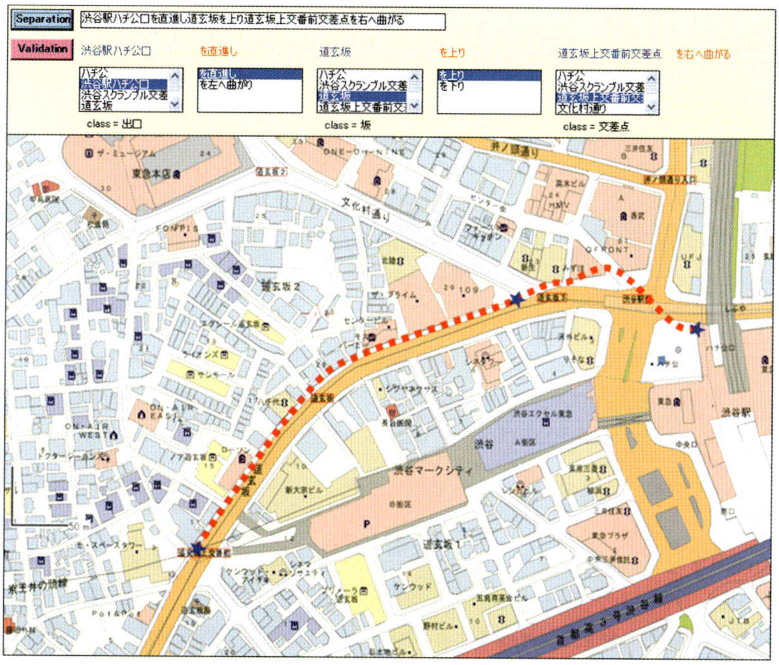

Figure 13B Example of geocoding a natural route description by a prototype system (Step 2)

The system searches the shortest path to the nearest one of nodes matched with *sa_desc*(2). This shows the result of processing *sa_desc*(0) + *sr_desc*(0) + *sa_desc*(1) + *sr_desc*(1) + *sa_desc*(2).

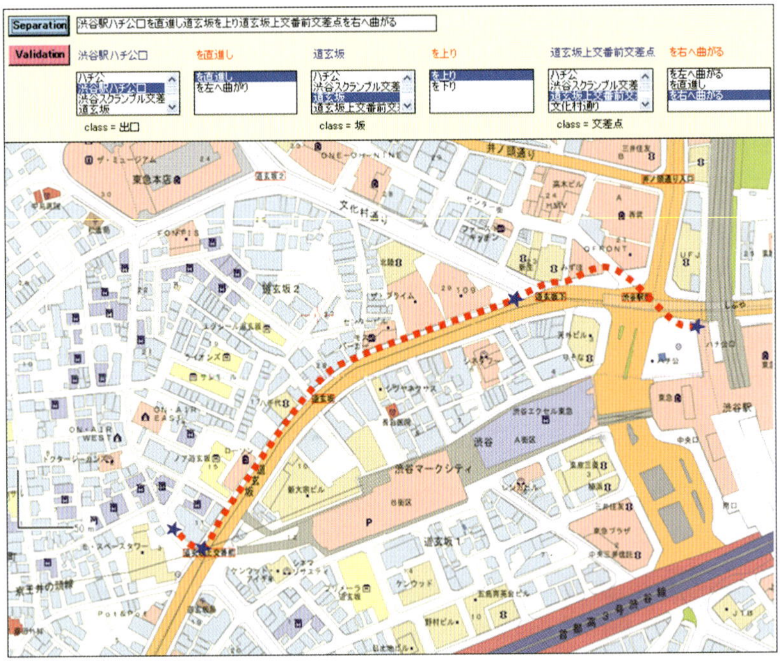

Figure 13C Example of geocoding a natural route description by a prototype system (Step 3)

The end node is deduced from the description of *sr_desc*(2) and the direction of the last matched link. This figure shows the result of processing *sa_desc*(0) + *sr_desc*(0) + *sa_desc*(1) + *sr_desc*(1) + *sa_desc*(2) + *sr_desc*(2).

5. Press the button "Validation" again to match the third word or the second noun, that is, the second spatial anchor, with the sidewalk network database. The second star shaped icon corresponding to the second spatial anchor appears at the opposite side of the dashed line on the map in Figure 13A. The interactive operation is almost the same as operation (3).
6. The fourth leftmost text box shows candidates of spatial relationships for the fourth word segment. It is almost the same as the operation (4) in Figure 13B.
7. Press the button "Validation" again to match the fifth word. It is almost the same as the operation (5) in Figure 13B.
8. The rightmost text box also shows candidates of spatial relationships for the sixth or last word segment. It is almost the same as the operation (4) in Figure 13C.

Viewpoint and Direction Geocoding for Scene Descriptions

Another example of an application of FSAS is geocoding method from a user's viewpoint and direction from scene descriptions with both text including scene descriptions and scene photographs with name labels. Figure 14 shows an overview of viewpoint and direction geocoding based on FSAS. Both scene description texts and scene photographs with name labels can be translated into FSAS composed of place names and "right" spatial relationships with rough or precise angles. If more than three of the place names can be matched

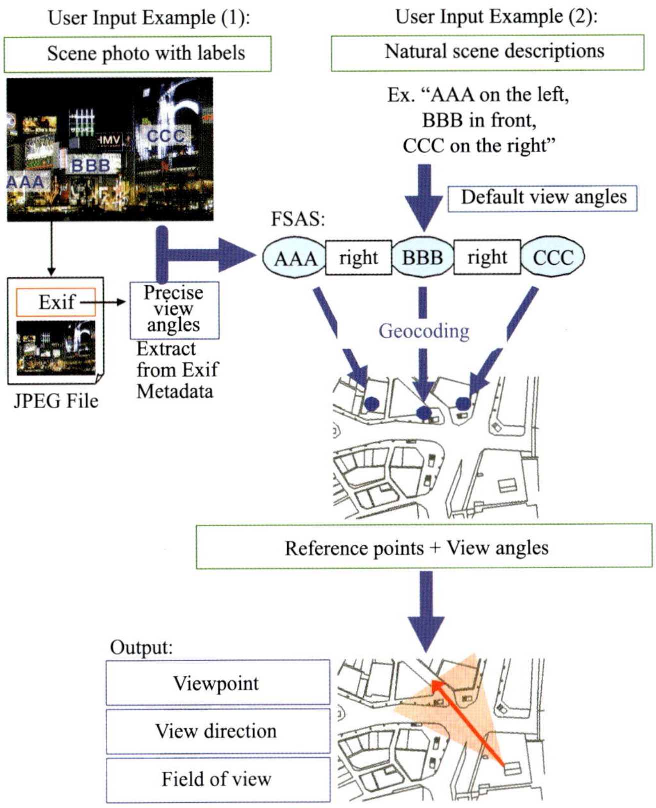

Figure 14 Overview of geocoding methods for detecting viewpoint and view direction from a scene photograph with labels and a natural scene description

with geographic objects on the sidewalk network database, the three reference points of the matched geographic objects and two view angles between them can be used to detect the point, direction and angle of the user's view.

This framework can be directly applied to an interactive system for deciding the orientation of a user with a digital compass and voice recognition function on a mobile device. For example, a user says "AAA" when pointing a billboard using a digital compass facing it, then repeats the point of the digital compass at two more billboards, and reads out the names of the billboards. After success of geocoding the three place names in voice using sidewalk networks, the three geocoded reference points and the two view angles between the reference points can be applied to the viewpoint and direction detection program. Then, the user can find her/his viewpoint and angle on the map immediately as we show in Figure 15. Many of current mobile phones in Japan have a high quality voice recognition function with the embedded DSP (digital signal processor) and some digital sensors including digital compasses, acceleration sensors, and thermometers. It is common for users to speak place names to fill in the field for geographic information retrieval and setting destinations of pedestrian navigation on mobile phones. Thus, the previous example of geocoding users' viewpoints and directions is not far from the current situation of use and infrastructure of modern mobile phones from the viewpoint of the hardware of such mobile devices.

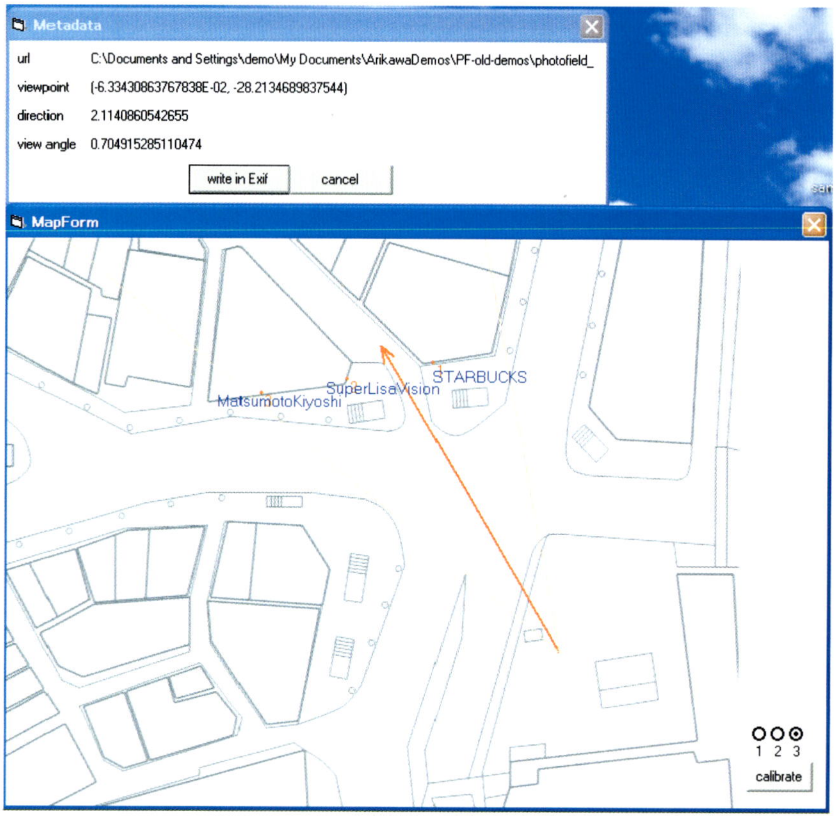

Figure 15 A prototype system of detecting the point and direction of a user's view from three reference points and two view angles between them

In the system, a user places three name labels on a scene photograph, then the name labels are geocoded into the geographic positions of geographic objects stored in the sidewalk network database and the view angles can be detected using the distances between labels on the photograph and the view angle of the camera's lens which is generally registered in Exif metadata of a JPEG file[8].

Detecting a camera's position and direction

Viewpoint and view direction can be detected by using user input as summarized in Figure 14. It is realized under the following conditions:

(1) A camera is directed in the horizontal direction.

(2) A view angle is known.

(3) A user indirectly inputs three sets of reference points on a map and the photo image by the direct interaction of placing name labels on a scene photograph on the computer screen.

The condition (1) limits view direction to two-dimensional. As for the condition (2), view angles of photographs generated by general digital cameras are known by using Exif information. Exif is a widely used standard for storage of information about cameras, images and shooting conditions within JPEG and TIFF image file headers[8]. Almost all digital cameras generate Exif information today. The condition (3) means that a user specifies points on a map and the corresponding three points on a photograph as shown in Figure 16. To be precise, a user needs to decide three vertical lines on the photograph because we use only x-coordinates. As reference points, those which are represented as points on a map such as corners of buildings, traffic lights, *etc.* are suitable.

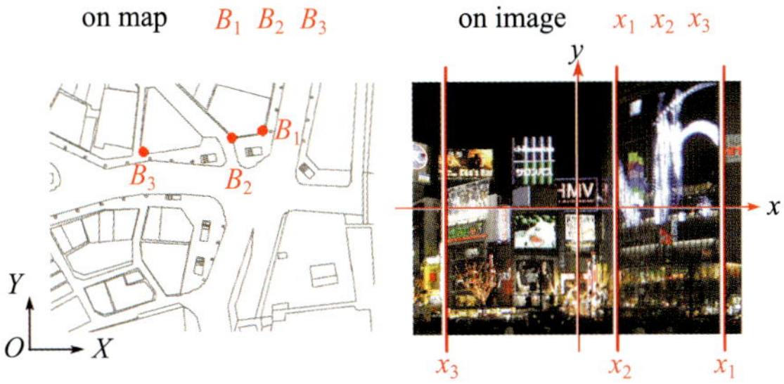

(X, Y, Z) Geographic coordinate system
(x, y) Image coordinate system

Figure 16 Reference points

We explain the procedure for deciding viewpoints and directions below. We use the following values which are first unknown values:

$S=(X_S, Y_S)$: Viewpoint (geographic coordinates)

$V=(X_V, Y_V)$: View direction (geographic coordinates)

And then known values:

θ_w: View angle

$B_i=(X_i, Y_i)$ $(i=1,2,3)$: Reference points on a map (geographic coordinates)

x_i $(i=1,2,3)$: x-coordinates of reference points on an image (image coordinates)

Figure 17 shows a relationship between these values. Angle θ_i from the view direction toward the reference point $B_i (i=1,2,3)$ are computed as follows:

$$\frac{\frac{w}{2}}{\tan\frac{\theta_w}{2}}=\frac{x_i}{\tan\theta_i} \tag{1}$$

where w is the pixel width of the image. θ_{21} and θ_{32} are computed as follows:

$$\theta_{21}=\theta_1-\theta_2 \tag{2}$$

$$\theta_{32}=\theta_2-\theta_3 \tag{3}$$

In focusing triangle SB_1B_2 and SB_2B_3, the following expressions are formed.

$$SB_1 \cdot SB_2=|SB_1||SB_2|\cos\theta_{21} \tag{4}$$

$$SB_2 \cdot SB_3=|SB_2||SB_3|\cos\theta_{32} \tag{5}$$

By using expressions (4) and (5), $S=(X_S, Y_S)$ is computed. View direction θ_V is computed as follows:

$$\theta_1' = \arctan\left(\frac{Y_1 - Y_s}{X_1 - X_s}\right) \tag{6}$$

$$\theta_V = \theta_1' + \theta_1 \tag{7}$$

where θ_1' is the angle between SB_1 and x-axis of the geographic coordinate system.

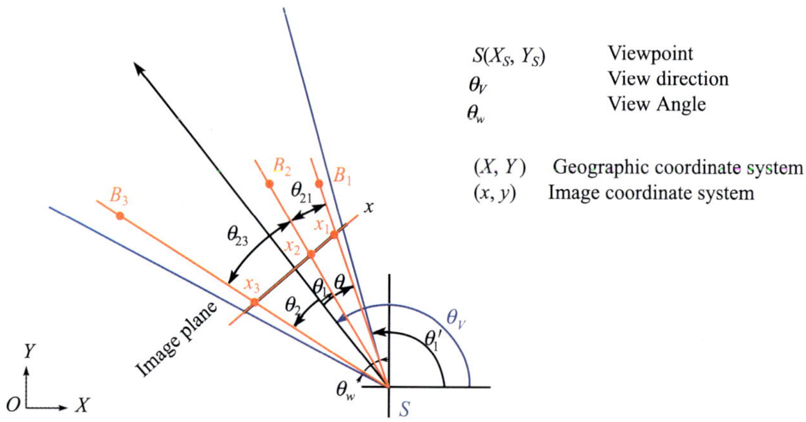

Figure 17 Deciding viewpoint and view direction

Displaying text labels on images

Text labels having geographic coordinates can be displayed on images by computing their image coordinates. Figure 18 shows a geometric model of geographic coordinates and image coordinates.

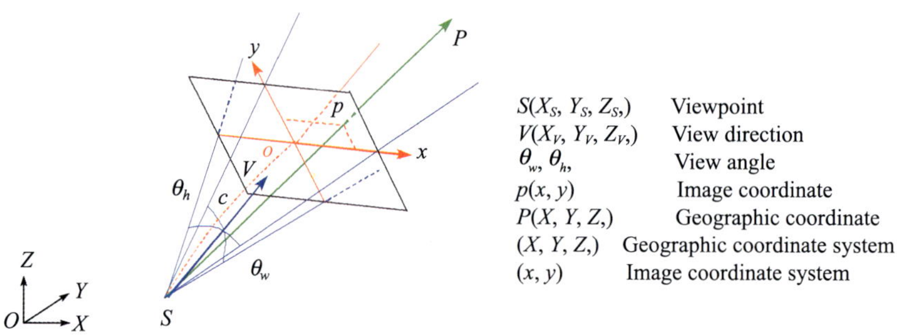

Figure 18 Collinearity equations

A relationship between them is expressed by collinearity equations[9]:

$$x = -c\frac{a_1(X - X_S) + a_4(Y - Y_S) + a_7(Z - Z_S)}{a_7(X - X_S) + a_8(Y - Y_S) + a_9(Z - Z_S)} \tag{8}$$

$$y = -c\frac{a_2(X - X_S) + a_5(Y - Y_S) + a_8(Z - Z_S)}{a_7(X - X_S) + a_8(Y - Y_S) + a_9(Z - Z_S)} \tag{9}$$

where $S = (X_S, Y_S, Z_S)$ is a viewpoint, *i.e.* camera's position, $(a_1, a_2, \cdots, a_9)$ are elements of a 3D rotation matrix, which rotates the view direction to the z-axis of geographic coordinate system. c is a focal length. With these equations, image coordinates (x, y) of a text label are computed from its geographic coordinates (X, Y, Z). Figure 19 shows two examples of a scene image with multiple text labels as results of geographic information retrieval.

A. With three-dimensional coordinates

B. With two-dimensional coordinates

Figure 19 Examples of augmenting geocoded labels on a scene image

Labels are displayed like Figure 19A when all substituted values to equations (8) and (9) (*i.e.* viewpoint, view direction and text label's geographic coordinates) are three-dimensional. When any of these values are two-dimensional, only the x-coordinate on an image is computed and the y-coordinate is not computed. Therefore we display labels as in Figure 19B.

Conclusions and Future Work

This chapter has introduced a new framework for advanced geocoding methods for natural spatial descriptions including route guide descriptions and scene descriptions using large scale sidewalk and POI databases, which can be used for geographic dictionaries to match candidate nouns in natural language with real place names. Our proposed methods are not always robust for irregular expressions which are common in our daily lives' communication.

However, they could become the first step for generalizing more advanced geocoding and geoparsing methods which are constructed on flexible structures, and can be made stronger for irregular expressions by increasing rules for adapting faults and changes in matching candidate noun and verbal parts with sidewalk network databases.

Our proposed FSAS represent linear ordered place lists which often correspond to route guide descriptions and scene descriptions. General graph structures of representing natural language descriptions are tree structures, but our proposed FSAS is based on list structures which are much simpler than the tree structures from the viewpoint of graph matching algorithms. The simplicity of the structures of intermediate representations makes it more feasible to be generated from digital documents and to be matched with geographic dictionaries based on sidewalk network databases. We have formalized the process of new geoparsing and geocoding using FSAS, and implemented test systems for proving if our proposed methods are feasible or not on large scale Japanese sidewalk network databases. Our test systems deal with only Japanese documents so far because sidewalk network databases are designed based on the Japanese language, but our proposed methods can be applied to other languages because they are simple and independent from the Japanese language.

We plan to extend the sidewalk network database and the processible expressions in the system, and implement other functions. In the sidewalk network database, the database needs to store z-coordinates. For example, spatial relationship descriptions can be anticipated by using the knowledge of not only classes of spatial objects but also geographic features stored in the sidewalk network databases. In walking starting from a lowest place, it is clear that we cannot go down there. Furthermore, with street grade, width and paving as attributes of a link, the system presents the route in consideration of user's preference or physical ability. The system can provide women who do not want to walk through poorly lit streets at night and elderly people who want to avoid steep slopes and long stairways with information about the best route for them. Also, we aim to realize a robust system. For example, the system corrects route descriptions when invalid route descriptions are detected. In addition, there is a problem about ambiguity of natural language. In that case, multiple solutions which are possible routes to a destination, should be ranked by some criteria of quality of the geocoding results.

Acknowledgements We would like to thank Mr. Kouzou Noaki for his contribution to implementing a prototype system and constructing a theory for analyzing natural route descriptions. We will thank Mr. Akihiro Ome and Dr. Takeshi Sagara for their collaborative support, advice and comments. Maps used in Figures 3, 8, 12 and 13 are provided by Shobunsha Publicatins Inc.[3], and maps in Figures 14, 15 and 16 are allowed to use by Zenrin Co. Ltd.[4].

References

[1] Arikawa M, Sagara T, Okamura K. Spatial Media Fusion Project. In: Proceedings of International

Conference on Digital Libraries: Research and Practice. Kyoto, Japan. IEEE Computer Society Press, 2000: 304–311.

[2] Japanese Morphological Analysis System ChaSen. http://chasen.aist-nara.ac.jp/hiki/ChaSen/

[3] Shobunsha Publications Inc. http://www.mapple.co.jp/

[4] Zenrin Co. Ltd. http://www.zenrin.co.jp/

[5] EZnaviwalk, KDDI. http://www.au.kddi.com/

[6] Herskovits A. Language and Spatial Cognition. Cambridge: Cambridge University Press, 1986.

[7] Nagao M, Sato S, Kurahashi S, Tsunoda T. Natural Language Processing. Tokyo: Iwanami Shoten, 1996 (In Japanese).

[8] JEIDA. Digital Still Camera Image File Format Standard (Exchangeable image file format for Digital Still Cameras: Exif). Japan Electronic Industry Development Association, Tokyo, 1998.

[9] Murai S. Spatial Information Engineering. Japan Association of Surveyors, Tokyo, 2001 (In Japanese).

Chapter 20 Multi-Agent-based Simulation of Pedestrian Crowds in Outdoor Events in Hong Kong

Lei Wu, Hui Lin and Jingjing Shi

Introduction

Public safety is a major issue especially when large outdoor events take place where thousands of pedestrians are gathered. Facing potential terrorist attacks and natural or technical disasters such as the September 11 tragedy in New York City in 2001, it is essential to be prepared to manage emergency situations to minimize loss of life and damage to property. Pedestrians and their movement patterns are thus one of the most important components of emergency management as many people have lost their lives in outdoor events which have become tragedies through crowding[1, 2]. In emergency management, the Fireworks Display over Hong Kong harbor at Tsim Sha Tsui is an important example where crowd managers are trained to handle possible accidents and panic and preview the action programs based on a conventional sand-box game. The sand-box game can only show static images of the target area and preset scenarios under pre-designed situations while it lacks cognition of the dynamic behavior of pedestrians. Meanwhile, it is required to make decisions according to the situations that change from minute to minute and in which pre-designed action programs may not work smoothly. Crowd safety in this situation is dependent upon the experience of field managers, which cannot always be effective. The model in this chapter is designed to address these concerns.

A considerable improvement to the traditional sand-box-game is provided by computer-based simulation approaches. These have been introduced recently to study pedestrian behavior by offering assistance in emergency management. Two schemes, cellular automation (CA) and agent-based modeling (ABM), which simulate pedestrians on a microscopic scale have shown better applicability than the flow dynamics model schema which functions at a macroscopic scale. It is believed that setting up pedestrian behavior models on a microscopic scale better represents the intelligent and self-organized behavior patterns of pedestrians[2–19]. In addition, the boundary of the two microscophic simulation schemes has become somewhat blurred in certain models of which pedestrians are treated as agents acting in a 2D cellular grid space[15]. These simulation models are useful in demonstrating the dynamics of pedestrian flow in a lattice from a physics perspective. However, the psychological impacts that inform the pedestrian behavior patterns cannot be represented in those simulations models, even including the social-force and random walker models[11, 20]. For practical application purpose, spatial complexity should be taken into account rather than operating these models on the idealized space of a homogeneous lattice.

The work presented in this chapter is the preliminary stage of a large project that aims to offer assistance for emergency management in the Tsim Sha Tsui (TST) fireworks display by simulating pedestrian behavior, integrated with the psychological impact of crowds and security issues on patterns of movement. Different roles involved in the event were modeled as independent agents acting in the study region based on their own intelligence. As a preliminary study, the arrival phase of the event has been simulated in setting up a platform for studying pedestrian panic behavior. In the following sections, the modeling scheme, the system framework, and the simulation sample will be presented. In addition, issues raised by the simulation will be discussed and panic behavior patterns for modeling will be suggested from a psychological perspective.

The Model Scheme

In this study, we modified the approach of Shi and Lin[21] identifying two principal classes of active agents: 'pedestrian' and 'crowd manager', involved in the event, and four classes of facilities including 'road', 'building', 'attractor' and 'subway station'. Each class of agents was defined with attributes and actions equivalent to an object-oriented scheme. The attributes contained the spatial and non-spatial information of agents, while the actions described the abilities of agents. In addition, the agents were built with intelligent rules, such as to make decisions where needed, as well as to represent interactions among them. The facilities are related to environmental factors affecting the agents' behaviors. The roads were the agents' walking space, the subway stations were the pedestrian entry points into the area, and the buildings determined the attractiveness of the space by their position in relation to the fireworks display, *i.e.* whether they blocked the view of the event or not.

Base board

For the sample case study, the study area covered all the buildings, roads and subway stations within the district, and the harbor, above which the fireworks explode. Therefore, a minimum rectangle 1,950 meters wide and 2,200 meters, long was specified. Ronald and Kirley[15] suggested that the typical area occupied by a pedestrian in a crowd that has gathered at such outdoor events is about 0.4 meter square. Our rectangle was therefore divided into approximately 2.7 million cells according to this typical size and each cell was coded to be one of the facilities as its cell type c_t.

Pedestrian

The pedestrian is the core component of the model. For this sample case, hundreds of thousands of participants were involved in the event. Multiple injuries and loss of life in case of any accident could thus result from this representation but it is a challenge to model such diverse behaviors on the microscopic scale. In this study, we attempted to model the behaviors of pedestrians more with respect to the psychological impacts that affect crowd behavior[22, 23].

Attributes

The attribute of a pedestrian agent contained both spatial and non-spatial information. The spatial coordinates and relative grid index to the area of the current location of the agent was represented as P_{ij}. The direction in which the pedestrian was heading (herein referred to as the 'heading direction') was also stored. Non-spatial information includes age, gender, education level, *etc.*, which could affect behavior patterns in a panic situation[23] and the initial walking speed was set to be 1.3m/s on average[4] with a random diversity of ±0.3m/s. This initial walking speed could be increased to a maximum of 2.2m/s according to Damien's experimental observations[24].

Actions

The action defined the abilities of movement and communication with others. The movements of a pedestrian agent consist of initializing its position, setting its heading direction and stepping towards its heading direction. Once an agent sets out from a randomly selected entrance, the model retrieves its initial position and spatial attributes. Within each time step, the agent sets its heading direction based on its behavioral logical rules set, and then it steps into the neighbor cell towards which it was headed. The neighbor cell was defined as one of the eight cells surrounding the central cell as shown in Figure 1. After stepping into the expected cell, the agent maintains and updates its attributes with respect to the latest situation.

The communications of a pedestrian agent consist of reporting its attributes on request and sending out requests for other agents' attributes. Following receipt of a request, the pedestrian could tell the age and gender, the current moving speed, the spatial coordination, and the attractor in response to the enquiry. Correspondingly, the pedestrian could send requests for attributes to the agents that are located in its accessible cognitive zone (ACZ) which is a topic we will discuss a little later. Moreover, commands from agents of crowd managers, such as prohibition of entering specific over-crowded streets, could be received and interpreted, affecting the decision making behavior of the agent.

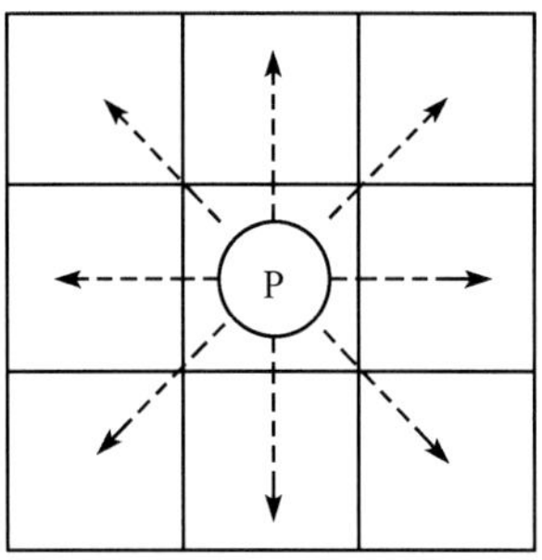

Figure 1 Neighbor Cells Definition where P represents a Pedestrian

Behavioral rules

When considering behavioral rules, three strategies for behaving were defined: the rule to find the attractor of the pedestrian, the rule to find the route toward the attractor, and the rule to assess satisfaction on reaching the attractor. Once a pedestrian was generated at any entrance in the study area, it keeps moving towards its attractor until it reaches a satisfied closest cell. In the arrival scenario, the attractor was the fireworks display, while in the evacuation scenario, the attractor was any one of the exits within the study area. Thus the first rule on finding the attractor was clearly defined. The pedestrian ascertained the nature of the current scenario and determined the attractor. The attractor could be changed by any one of three conditions: a command from crowd managers, discovering an accident in its ACZ, or receiving information that other pedestrians had changed their attractors in the ACZ in question.

With the attractor determined, the pedestrian tried to find a route to reach the attractor. Shi[25] suggested selecting a route from the valid routes set. However, a classical path finding algorithm on vector networks was not suitable here as the road was represented on the raster base board due to the two-dimensional characteristic of the pedestrian[10, 13]. In contrast to vehicle routing on a one-dimensional linear road, the pedestrians could walk on the roads in two dimensional form and direction, meaning that, for example, in the open space, pedestrians could not only walk in a line along the roads, but they could also walk in any direction from the center lines of the roads.

Although a vector-based path finding algorithm can be applied in this scenario, (with available commercial software to do the job), the solving process needs to be much quicker in order to make the simulation practical. Based on the case data, a sample test for finding the shortest path from one of the entrances to the fireworks has been carried out using the Spatial Analyst Extension of the ArcGIS 9.2. The hardware platform was an ordinary desktop PC with Intel P4 2.4 GHz CPU, and 1 G RAM. The test took up to 10 seconds for each individual search, which was unacceptably long considering that there were tens of thousands of agents involved in the simulation. Furthermore, heuristic algorithms such as swarm intelligence were also considered[3] in this kind of simulation. However, limitations still remained in terms of the number of total cells and the number of pedestrian agents that the system could handle. Another kind of algorithm using a probability-based method has been applied to such lattice spaces[4, 9, 13, 18]. Each cell was given a probability according to the distance between the cell and the attractor. This algorithm was effective and efficient in a lattice space. But in the context of real spatial data involving roads and paths where the connectivity and the crowd dynamics should be taken into account, pre-calculated possibilities based on distances were inadequate.

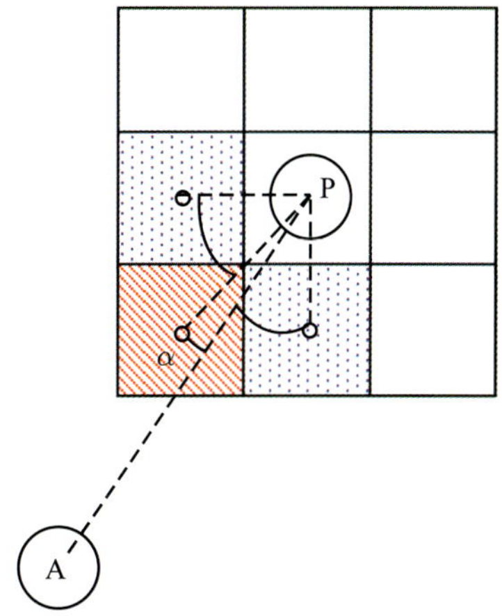

Figure 2 Sample of selecting heading direction, where A represents an attractor

In this study, we used a dynamic attractor-oriented method to find the path. Following identification of the attractor, the pedestrian calculated the direction from the current cell to the attractor, then selected the nearest unoccupied neighbor cell. The surrounding circle of a cell was divided into eight equal parts, each part being at 45 degrees to adjacent parts The part that contained the direction of the attractor was the preferred choice as seen in Figure 2 where the eight neighbor cells could be recoded with respect to the heading cell, left and right heading, left and right, left and right back, and back. If the heading cell was a road cell and unoccupied, the pedestrian would step into the cell and flag it as occupied. Otherwise, the pedestrian would check the cells according to its priority which was dynamically recoded and then select one from the alternatives. If there was no unoccupied road cell, the pedestrian would wait at this time step. In Figure 2, the red hatched cell would be the heading cell with first priority. If the heading cell was not available, the blue doted cells would be the left heading, which located lower, and the right heading, which located higher, with second priority.

Mathematically, this algorithm could be represented as follows:

$$V_i = \begin{cases} 1, & C_i : Road \,\&\, Unoccupied \\ 0, & Otherwise \end{cases}$$

where V_i is defined as the availability of the cell i. Then

$$C_t = \begin{cases} Null, & \prod_i V_i = 0 \\ C_i, & \min(\alpha_i), V_i \neq 0 \end{cases}$$

where the C_t was the target cell, cell i would be selected while α_i, which is the angle of the line from P to A and the line from P to centroid of cell i, reached a minimum.

Figure 3 describes the framework as a sequence of operations for finding a cell to move, where "available" means the cell is a road cell and unoccupied by any other agent.

We considered the rule to assess satisfaction on reaching the attractor to be unstable. The pedestrians kept moving to get closer to the attractor. They only stopped to wait when all the neighbor cells were unavailable. Whenever there was a cell available which was closer to the attractor than the cell the pedestrian currently occupied, the pedestrian would move into that cell. With such a rule, a dynamical balance would be reached, where that all the pedestrians would take the cells that were as close to the attractor as possible.

Crowd manager

The crowd manager was the agent that moved and inspected the area to maintain crowd safety. The manager had three objectives. The first was to synchronize information with other managers and close the road in case of any accident. The second was to investigate the cell where pedestrians reported an accident. The third was to direct pedestrian flows to less dense roads if the manager found the zone was too crowded.

In the model, manager agents were evenly and randomly distributed across the area. When the simulation started, the managers used random walking behavior to freely move into any unoccupied cells. When a manager agent found an accident in its ACZ, it would send information to all other managers. Then the manager would mark the cell with 'accident' and the cells around it as 'unavailable'. Sometimes the manager received information from the pedestrians saying that there had been an accident somewhere; the manager then

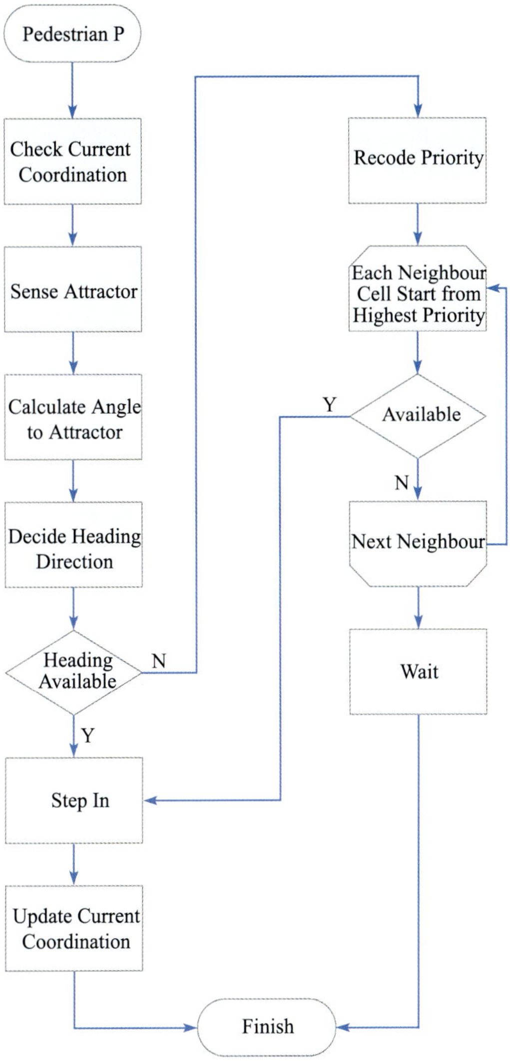

Figure 3 Flow chart for finding a path

had a special attractor and it would find a route to achieve the action. If the manager found that the information was accurate, it would behave in the same manner as if it had found an accident. If not, then the manager would spread a confirmation that the attractors of the pedestrians need not be changed. Also as the manager patrolled the area, it would mark certain cells of roads as unavailable if it found that too many pedestrians stayed in the zone and were not moving ahead. When the zone was no longer crowded, the manager would mark the cells as available again.

The Simulation System

After we modeled the agents involved in the event, we developed a simulation system to test our model and analyze crowd phenomena in this kind of outdoor event. Also our platform was built for further study of modeled pedestrian behaviors in depth, in terms of their psychological impact and thus the behavior set of the agents was designed to be expandable.

Developing tools

The system was built in the object-oriented Visual C++ 6.0 with the ArcObject development kit 9.2. Object-oriented VC++ offered the encapsulation framework, in which the attributes and actions of agents were implemented. The ArcObject development kit 9.2 offered support for accessing and visualization spatial data. All the programs were written to an object-oriented standard, and much effort was put into optimizing the efficiency of the system so that the simulation could be run as closely as possible to real time to obtain practical results.

Data

In this study, we obtained four data layers as the foundation of the system, including three dimensional buildings, two dimensional roads, point shapes at the location where the fireworks flashed, and points for the entrances to the study area. The three dimensional buildings were digitized from maps in a former project and the height value of each building was stored in the database. The vector building layer was converted to a raster format based on the height value field and clipped by the base board with a resolution of 0.4 m×0.4 m for visibility analysis of the study area with respect to the fireworks.

The road layer was a two-dimensional polygon layer, which resembled strips. Pedestrians could walk on the roads in any direction until they reached the boundary of one road. This vector layer was also converted to raster. There was also a point layer indicating the place where the fireworks flashed in the sky. The visibility analysis was carried out from this point to each cell in the study area. Also, this point was the attractor to all pedestrians in the arrival phase of the event if there was no accident or emergency. The entrance layer indicated all possible sources through which the pedestrians came into this area where the subway stations, bus stops, and start points of roads were the entrances.

Framework flow

When the system began to carry out the simulation, four data layers were loaded and converted to raster formats. A huge matrix was created to represent the base board of the study area. Each element was coded to one of the facilities based on spatial locations including roads, buildings, entrances or not-at-all if no other data overlayed. A pedestrian factory was constructed to take charge of generating pedestrian agents. Randomly selecting an entrance as the start point, the factory generated pedestrian agents at a user defined rate and initialized the agent attributes. Then, within each time step, existing pedestrian agents searched a route to the attractor and chose a neighbor cell to move into, as described in the previous section on pedestrian behavior rules. The flow chart of arrival scenario is described in Figure 4.

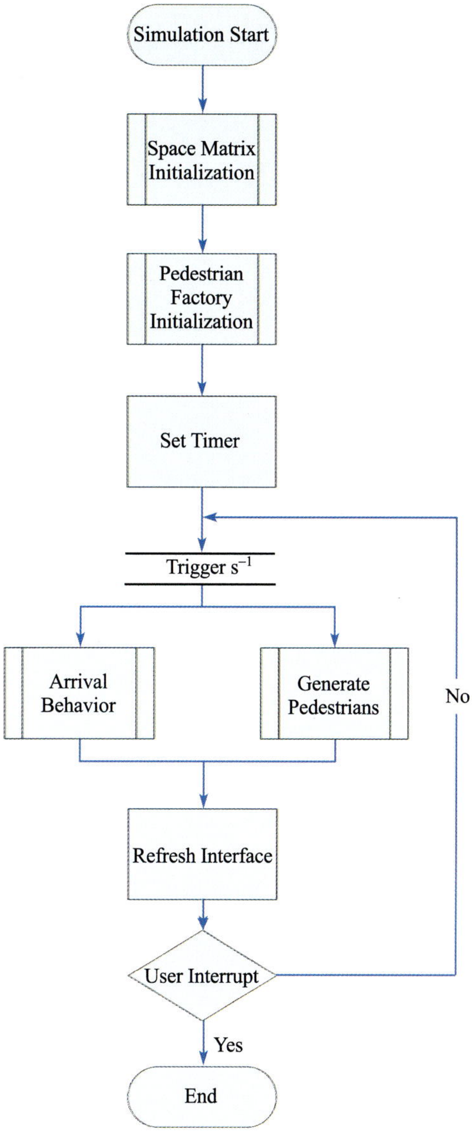

Figure 4 Flow chart of the arrival simulation

System interface

Figure 5 is a screen capture shot which shows the interface to the system. The main part of the dialog-based system was the data view window in which all basement data layers were loaded. The light yellow polygon represents the administration boundary of the study area while buildings in the area are pink blocks. The red stars indicate the locations where the fireworks flashed, and the blue symbols are the entrances, including subway stations, bus stops and road start points. On the right panel, three kinds of simulation scenarios were provided while modeling the arrival behavior. The slide bar with the time scale caption was used to adjust the time period of one time step so that the simulation could be accelerated if the computer had sufficient capacity. Four buttons were used to control the progress of the

Figure 5 Interface to the simulation system

Figure 6 Snap-shot of the simulation in its early time steps

simulation. The start button was enabled by default while others were disabled. The simulation was activated by clicking the start button. Simulation could be temporarily halted while all data was kept in memory by clicking the pause button and could be continued by clicking the pause button a second time, where the operator would see the Go On sign. Clicking the stop button terminated the simulation, discarded all the results and reset the system to its initial conditions. Once the finish button was clicked, the simulation was finished and all the results were written to database.

Simulation results

When the arrival behavior model was finished, we took the simulation of arrival scenario as our example. Figure 6 is a snap-shot from this simulation at time step No. 322 on the real time scale. Figure 7 is another snap-shot after a long run-time. The black dots, which resemble lines due to their heavy density, represent the pedestrians. The simulation demonstrated how the pedestrians come from different entrances, searched their routes toward their attractors, which were the fireworks in this phase, and then how the pedestrians avoided collisions if possible, by choosing their route. Through the latter figure, we find that pedestrians gathered at the southern square which was the closest place to the attractive firework displays.

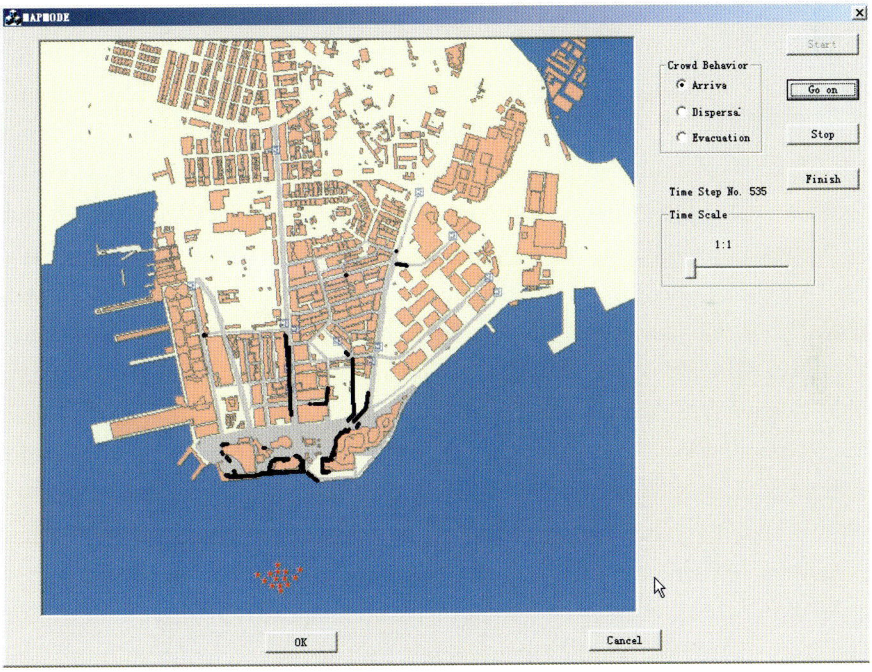

Figure 7 Snap-shot of the simulation in its later time steps

Discussion

Through the simulation results seen in Figure 7, we found that although most pedestrians managed to find their way to the southern square to enjoy the fireworks display, some pedestrians were trapped at road crossings, and this did not seem realistic. This was due to the spatial complexity of actual road data, which is different from the flat lattice board. In a lattice board, there are no convex corners to block the agent's path toward the attractor but this is the case in reality. Figure 8 illustrates how a pedestrian was trapped due to spatial complexity by using the dynamic attractor-oriented method to find its way to the attractor. In the illustrated situation, pedestrian P wanted to find a way to the attractor A. Since the attractor was located at the left bottom corner of the pedestrian's location, the pedestrian would head in this direction to southeast. But due to the spatial constraints posed by the road, P could only choose to step into the two blue dotted cells. After it stepped into either one of the two cells, it still found that the attractor lay to the southwest. Its decision in the next time step was to move into the original cell, which placed it in exactly the same situation as in the previous time step. Therefore an endless loop between the original cell and the two blue dotted cells was formed, thus trapping the pedestrian.

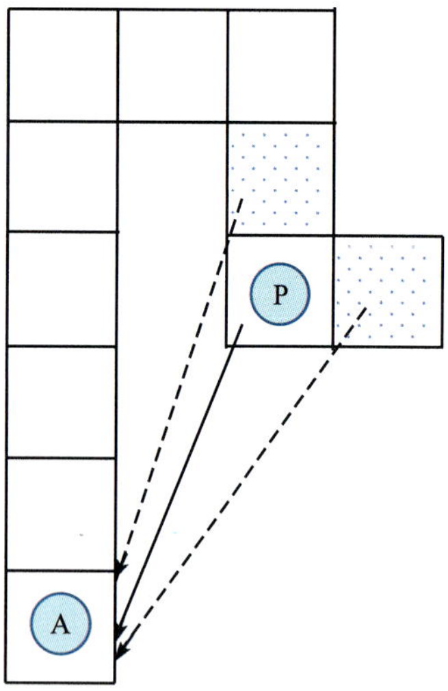

Figure 8 Illustration of a trapped pedestrian

Thus at the macroscopic scale, the path finding algorithm based on the direction to the attractor sometimes failed to find a route to a better cell leading toward this attractor. The classical algorithm for path finding could solve this issue although its time cost is likely to be very high. However, the author would argue that even if the time cost could be ignored by, for example, performing simulation on a high performance computer (HPC), the classical algorithm was still not practical as it could not able to take into account various human factors; we cannot assume that pedestrians all have the same level of infinitive intelligence or prefect information, nor that they can search each cell to find a "best" route.

It would be more realistic to assume that a large number of participants in this kind of event are not familiar with the district, as they may be tourists or visitors. Therefore, only some of the roads could be counted in while finding the path, which relates to the topic of the accessible cognitive zone (ACZ). The ACZ is the zone in which the pedestrian has full spatial cognition for sending and receiving information on most common communication methods. In other words, only those spatial objects that are present within the ACZ are considered by the pedestrian. Therefore, we suggest further investigations into this kind of human behavior to show how a pedestrian could find a reasonable route to its target.

Another problem to consider in making the simulation of crowd behavior during an evacuation more realistic, is that the psychological impacts on modeling behavior in depth. As we have done in this chapter, the behavior of pedestrians in normal situations is rational and logical. During an emergency, however, the actions and mental states of crowds are highly emotional and irrational[23] which makes the model much more complicated than its simulation of arrival behavior. Psychologists have set up various experiments to investigate the mental process of crowds, and we plan to implement some of these with respect to the spatial data in future studies.

Conclusion

This chapter reports the preliminary steps taken to model pedestrian behaviors in an outdoor fireworks display at a microscopic scale integrating elements of psychological impact. A multi-agent-based simulation system was built based on a model to simulate crowds in a fireworks display. Sample cases demonstrated that the pedestrian flow dynamics consisted first of the movement and location of individual agents during the arrival phase of the event. This preliminary model needs improvement as mentioned. The practical path towards finding out how logic of human beings could be studied and applied to the agent behavior rules, needs to be further developed. Also future studies should focus further on the impact of psychological factors of different scenarios, for example, by simulating the evacuation behaviors of pedestrians, which would be much more complicated and diverse than the behaviors of pedestrians in non-threatening situations. Other conditions involved in the simulation not considered in this chapter include vehicle flow on the roads. The assumption in this chapter was that no other agent occupied the road but the reality is that sometimes vehicles including ambulances and other emergency vehicles take up much of the road space. To produce a more realistic simulation of pedestrian behaviors, this would have to be investigated in future studies.

References

[1] Helbing D, Buzna L, Johansson A, Werner T. Self-Organized Pedestrian Crowd Dynamics: Experiments, Simulations, and Design Solutions. Transportation Science, 2005, 39: 1–24.

[2] Kerridge J, Hine J, Wigan M. Agent-Based Modelling of Pedestrian Movements: The Questions that need to be Asked and Answered. Environment and Planning B: Planning & Design, 2001, 28: 327–341.

[3] Batty M, Desyllas J, Duxbury E. The Discrete Dynamics of Small-Scale Spatial Events: Agent-Based Models of Mobility in Carnivals and Street Parades. International Journal of Geographical Information Science, 2003, 17: 673.

[4] Burstedde C, Klauck K, Schadschneider A, Zittartz J. Simulation of Pedestrian Dynamics using a Two-Dimensional Cellular Automaton. Physica A, 2001, 295: 507–525.

[5] Cunningham P P and Cullen D A. Pedestrian Flow Data-Collection and Analysis. Proceedings of the Institution of Civil Engineers-Transport, 1993, 100: 59–69.

[6] Bing Q, Huili T, Chaoying Z, Lingjiang K, Muren L. Cellular Automaton Simulation of the Escaping Pedestrian Flow in Corridor. International Journal of Modern Physics C: Computational Physics & Physical Computation, 2005, 16: 225–235.

[7] Georgoudas I G, Sirakoulis G C, Andreadis I T. A Cellular Automaton Crowd Tracking System for Modelling Evacuation Processes. Cellular Automata, Proceedings, 4173, 699–702.

[8] Golledge R G and Stimson R J. Spatial Behavior: A Geographic Perspective. New York: Guilford Press, 1997.

[9] Nishinari K, Sugawara K, Kazama T, Schadschneider A, Chowdhury D. Modelling of Self-Driven

Particles: Foraging Ants and Pedestrians. Physica A, 2006, 372: 132–141.

[10] Helbing D, Farkas I, Vicsek T. Simulating Dynamical Features of Escape Panic. Nature, 2000, 407: 487–490.

[11] Helbing D and Molnar P. Social Force Model for Pedestrian Dynamics. Physical Review E, 1995, 51: 4282–4286.

[12] Kirchner A, Nishinari K, Schadschneider A. Friction Effects and Clogging in a Cellular Automaton Model for Pedestrian Dynamics. Physical Review E, 2003, 67: 56–122.

[13] Kirchner A and Schadschneider A. Simulation of Evacuation Processes using a Bionics-Inspired Cellular Automaton Model for Pedestrian Dynamics. Physica A, 2002, 312: 260–276.

[14] Lovas G G. Modeling and Simulation of Pedestrian Traffic Flow. Transportation Research Part B-Methodological, 1994, 28: 429–443.

[15] Ronald N and Kirley M. Pedestrian Modelling: A Comparative Study using Agent-Based Cellular Automata. Computational Science–ICCS 2006, Part 3, Proceedings, 2006, 3993: 248–255.

[16] Song W G, Yu Y F, Fan W C, Zhang H P. A Cellular Automata Evacuation Model Considering Friction and Repulsion. Science in China Series E-Engineering & Materials Science, 2005, 48: 403–413.

[17] Weng W G, Yuan H Y, Fan W C. A Cellular Automaton Evacuation Model Based on Mobile Robot' s Behaviors. Chinese Science Bulletin, 2007, 52: 680–684.

[18] Yamamoto K, Kokubo S, Nishinari K. Simulation For Pedestrian Dynamics By Real-Coded Cellular Automata (RCA). Physica A: Statistical Mechanics and Its Applications, 2007, 379: 654–660.

[19] Yang L Z, Fang W F, Fan W C. Modeling Occupant Evacuation using Cellular Automata—Effect of Human Behavior and Building Characteristics on Evacuation. Journal of Fire Sciences, 2003, 21: 227–240.

[20] Hoogendoorn S P and Bovy P H L. Pedestrian Route-Choice and Activity Scheduling Theory and Models, Transportation Research: Part B, 2004, 38: 169.

[21] Shi J and Lin H. Developing Pedestrian Behavioral Models for Crowd Management in Outdoor Recreational Events. In: 13th International Conference on Geoinformatics. Toronto, Canada. August 17–19, 2005.

[22] Brown R W. Social Psychology. New York: Free Press, 1965.

[23] Smelser N J. Theory of Collective Behavior. London: Routledge and Kegan Paul, 1962.

[24] Daamen W and Hoogendoorn S P. Experimental Research of Pedestrian Walking Behavior. Pedestrians and Bicycles 2003, 2003: 20–30.

[25] Shi J. Crowd Management for Large-scale Outdoor Events–Multi-agent-based Modeling and Simulation of Crowd Behaviors. PhD Dissertation, The Chinese University of Hong Kong, Sha Tin, Hong Kong, 2006.

Chapter 21 Visualizing Children's Walking Behavior Using Portable Global Positioning (GPS) Units and Activity Monitors*

Yi Gong and Roger Mackett

Introduction

Today, children enjoy less freedom when walking and playing in urban settings than they did in previous eras. Evidence shows that such movement in current generations of children has become more and more curtailed compared to past generations. A story from a UK national newspaper *The Daily Mail* (15th June 2007) reflects how children's liberty has shrunk over the years. Great-grandfather 'George' (see Figure 1) used to walk six miles to go fishing when aged 8 in 1919, grandfather 'Jack' aged 8 in 1950 walked about one mile on his own to the nearby woods. 'Vicky', the mother who was aged 8 in 1979, was able to walk to the swimming pool alone half a mile away. In 2007, George's great-grandson 'Edward' is driven the few minutes to school, and is taken by car to a safe place to ride his bike. He can roam no more than 300 yards from his home.

While movement on foot has declined over the years, more motorized trips are now generated by children than ever before, particularly on the so-called 'school run' where children are ferried by their parents and guardians from home to school. British children aged 16 and under make 30 percent of their trips to school as car passengers with most of the rest on foot[1]. The percentage of cars taking children to school in the morning peak period (8–9 am.) in urban areas during term-time increased from 7 per cent in 1991/1993 to 12 per cent in 2006. However, this proportion peaked at 15 per cent in 2004 and has fallen slightly since then. The average length of the trip to school for children aged 5–10 has increased from 1.3 to 1.5 miles (2.1–2.4 km) in the last 15 years, while the proportion of trips to or from primary school for which these children traveled alone (with no other child or adult) fell from 21 percent of trips in 1985/1986 to 5 percent in 2006.

This loss of freedom and more motorized travel have resulted in an increase in sedentary activities and a decrease in daily energy expenditures[2–4] is affecting the rapid increase in the prevalence of childhood obesity. Andersen *et al.*[5] found that the amount of time spent by children sitting in front of a computer or TV screen is negatively correlated with their level of physical activity. The children who watched TV more than four hours a day, statis-

* Supported by the UK Engineering and Physical Sciences Research Council (EPSRC) (Grant No. GR/ T09378/01).

tically, had more body fat than those who watched for less than two hours a day in 1998. In England, nearly a quarter of adults and about 10% of children are now obese, with a further 20%–25% of children overweight[6]. The latest UK Government Foresight modeling approach suggests that about 25% of all children under 16 in the UK could be obese by 2050[7]. As childhood obesity restricts children's body activity, it is more likely to lead to adulthood obesity, which has a direct link to heart disease, stroke, and type 2 diabetes in the later life. Physical inactivity is thus a crucial issue in both child and adult health.

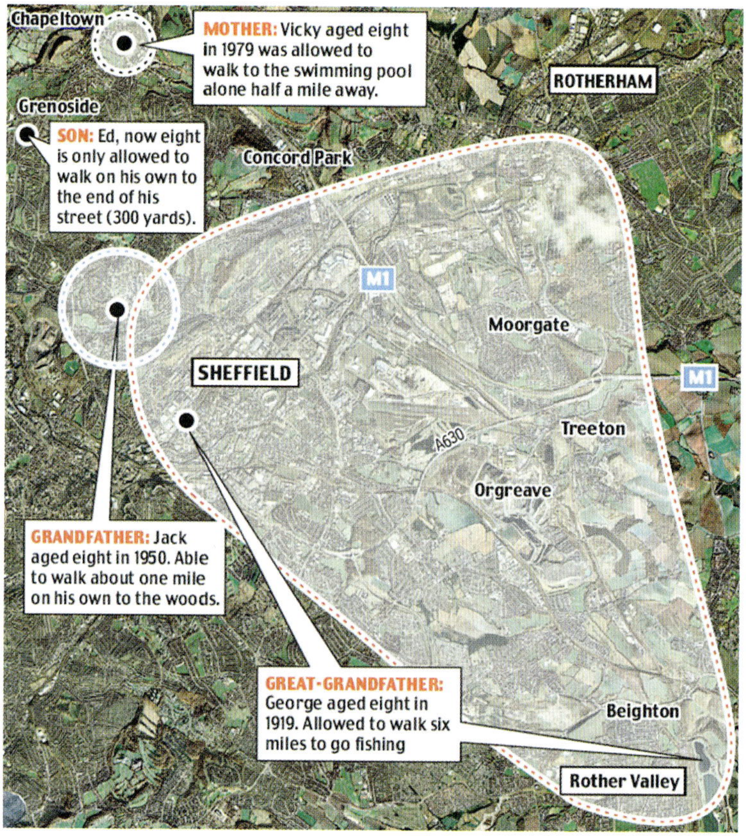

Figure 1 Ever decreasing circles: the changing shape of playtime

Source: Daily Mail[1)] (15th June 2007)

Recent studies have increasingly argued that the built environment in which people are located, has strong impacts on their physical activity level in the form of walking, and therefore on their health[8, 9]. Because walking itself is a distinct type of physical activity, walking, either for transportation or recreation purpose provides the opportunity to accumulate personal physical activity. Particularly for children and the elderly, an individual's requirements for physical activity can be naturally built into daily life, and it is widely recognized that walking has important beneficial effects on health[10, 11].

---

1) http://www.dailymail.co.uk/pages/live/articles/news/news.html?in_article_id=462091&in_page_id=1770.

Broadly defined, the built environment refers to those man-made surroundings that provide the setting for human activity, including issues from the design and integrity of housing to land use, and urban planning[12]. Interactions with the built environment are complex, and do not always lead to healthy modes of living and travel. For example, the original *American Zoning Enabling Act* was predicated on health, safety and welfare impacts on land use, but modern exclusionary zoning practices have actually resulted in more motorized travel, with greater travel distance between where people live, work and play. The high level of car use prompted by such design considerations is now believed to contribute to obesity through decreasing physical activity.

Although evidence suggests that urban form has a powerful influence on walking including land use patterns, transportation systems and density[13, 14], its implications for health are not fully understood. From limited research to date, surprisingly little agreement exists on how and to what extent the built environment affects walking. There are many uncertainties and complexities which can affect walking behavior. Current knowledge and research methods in this area are poor due to the lack of objective measurements of human behavior in the built environment. Until now, most studies which link physical activity with built environment have been consistently 'aspatial', and more qualitative than quantitative[15]. Thus, there is a need for much better understanding and measurement. In this exploratory study, possible ways to determine how children's daily energy is expended during their outdoor activities are examined with an emphasis on methods used to explore to spatial behavioral data through geo-visualization techniques. By displaying and exploring the various results, it is hoped that increased understanding of these environmental characteristics will be obtained in terms of those features that can be used to motivate children to be active and provide the opportunities that they need to burn calories.

The data in this chapter are from an EPSRC project entitled CAPABLE (Children's Activities, Perceptions and Behavior in the Local Environment) carried out at University College London (see http://www.casa.ucl.ac.uk/capableproject/) across four academic departments—the Centre for Transport Studies, the Centre for Advanced Spatial Analysis, the Bartlett School of Planning and the Department of Psychology. In this work, 162 children aged 8–11 years old from primary schools in the county of Hertfordshire, to the north of London, were fitted with GPS (Global Positioning System) units as well as activity monitors and diaries. The fieldwork was conducted from October 2005 to March 2006 and was supplemented with detailed questionnaires, mental map drawing and travel diaries all completed by the children involved. All this data allowed the establishment of when, where and with whom the children had been in relation to their various activities which they engaged in within their local environments over a 96 hour period.

The Methods Used

Survey methods

The questionnaires completed by the children contained questions about their personal characteristics. The children were invited to volunteer within their primary school classes and those who did, were asked to wear both activity monitors (RT3s) and portable GPS units for fours days—Thursday, Friday, Saturday and Sunday. The activity data were collected by RT3 tri-axial accelerometers produced by Stayhealthy, USA (see Figure 2A). These activity monitors measure movements in three dimensions, and generate an overall vector magnitude (VM) expressed in terms of activity calories. By using data on the age, gender, weight and height of the child, VM was then converted into activity calories (the calories used in undertaking physical activity) per minute per kg of body weight.

The latest developments in GPS technology enable the data gathering process to move out of the laboratory environment into the 'wilds' of urban space. In the past, monitoring walking using GPS was not easy because of the size of GPS equipment and accuracy they provide and thus a large proportion of the studies have been vehicle-based transportation studies[16–18]. The size and weight of GPS receivers have been found to be extremely important when using GPS to track the trips made by bicycle and foot[17] and around 41 per cent of a 150 adult participant survey reported negative feedback on the size and weight of the 2-kilogram GPS receivers in a study in the Netherlands[19]. More recently, GPS receivers have become smaller and cheaper. Rodriguez *et al.*[20] carried out a pilot experiment which suggested that data recorded by portable GPS units are sufficiently precise to track pedestrian movements. Using portable GPS allows the tracking of pedestrian movement in almost all types of outdoor space, so that the monitoring is independent of scale or area delimitation[21]. Locations can be closely monitored at a small scale (*e.g.* park or road) as well as at the urban or regional level. It presents a more precise tracking of the interactions between people and built environment. For example, portable GPS devices can now show which side of the street the person is walking on.

Here, two things have to be stressed. First, only a few of studies have employed GPS for tracking children, often with small samples. For example, Elgethun *et al.*'s GPS-PAL vests have been tested on only 11 children aged from 2 to 8 years old[22]. Secondly, although GPS technology has been used widely with a PDA or diary in travel surveys to collect more data on the participant's behavior, GPS technology is rarely integrated with the other research instruments. One example is Lipp *et al.*'s study, which combined GPS with brain activity recorder on the homing pigeons[23]. This opened up new analytical perspectives and Lipp *et al*[23] showed that pigeons use highways and their exits as landmarks during their homeward journey. Once the goal moves beyond mere observation towards an understanding of the underlying reasons and more detailed analysis, a hyper-approach involving GPS integrated with other methods and supplemental information offers significant new opportunities. This is where integrating GPS with activity monitors could prove to be a useful tool. GPS is a promising tool for complementing accelerometer-based physical activity measures. This combination facilitates the simultaneous collection of positional data and energy expenditure data of a moving person. This permits the examination of the relationship between where physical activity actually occurs and the nature of the local environment[20].

The GPS receiver used in this work was the Garmin Foretrex 201 portable monitor illustrated in Figure 2B, which offered the best combination of capacity, weight, battery life and cost that could be obtained from the market at the time of the survey. It weighs 78 g, measures 8.4 cm×4.3 cm×1.8 cm, with a battery that lasts 15 hours and has a capacity for storing 10000 points. Children can easily wear it on their wrists all day long and charge it overnight. The location data is recorded as latitude and longitude in degrees and minutes (HD° MM′) in the GPS unit's memory in chronological order in the World Geodetic Survey 1984 projection. This was then downloaded into a PC using Garmin's MapSource Trip and Waypoint Manager which enabled further spatial analysis. As the GPS units were set up to obtain every possible reading during the survey, GPS positioning readings were not fixed at regular intervals. Occasionally a large number of location points were recorded within a short interval of time. After the data for each child had been collected, a computer program was developed to remove the points which were unlikely to be correct based on the maximum feasible traveling distance within a given time.

Besides all this, the participating children also were asked to complete a travel and activity diary (see Figure 2C) for the four days of tracking and this supplemented the GPS and energy information with information on the activities and trips they participated in, and with whom. All data were initially input into Excel spreadsheets, and programs written in Visual Basic were used to integrate the data from the GPS and RT3 monitors into an Access database which provided greater flexibility for analysis. Temporal sequencing was used to reconcile all these three sources — the GPS, RT3 and travel diaries data. Because of difficulties with the GPS equipment resulting from signal loss or poor of battery life due to the failure of the charger, as well as children forgetting to complete their diaries, not every child provided a complete set of data. The events recorded in the children's activity diaries were classified, using a typology developed previously[11].

A. RT3 monitor B. GPS receiver C. Travel diary

Figure 2 RT3, GPS and the travel diary

As shown in Table 1, 162 children participated in the survey and completed a diary. All children in Years 4, 5 and 6 from two schools in Cheshunt, Hertfordshire were invited to take part in the survey. All the children who wished to take part, with written permission from parents, were included in the study. Only twenty children could complete the survey at any one time due to availability of the equipment and their need for the team and the teachers to monitor the exercise quite closely during the 4-day period.

Table 1 Children samples from the two schools

	Boys	Girls	Total
Year 4 (age 8–9)	25	32	57
Year 5 (age 9–10)	33	36	69
Year 6 (age 10–11)	16	20	36
Total	74	88	162

The results in this chapter are drawn from the two schools in Cheshunt, Hertfordshire (identified in Figure 3), an area immediately north of Greater London. It is a predominantly residential suburb with a large park and a golf course in the north east part (where the school A is located), and industrial factories and busy motorways in the south east (where school B is located). Field visits along with Google Maps, Microsoft Live Map and the Ordnance Survey (OS) MasterMap (Topology Layer and Address Layer 2) were used to establish the nature of the places where children moved about in this study. The OS MasterMap topology layer classified all the land parcels into different types, such as buildings, natural environment, general surface, path, road or tracks, roadside, structure and so on. Address layer 2 categorizes the address as residential, office, retail, services, industrial, educational, hotel, transport and general-commercial land use.

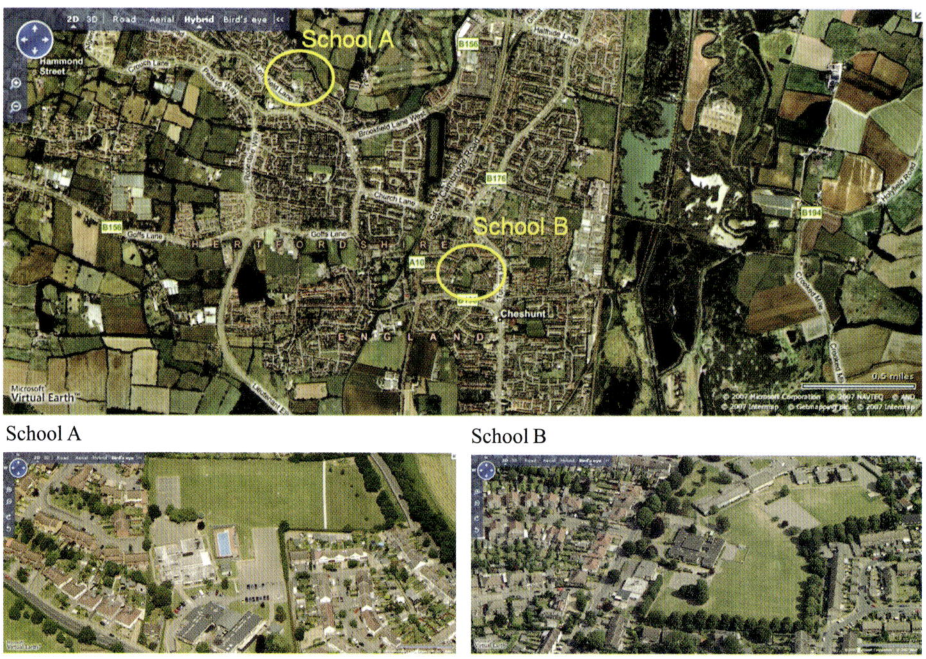

Figure 3 Aerial map of Cheshunt, Hertfordshire where the two schools are located

Results

Visualizing children walking

Compared with motorized travel, the physical locations of walking are less constrained by the transportation network and have much more complexity in terms of where they take place. In reality, people not only walk on the streets: they walk across parks, fields and other open space, which is quite free in comparison with driving where trips are constrained to certain types of streets and dominated by traffic regulations. The GPS trajectories from the children demonstrate this complexity. For example, Figure 4 shows the walking trip from home to school on two mornings for a boy aged 10. The major street network from OS MasterMap ITN data is colored as purple and green. According to the street network, the shortest path is route 1 which is colored in turquoise. However, instead of walking up the shortest path in the street network, this boy actually chose to zigzag using a less direct route (the blue route on Thursday and yellow route on Friday) to school. He walked across a little alley-way between houses (as shown in Figure 4A), then across a roundabout and took the path between the houses and open fields in Figure 4B. In the final part of the journey, he chose a long alley-way instead of a busy road (see Figure 4C). Not all these small alley-ways and open spaces which he walked past are included in the OS street network. In terms of locations, this type of walking behavior thus becomes much more complex than would be revealed if the analysis were based just on streets. To some extent, the calculation of the shortest routes for walking between locations on the street network thus becomes unrealistic and meaningless.

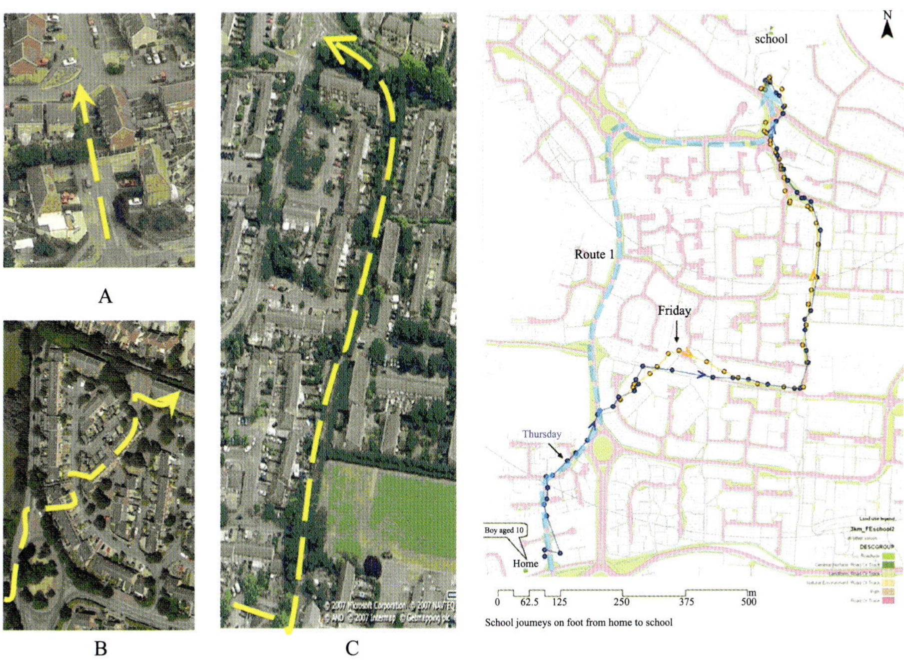

Figure 4 Trips to school from a boy

Mapping movement is different from mapping a stationary object. The development of methods to record people in different locations at different times has stimulated a huge interest amongst geographers and urban planners. Space-time paths have been largely adopted in behavioral studies since Swedish geographer Torsten Hägerstrand firstly introduced time geography in 1970[24]. New geo-visualization of spatial data is not only effective in revealing the complex interaction between the spatial and temporal dimensions in structuring human spatial behavior, but also provides an effective set of tools for exploratory spatial data analysis that can help in the formulation of more realistic computational or behavioral models[25].

Figure 5 presents a child's daily activity pattern in a two dimensional fashion. This child walked to school in the morning before 9 am. (pictured in Figure 5A), and stayed in the school till 3:30 pm., then walked back to home with friends (see Figure 5B). After arriving home, the child went out and played with friends in the neighborhood ending up at a friend's house (shown in Figure 5C). After this, he went out from the friend's house and then returned there later to stay overnight (Figure 5D). This child walked on two different paths during the school journeys. The trip from school to home in the afternoon was longer than the journey to school in the morning. All the walking and playing happened within a small residential neighborhood (less than 1 km^2). Figure 5E shows this child's calories expenditure. By linking with this his spatial movement pictured above, it can be seen that he spent a cluster of high active calories per kg on the way to school in the morning, during two school breaks (the morning and lunch breaks), playing out with his friends after school and on the way to his friend's house.

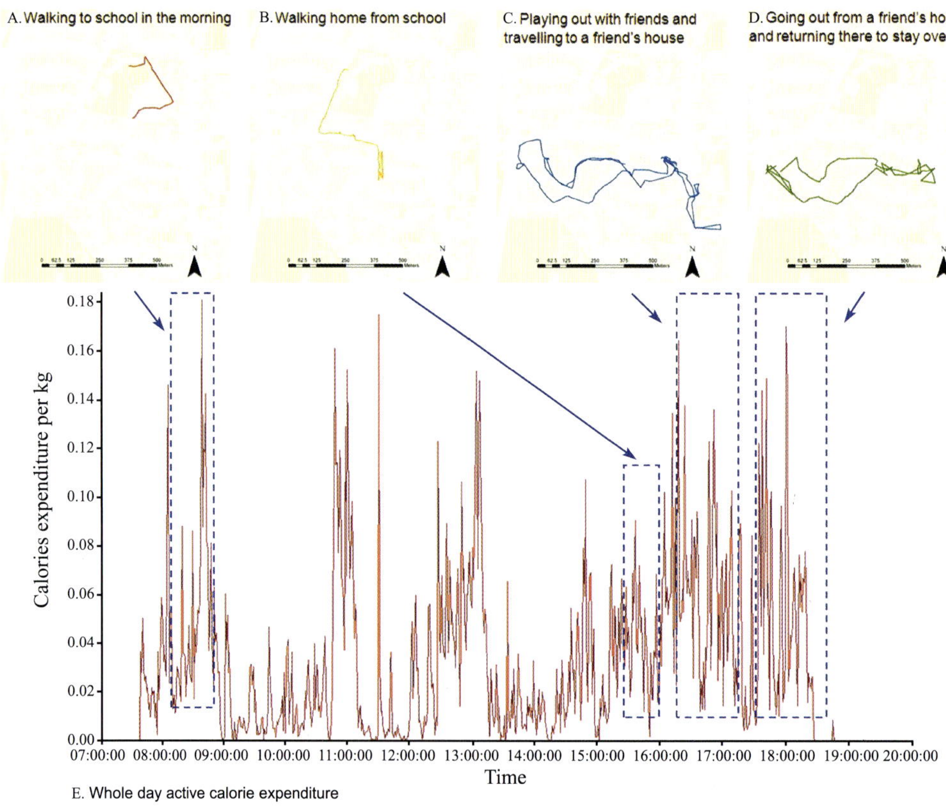

Figure 5 A child's trajectories and energy expenditure in a day on a 2D Map

By applying space-time paths, this child's daily behavior between 3 key locations can be presented visually in a three dimensions as shown in Figure 6. Time is presented as the third dimension, and the horizontal coordinates represent the position on the Easting-Northing plane. The behavior therefore can be traced in both space and time as a 3D trajectory. The trajectory is divided into 5 different sections to differentiate the different activities this child took part in. An almost vertical path between 9:00 and 15:00 locates where the school is, as the child spent most of his time in school. After finishing at school, the child moved around through space, so the path becomes more horizontal. The speed of movement can be seen as the slope of the line on the vertical direction. The faster the movement, the flatter the line. Section 1, presenting the walk to school, shows higher speed than Section 3, illustrating walking home from school.

Coloring the space-time path according to different attribute data, further exposes this child's walking behavior. Figure 7 to Figure 10 zoom into the details of space-time paths of this child's four different trips (Sections 1, 3, 4 and 5 in Figure 6). The color of the trajectories shows the calories per kg the child spent (on the left), and the walking speed (on the right). According to the space-time path, the redder the color, the higher the energy expenditure, and the faster the walking. Taking walking to school as an example, different levels

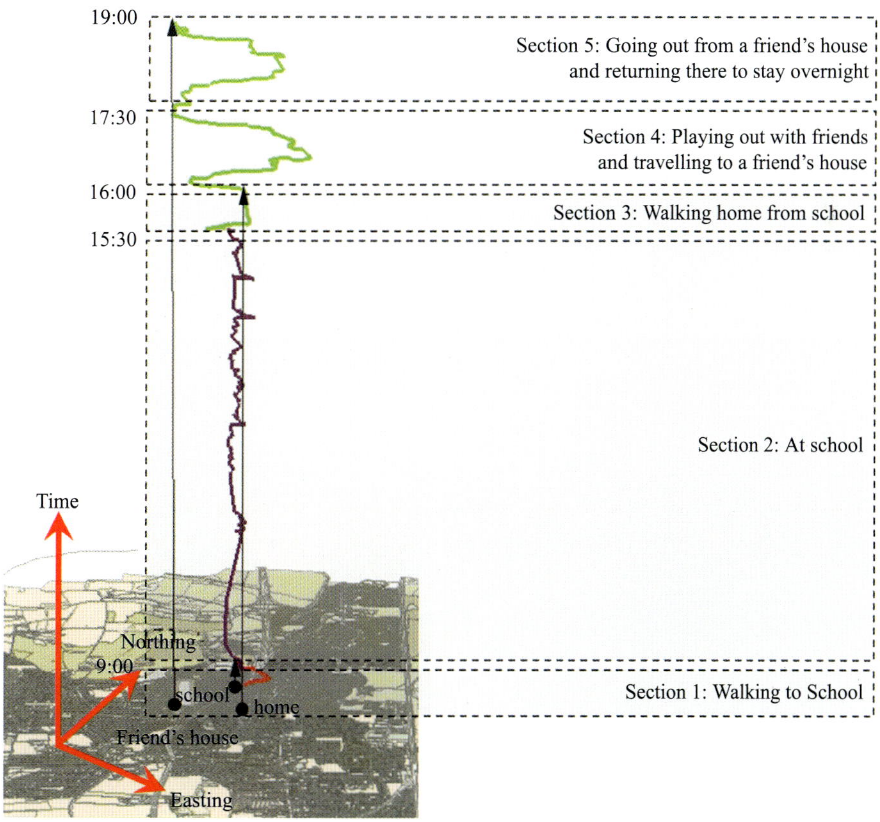

Figure 6 Space-time path of a child's activity over a day

of calorie consumption took place during the whole walking journey. A period of continuous higher calorie expenditure can be seen when approaching the school, compared with the child being less active at the beginning of the trip (pictured in Figure 7 and Figure 8). This was the result of the increase in the speed of walking when approaching the school after leaving home. Longer journeys in Figure 9 and Figure 10 show clearer differences in the slopes. The space-time paths show that this child walked and stopped to wander around during those two journeys. Higher calories were expended when this child walked faster (see the ellipse marked 2 in Figure 9), as well as when slow walking speeds were recorded by GPS (see ellipse 1 in Figure 9 and ellipse 3 in Figure 10). This indicates that the child jumped up and down or moved around in those two locations rather than walking straight.

Section 1: Walked to school

A. Coloured by activity calories per kg   B. Coloured by speed (meter/second)

Figure 7 Space-time path of walking to school

Section 3: Walked home from school

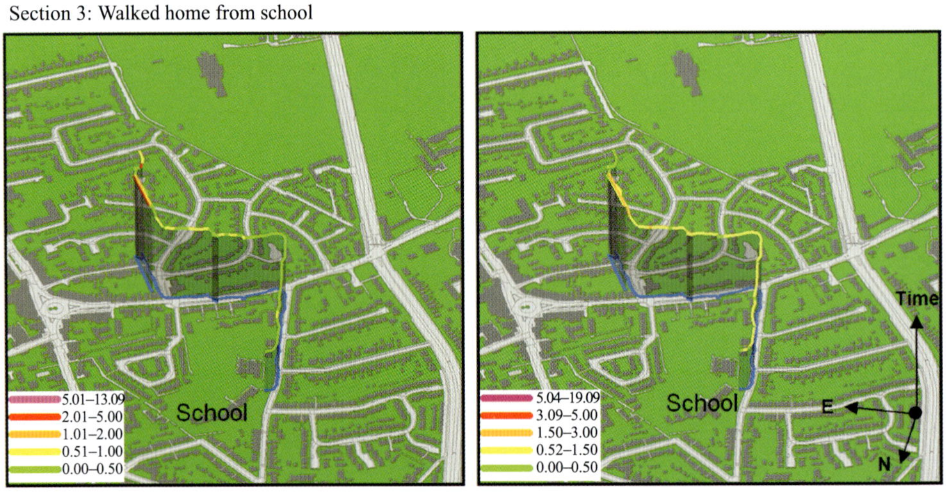

A. Coloured by activity calories per kg   B. Coloured by speed (meter/second)

Figure 8 Space-time path of walking home from school

Section 4: Playing out with friends and travelling to a friend's house

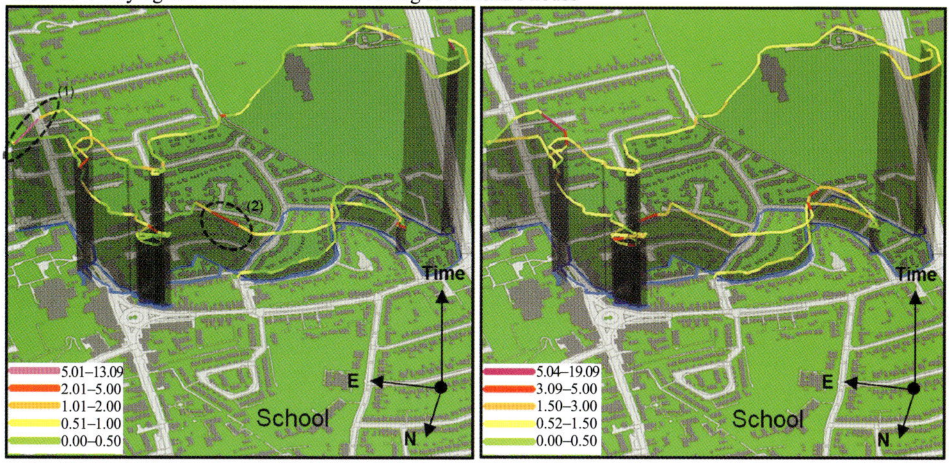

A. Coloured by activity calories per kg B. Coloured by speed (meter/second)

Figure 9 Space-time path of playing out with friends and traveling to a friend's house

Section 5: Going out from a friend's house and returning there to stay overnight

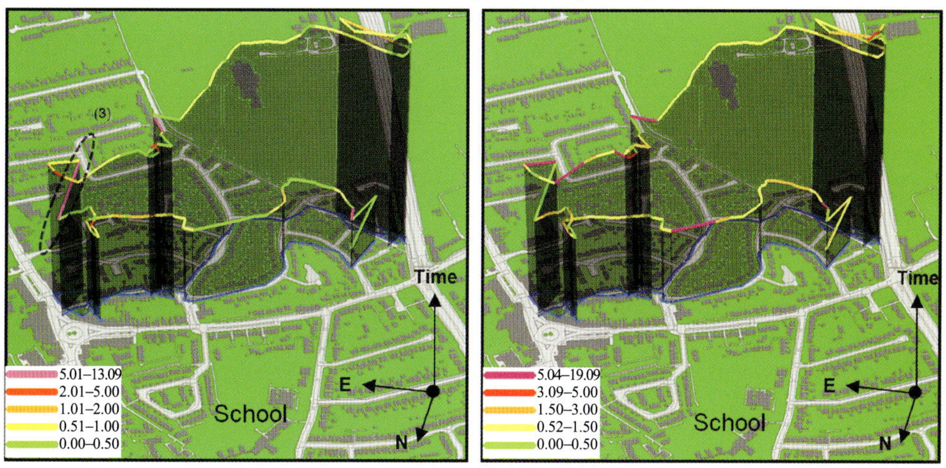

A. Coloured by activity calories per kg B. Coloured by speed (meter/second)

Figure 10 Space-time paths of going out from a friend's house and returning there to stay overnight

Alternatively, the fine-scale data collected through GPS, RT3 and the travel diaries can be used to project children's walking behavior in a real time sequence which can be animated to illustrate the different spatial patters that children generate. By using Google Maps, children's walking trajectories and corresponding energy expenditure data at the same time can be displayed in the web browser[1]. This geo-visualization technique shows more detailed information on how children move around on foot in their local environment. Figure 11 displays two school journey trips from two different children. Figure 11A is the journey from home to school, while 11B is the journey from school to home. The color of the dots

1) More details see http://www.casa.ucl.ac.uk/news/newsStory.asp?ID=150.

represents the energy expenditure per minute. The higher energy consumed, the redder the dots are (on a scale from blue through yellow to red). Both children walked across a recreation area which is located in a large open green space. The energy expenditure became notably higher here compared with other parts of the journey. By applying these different geo-visualization techniques to our fine scale data, a better understanding can be gained visually of children's walking behavior in their local environment.

A. A child walked home from school   B. A child walked to school

Figure 11 Visualizing children's walking using a Google Map

Analyzing walking in the local environment

Whilst analyzing the spatial behavior of individual students is useful, it is also possible to carry out statistical analysis of the behavior of groups of children. Summary variables need to be defined in order to do this. The average turning angles and speed of walking in meters per second over the ground obtained from GPS data, as well as the intensity of movement in 10^{-2} activity calories per minute per kg from RT3 monitors, were used to measure children's walking behavior. Turning angles represent the difference in the direction between two successive locations for children walking, based on 1 minute intervals. The changes in walking directions have been linked with the geographic locations and presented by five different colors in Figure 12 and Figure 13. Yellow and orange color indicate larger turning angles (over 90 but less than 180 degrees), while smaller turning angles (less than 90 degrees) are colored by red and purple. The walking patterns of the children can be seen in Figure 12. A comparison of the differences between boys and girls' walking in terms of speed and energy expenditure per minute per kg and turning angle has been made (see Table 2). It can be seen that boys move faster than girls and use more energy while girls tend to meander in a more relaxed way. The differences for speed and intensity are statistically significant, but those for the mean angles are not.

It is possible to establish whether children behave differently in different types of environment, using the GPS trace superimposed on the OS MasterMap topology. For illustrative purposes, two types of environment are identified: roads and open space. Figure 13 shows the spatial patterns of walking on roads and open space. Table 3 shows the summary statistics associated with these patterns. It can be seen that children were found to walk faster, straighter and more intensely on roads compared to when they were walking in open spaces (parks, fields, woods, and so on). All these differences are statistically significant. Open space offers more opportunity to play and meander about as shown in Figure 9. Related work on the relationship between children's walking movement and local environment have been published elsewhere[26]. In summary, open space offers the opportunity for children to explore and meander about, alone or with friends, which are healthy and sociable aspects of growing up.

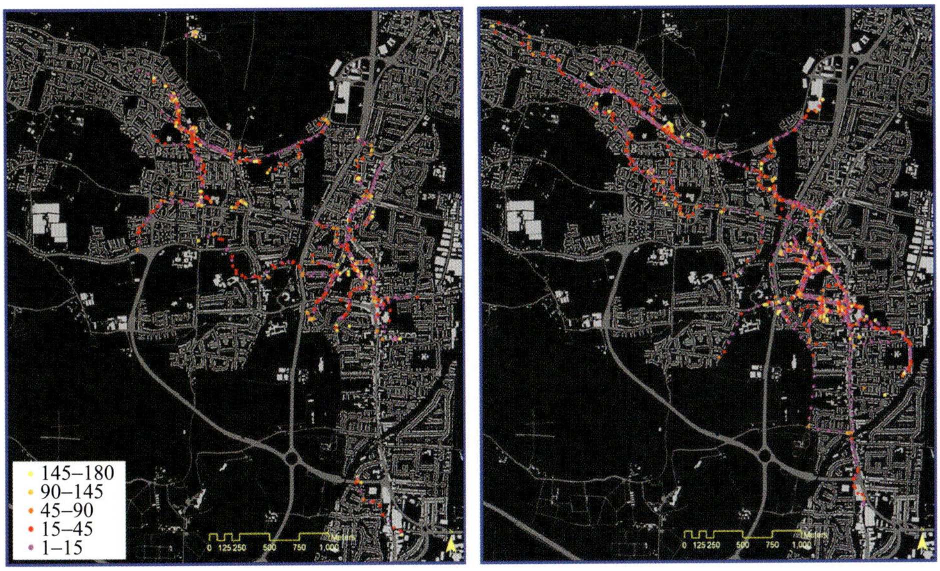

A. Walking boys   B. Walking girls

Figure 12 Turning angles in degrees when walking, at one minute intervals

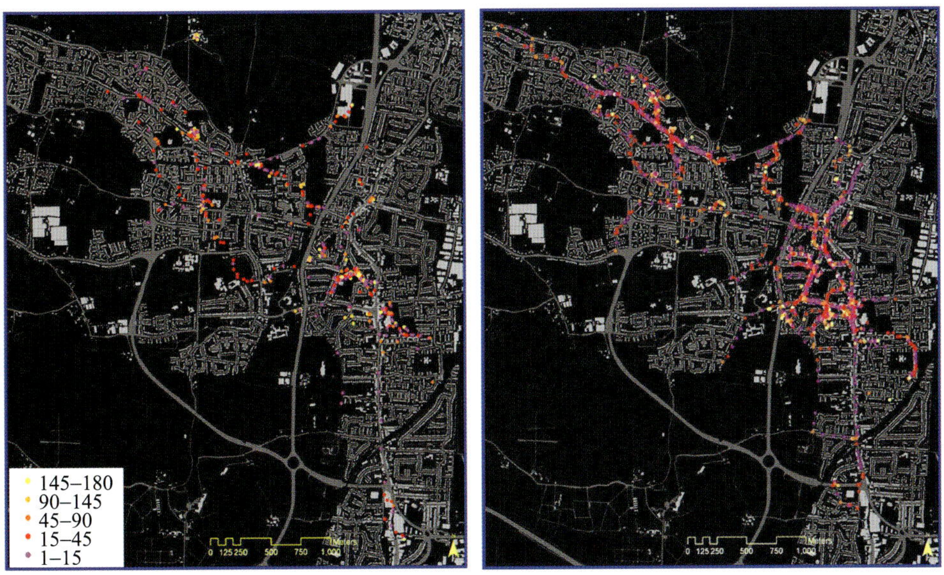

A. Open space   B. Road

Figure 13 Turning angles in degrees when walking, at one minute intervals

Table 2 Gender differences in children's walking patterns

	Boy	Girl	t	df	p
Speed	0.9	0.8	2.48	820.42*	0.013
Intensity	7.3	6.1	5.81	844.14*	<0.0005
Angle	38	46	2.52	855.13*	0.012

Note: * Equal variance not assumed.

Table 3 Children's walking patterns in different types of environment

	Roads	Open space	t	df	p
Speed	0.8	0.7	4.05	1498	<0.0005
Intensity	6.7	5.8	4.66	1614	<0.0005
Angle	41	49	2.34	570.97*	0.019

Note: * Equal variance not assumed.

Conclusion

This chapter has presented a preliminary study of how geo-visualization techniques can be used to explore the uncertainty and complexity of children's walking behavior in the local environment. By using geo-visualization, children's walking behavior becomes more tangible and easily understood when related to the different land uses that they encounter as they move. The study suggests the need for the development of a much stronger approach based on space-time geography involving collapsing several variables such as the energy expended, location, gender, speed, acceleration and a host of related variables into a framework that allows seamless visualization of these patterns. This is the intention for further analysis of the data which will be supplemented with material associated with mental maps and local perceptions of the environment.

Acknowledgements This research was funded by the UK Engineering and Physical Sciences Research Council (EPSRC) on a project entitled 'Children's Activities Perceptions and Behavior in the Local Environment (CAPABLE)' (Grant GR/T09378/01). The co-operation of the children who took part in this exercise and their teachers and parents is greatly appreciated, as is assistance from Hertfordshire County Council. We are grateful to Belinda Brown, Kay Kitazawa and James Paskins for their contribution to the fieldwork and to Richard Milton for helping to develop the Google map animation.

References

[1] Department of Transport. National Travel Survey 2006 Transport Statistics Bulletin. London, 2007.

[2] Klesges R, Eck L, Hanson C. Effects of Obesity, Social Interactions, and Physical Environment on Physical Activity in Preschoolers. Health Psychology, 1990, 9: 435–449.

[3] Sallis J F, Nader P R, Broyles S L, Berry C C. Correlates of Physical Activity at Home in Mexi-

can-American and Anglo-American Preschool Children. Health Psychology, 1993, 12: 390–398.

[4] Baranowski T, Wo T, Durant RH. Observations on Physical Activity in Physical Locations: Age, Gender, Ethnicity, and Month Effects. Research Quarterly Exercise Sport, 1993, 64: 127–133.

[5] Andersen R E, Crespo C J, Bartlett S J, Cheskin L C, Pratt M. Relationship of Physical Activity and Television Watching with Body Weight and Level of Fatness among Children: Results from the Third National Health and Nutrition Examination Survey. Journal of the American Medical Association, 1998, 279: 938–942.

[6] Canoy D and Buchan I. Challenges in Obesity Epidemiology. Obesity Reviews, 2007, 8: 1–11.

[7] McPherson K, Brown M, Associates M B. Tackling Obesities: Future Choices—Modelling Future Trends in Obesity and the Impact on Health. London: UK Government Office for Science, 2007.

[8] Hill J O, Wyatt H R, Reed G W, Peters J C. Obesity and the Environment: Where Do We Go from Here? Science, 2003, 5608: 853–855.

[9] Papas M A, Alberg A J, Ewing R, Helzlsouer K J, Gary T L, Klassen A C. The Built Environment and Obesity. Epidemiologic Reviews, 2007, 29(1): 129–143.

[10] Frank L D and Engelke P O. The Built Environment and Human Activity Patterns: Exploring the Impacts of Urban Form on Public Health. Journal of Planning Literature, 2001, 16: 202–218.

[11] Mackett R L, Lucas L, Paskins J, Turbin J. The Therapeutic Value of Children' s Everyday Travel. Transportation Research Part A, 2005, 39: 205–219.

[12] Cummins S K and Jackson R J. The Built Environment and Children' s Health. Pediatric Clinics of North America, 2001, 48: 1241–1252.

[13] Frank L D and Engelke P. How Land Use and Transportation Systems Impact Public Health: A Literature Review of the Relationship between Physical Activity and Built Form. Atlanta, GA: Georgia Institute of Technology, 2000.

[14] Boarnet M G. The Built Environment and Physical Activity: Empirical Methods and Data Resources. Prepared for the Committee on Physical Activity, Health, Transportation, and Land Use, Washington DC, 2004.

[15] Jones A and Panter J. Tackling Obesities: Future Choices—Obesogenic Environments—Evidence Review. London: UK Government Office for Science, 2007.

[16] Wagner D P. Lexington Area Travel Data Collection Test: GPS for Personal Travel Surveys. In: Final Report for OHIM, OTA, and FHWA. Columbus, OH: Battelle, 1997.

[17] Wolf J L. Using GPS Data Loggers to Replace Travel Diaries in the Collection of Travel Data. Atlanta, GA: School of Civil and Environmental Engineering, Georgia Institute of Technology, 2000.

[18] Chung E H and Shalaby A. A Trip Reconstruction Tool for GPS-Based Personal Travel Surveys. Transportation Planning and Technology, 2005, 28: 381–401.

[19] Draijer G, Kalfs N, Perdok J. Global Positioning System as Data Collection Method for Travel Research. Transportation Research Record, 2000, 1719:147–153.

[20] Rodriguez D A, Brown A L, Troped P J. Portable Global Positioning Units to Complement Accelerometry-Based Physical Activity Monitors. Medicine and Science in Sports and Exercise, 2005, 37(Supp): S572–S581.

[21] Nielsen T and Hovgesen H. GPS in Pedestrian and Spatial Behavior Surveys, paper for Cities for

People. In: The Fifth International Conference on Walking in the 21st Century. June 9–11, 2004, Copenhagen, Denmark.

[22] Elgethun K, Fenske R A, Yost M G, Palcisko G J. Time-Location Analysis for Exposure Assessment Studies of Children Using a Novel Global Positioning System Instrument. Environmental Health Perspectives, 2003, 111: 115–123.

[23] Lipp H-P, Vyssotski A L, Wolfer D P, Renaudineau S, Savini M, Tröster G, DellOmo G. Pigeon Homing Along Highways and Exits. Current Biology, 2004, 14: 1239–1249.

[24] Hägerstrand T. What About People in Regional Science? Papers in Regional Science, 1970, 24: 6–21.

[25] Kwan M P and Lee J. Geovisualization of Human Activity Patterns Using 3D GIS: A Time-Geographic Approach. In: Goodchild M and Janelle D (Editors). Spatially Integrated Social Science. New York: Oxford University Press, 2004: 48–66.

[26] Mackett R, Belinda B, Gong Y, Kitazawa K, Paskins J. Children' s Independent Movement in the Local Environment. Built Environment, 2007, 33: 454–468.

Chapter 22 Videogames: The New GIS?

Ifan D. H. Shepherd and Iestyn D. Bleasdale-Shepherd

"Comparisons are odious." (Gilbert of Hay, 1456)

Introduction

The positioning of one technology or medium relative to others is a common academic pastime. For example, numerous studies have compared GIS with other spatial and non-spatial digital technologies, including: remote sensing[1, 2], mapping[2, 3], databases[4], and CAD[4, 5]. For their part, videogames have been compared and contrasted with novels, drama, comics and film[6]. Until comparatively recently, however, the subject matter and technologies of most videogames have not been similar enough to warrant detailed comparison with GIS. However, this is no longer the case. Not only do many games present vibrant and realistic worlds on screen, as in Figure 1, but the spatiality of games is increasingly recognized as being a key property of these distinctive digital environments[7, 8]. For this reason, a detailed head-to-head comparison with GIS is long overdue. In undertaking such an analysis, we advance two broad propositions: that the two technologies and their uses are far more similar than might be expected; and that GIS have a great deal to learn from videogame technology and design. Although the central focus of this chapter is on videogames and what they have to offer, they will also be used as a lens through which to better understand GIS, and to identify areas for further development.

Not every kind of videogame is germane to our analysis. There are numerous genres and sub-genres of this form of interactive multimedia[8, 9], but the main focus here is on games which have recognizable geographical content. Among these are most action games (and especially first-person and third-person shooters), many role-playing and adventure games, several kinds of simulation game (including driving and racing games), and strategy games which involve the evolution of settlements, cities and civilisations. There are many videogames with a less recognizable geographical content, including: puzzle games (especially board games, quizzes and classic titles such as *PacMan* and *Tetris*), music games (such as *Guitar Hero* and *Jam Sessions*), and many children's games. Although a large number of sports games (*e.g.* tennis, soccer, snooker, American football, and Olympic events) have little intrinsic geographical content, others (such as golf and car racing games) are of considerable interest to GIS, because of their often fastidious and innovative approaches to micro-terrain modelling and townscape representation. These, and other less geographically relevant videogames, will be discussed where aspects of their design, role and usage are relevant in a GIS context.

Figure 1 Realistically rendered synthetic landscape (*Half-life 2*)

Although a broad view is taken of what constitutes a videogame, it was decided to exclude virtual world software whose primary goal is to enable users (typically in online environments) to act out alternative lives in relation to one another, albeit in specially constructed spatial environments. (These include virtual social worlds such as *Second Life* and *Habbo*.) Although some of the functionality of such virtual environments is also found in multi-player role-playing games, such as *World of Warcraft*, the primary objective of these sites is social networking rather than game playing. It is, however, recognized that this distinction is not watertight, and that at the time of writing some social networking sites are beginning to incorporate games-like activities as a means of attracting and retaining subscribers. It should also be noted that we use the term GIS throughout as a shorthand for the relatively close-knit family of software that enables analysts to manage, display, analyse and report spatial information in an integrated environment. This term therefore should be understood to include full-service GIS, desktop mapping software, spatial data visualization programs, and related software. The focus will be largely on software that models aspects of the real world for users driven by relatively serious analytical intent.

Throughout this chapter, the acronyms VG and GIS will be used respectively to refer to videogames and geographical information systems. Assiduous readers will note that while numerous videogame titles are mentioned in the text, few GIS are identified by name. This is partly because of the videogame focus of this chapter, but it also reflects the fact that there are relatively few GIS and, unlike individual games, most full-featured GIS contain a relatively wide range of functionalities characteristic of this kind of spatial analysis and display software. In contrast, there are thousands of individual VGs, and more than a dozen VG genres, and there is consequently a far greater diversity between individual products than among GIS. Where distinctive features are found in specific GIS, individual software will be identified by name.

Similarities Between Videogames and GIS

A common definition of GIS is that it is a system (part hardware, part software, and part human) that is used to solve problems with a spatial dimension, by providing tools for the acquisition, storage, integration, update, retrieval, manipulation, analysis and display of spatial information. In this section, key GIS characteristics and functionalities will be critically examined in relation to VGs, with the following three aims in mind: to evaluate how far VGs provide the capabilities of GIS; to determine limitations in the functionality of VGs in terms of spatial problem solving; and to indicate ways in which GIS might benefit from adopting VG functionality.

Spatial representation

Videogames may be classified on the basis of two aspects of their spatiality: the extent to which they attempt to model the real world, and the kind of representational system they use in portraying elements within their worlds. (In terms of the second of these dimensions, only visual representation is considered here, for reasons that will become clear in a later section). For the purposes of the current discussion, these two aspects will be referred to as reality and realism respectively. Figure 2 maps out this reality—realism space, and indicates where three related spatial technologies (VGs, GIS, and social networking virtual worlds) are positioned in relation to these twin characteristics. It is suggested that VGs occupy a far greater territory in this space than either GIS or social networking virtual worlds. Within the area of the diagram occupied by VGs, a selection of videogames has been located. (For readers less familiar with contemporary VGs, a lookup table is provided in the Notes.) These locations

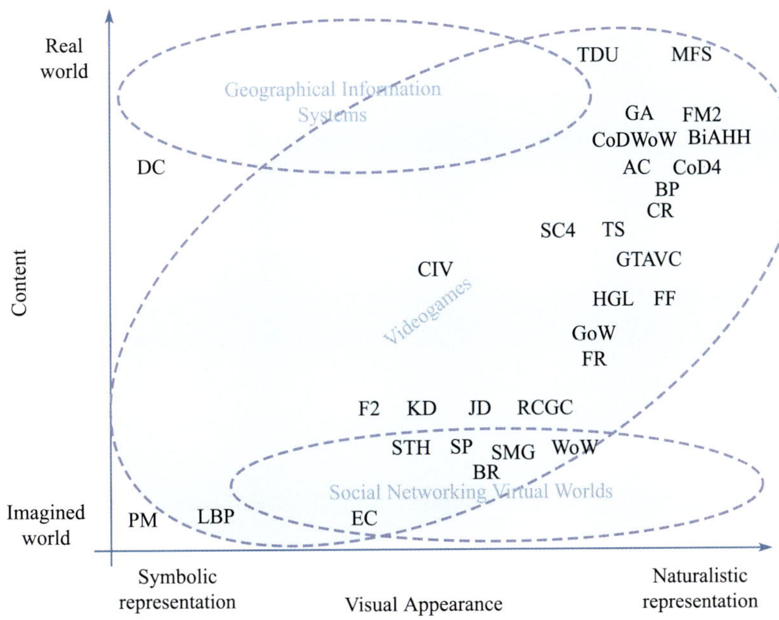

Figure 2 Videogames, GIS and networked virtual worlds in reality—realism space

are best thought of as modal locations, because a number of VGs present several contrasting virtual environments to the player.

Visual realism

The rich visual realism which many contemporary VGs present in their virtual worlds is immediately apparent to any spatial analyst who looks over their children's shoulders. Landscapes extend across extensive areas, terrain is draped with plausible vegetation patterns and the infrastructure of human habitation, building interiors are designed with considerable attention to architectural and engineering detail, and many of these virtual environments are enlivened with the dynamic trappings of modern life, including vehicles and people in motion. Some of this visual realism is undeniably a showcase for VG talent and technology, and especially for graphic artists and graphics hardware. However, a great deal of it reflects the way in which many modern videogames place the player at ground level, and therefore represent the world as seen through the eyes of the pedestrian or the driver, rather than from the air as in most GIS. However, much of the visual realism is wrapped around a world model that is largely or entirely fictional.

Representing the real world

The environments encountered by players in VGs run the gamut, from authentic reproduction of real places using real spatial data to entirely fictional places using synthetic data. The majority of VG environments, and especially those built for role-playing games, are imagined worlds, created from the imagination of the games designer. At this lower end of the reality spectrum (as depicted by the vertical axis in Figure 2), are to be found innumerable fantasy worlds (*e.g. Final Fantasy, Jak & Daxter*, the *God of War* series and *Little Big Planet*) and several impossible ones (*e.g. Echochrome*, with its Escher-like 3D geometry, and *Braid*, with its distortions of time). There are also some casual and passive games which visualize user data to create an abstract 3D scene (*e.g. Packet Garden*, which uses 3D glyphs to construct virtual landscapes from multivariate data describing the player's online surfing habits).

Also at a low level on the reality scale are worlds which may appear to be based on some form of geography or cosmology (as in role-playing games such as *Fable II*), but where the reality is only skin deep. The spherical planetoids in *Super Mario Galaxy*, for example, bear little or no resemblance to real world exemplars; and the 3D globes that may be explored in *Civilization: Revolution*, *Katamari Damacy*, *Spore*, and *Ratchet & Clank: Going Commando* lack either the dimensions or the realism of the earth as seen from space. However, numerous VGs at this relatively low point on the reality scale attempt to mimic the visual appearance of the real world. These include a number of first-person games (*e.g. Crysis*), and many racing games (*e.g. Burnout: Paradise*), while many spatial strategy games (*e.g.* various versions of *Civilization*, *Sim City* and *The Settlers*) also provide recognizably human landscapes and elements of real-world behavior. In creating the level of visual fidelity evident in these games, designers are increasingly aided by software that procedurally generates terrain and buildings, often using architectural templates taken from real cities.

Many games with this reality: realism mix are set in historic locations, but designers typically model the environments with a mix of artistic licence and architectural and geographical authenticity. Examples include: *Assassin's Creed* (set in the cities of Acre, Damascus and Jerusalem at the time of the Third Crusade in the late 12th century), and *Hellgate: London* (set in a futuristic version of London). Although most videogames based on past wars (the military shooter genre) appear to recreate the look and feel of a battlefield, they tend to use a particular conflict merely as an 'inspiration' or hook; the gameplay itself often bears only passing resemblance to the battles fought. Examples include *Medal of Honor: Allied Assault*, based loosely on World War II, *Call of Duty 4: Modern Warfare*, based on modern battlezones, and the squad shooter *Conflict: Desert Storm*, based loosely on Desert Storm. Some titles (*e.g. Rainbow Six Vegas 2* and *Call of Duty 4: Modern Warfare*) take considerable pains to visualize weaponry, vehicles, soldiers clothing and insignia in true detail, but their representation of geography and historical events is much less authentic. Among VGs with a greater purchase on reality are *Call of Duty: World at War*, which aims for both geographical and historical authenticity (including the use of audio recordings of World War II weaponry), and *Brothers in Arms: Hell's Highway*, which not only models actual theatres of conflict (the Dutch countryside during in World War II), but also enables the player to participate in reasonably authentic forms of fighting. But for every title that attempts to be true to the battle that inspired the game, there are many others which make only a token effort to replicate the geographical environment of the war in which their virtual fighting is staged.

Much the same mixture of relatively low reality and high visual realism may be found in several VGs set in contemporary locations. In some examples (*e.g. Fracture*, set in the San Francisco Bay Area), the cityscape is 're-imagined', using a mixture of familiar and unfamiliar elements. In other games, locales are described as 'inspirations' rather than replicas. Examples include the skateboard game *Proving Ground*, which is based loosely on Philadelphia, Baltimore and Washington DC, and the car racer *Project Gotham Racing*, which has been set variously—and in most cases somewhat symbolically—in New York, London, Tokyo, Las Vegas, Macau, Shanghai, St Petersburg and Quebec.

However, some VGs are more closely based on real world locations, altering the real geography only where it is dictated by the needs of the gameplay. An example of this approach is the car racer *Getaway*, whose designers took considerable efforts in collecting detailed streetscape information (including building textures captured through extensive field photography). Although the street layout is simplified and stylised, it bears more than a passing resemblance to the major thoroughfares in those parts of central London in which the game is set. Similarly, *Grand Theft Auto IV* involved considerable research into New York, including the taking of tens of thousands of photographs[10]. (See Ref. [11] for further examples of VGs set in replicas of real cities.) This mixture of real-world and synthetic data is also found in largely fictional VGs (*e.g. WRC: Rally Evolved* and *Forza Motorsport 2*), which use real-world data in the same way that some GIS worlds incorporate imaginary (or at least synthetically derived) data[12,13].

There is only a small number of VGs whose virtual environments are based on real geography and real spatial data, and therefore occupy the upper end of the vertical axis in Figure 2. One example is the car racer *Test Drive Unlimited*, which models the terrain and route network of the Hawaiian island of Oahu by integrating multiple sources of geospatial data for the island. (This permits the display of a realistic in-car SatNav.) A second example is *Microsoft Flight Simulator*, which drapes aerial photo data onto digital terrain models, and uses accurate 3D digital models of actual airplanes and detailed airfield and airport layouts, to provide a realistic simulation of flying that unfolds in real time. (It should be noted that much of the functionality of these two games, including the virtual SatNav, is available in Google Earth via user mashups[14].) A third example of a game based on real world geography is the multi-player world strategy game, *Defcon*, which charts player progress on a cylindrically projected world map as in Figure 3.

Figure 3 World strategy game indicating buffered radar installations (*Defcon*)

Although these three examples are located towards the top of the vertical reality scale in Figure 2, they are spread across the horizontal realism scale. In terms of its real-world visual appearance, *Microsoft Flight Simulator* is closest to the naturalistic end of the horizontal scale, although for most of the time the player has a relatively high-altitude bird's eye view. A little way to the left of this is *Test Drive Unlimited,* whose visual representation of island scenery is frequently rather stylised. Finally, *Defcon* is over at the symbolic end of the horizontal scale, with gameplay being organized around an atlas-like world map. It should be noted, however, that VGs evolve through time, and many titles shift their position and/or modify the area of the reality:realism space they occupy as new versions appear on the market. In this context, it is interesting to note how Google Maps, originally developed to represent the real world, is increasingly being used to display virtual worlds (*e.g. GTA IV*'s Liberty City at grandtheftauto.ign.com/maps/1/Liberty-City-Map and Second Life at: slurl.com).

Fitness for purpose

For VGs and GISs alike, the level of ground truth provided by the software is only useful to the extent that the player or the spatial analyst requires it. Indeed, the reaction from some quarters of the gaming community to increased realism in VGs has been extremely negative. One commentator[15] rails against the obsessive pursuit of realism by some games producers, suggesting that the essence of some classic games (such as *Sonic the Hedgehog* or *Pac-Man*) would be destroyed by using realistic rendering and physics. For Biffo, "one of the big appeals of games is that they are a window on to fantasy … in the more general sense of being transported to somewhere that isn't real". In a statement that serves to maintain clear water between videogames and GIS technology, he goes on to argue that: "If you're going to create a CGI world that's only one step away from our real world, then why not just make a live-action movie?"—or, one might add, use a GIS? What GIS analysts know only too well is that unrealism, in the form of map symbols and visualization glyphs, can often serve to deliver greater understanding than photorealistic displays.

A considerable tension has been introduced into VGs by the desire to create truthful representations of the real world while at the same time designing virtual environments which provide a compelling game experience. This tension is central to the game designer's craft, as is evidenced by the critical view of a highly influential games designer[16]:

"A game creates a subjective and deliberately simplified representation of emotional reality. A game is not an objectively accurate representation of reality; objective accuracy is only necessary to the extent required to support the player's fantasy. The player's fantasy is the key agent in making the game real."

Using modern VG technology it is now just as easy to model the geography of an actual racing circuit or a battlefield as it is to recreate an actual race or battle. However, the faithful recreation of real events runs the risk of suppressing the freedom associated with player decision-making behavior that is the central feature of interactive games. Instead of interaction, the player would confront a digital version of the kind of event reported on a sports TV channel or staged by a battle recreation society. As one war simulation game reviewer put it[17], the challenge for developers is to tread the fine line between testament and entertainment. This important dilemma is largely absent in GIS, whose primary goal is the representation of geographical reality.

The foregoing discussion reveals as illusory the presumption that VG environments are always synthetic while GIS worlds are always real. Indeed, GIS have until very recently provided low-fidelity visual realism (in the form of cartographic symbolism), minimal fidelity in terms of non-visual data representation[18, 19], and only medium fidelity in terms of dynamic modelling (a reflection of the static map inheritance of GIS). Moreover, the central problem of spatial data uncertainty further underscores the fact that even though GIS attempt to model the real world as accurately as possible, their representations are frequently significantly flawed[20, 21]. It is therefore a crude over-simplification to suggest that while videogames have been high on realism but low on reality, GIS have been low on realism but high on reality. Each in its own way adopts blends of the real and the imagined, the naturalistic and the symbolic, as needs dictate.

Spatial modelling technologies

The technologies used by VGs and GIS in building their respective digital worlds are increasingly common to both domains, though with perhaps different emphases. Vector data models are widely used, though in VGs they are almost exclusively applied to 2.5D and 3D objects. Digital terrain models are also commonplace, with both VGs and GIS using multiple resolution terrain data. The grid-cell data model is largely absent from VGs, though a variation of it appears in the grid-based terrain models used in a number of strategy games (*e.g. Sim City*). On the other hand, image and texture data are widely used, and recent innovations in texture mapping from the VG and CGI communities are now routinely used by GIS to drape aerial photos and satellite images over terrain models. Additional realism is provided in VGs through surface rendering that uses a range of sophisticated computer graphics techniques, including: bump mapping, environment mapping, radiosity, dynamic lighting, and dirt mapping (which is put to good effect in the car rally game, *Colin McRae: DIRT*). Other technologies used in both camps include: the use of hierarchical data structures (and especially quadtrees), spatial data integration (especially of various kinds of surface data), techniques which minimise processor loads in complex scenes (*e.g.* tiling and level of detail or LOD), and layering (which in VGs is mainly used to model dynamic accretions on surfaces).

Various algorithms are also similar in VGs and GIS. These include: spatial search (using sophisticated spatial indexing), object buffering (used for object collision detection), and path planning algorithms (used to transport the user from one location to another). (The artificial intelligence developed in VGs for automatic camera control in third-person games is perhaps unique to VGs, and will be discussed further below.) However, this is where VGs and GIS begin to part company. For their part, VGs have focused more on innovative and flexible methods of user interaction and navigation in 3D environments, and in developing dynamic models of environmental and human processes. Several specialist data structures are used to support fast surface rendering (*e.g.* indexed vertex arrays), dynamic lighting (*e.g.* normal maps), rapid search (*e.g.* BSP trees and nested voxel grids) and navigation (*e.g.* traversability graphs) in 3D environments, and related structures for rapid collision detection (*e.g.* convex hulls). In contrast, GIS have placed more emphasis on developing tools for spatial analysis, and perhaps most notably in terms of 2D spatial overlay. In general, VGs tend to adopt spatial models and data structures that ensure optimal real-time rendering and interaction, realistic sound and haptic responses, and continuous data updating during gameplay. (In *Splinter Cell: Conviction*, for example, the dynamic navigation mesh which enables realistic navigation behavior is updated as objects in the world move or change.) In contrast,

GIS have tended to adopt spatial models and data structures which are primarily optimised to support analytical operations on data that represent relatively static worlds.

Attribute data

One of the more distinctive features of GIS technology is its integration of attribute and locational data[22, 23]. On the surface, VGs appear to be all visual representation and action, with little obvious evidence of the attribute data associated with real-world features and objects. This perception, however, is misleading. Although VG software may run to millions of lines of code, most games also contain very large volumes of 'content' or 'assets'. Like GIS, VGs also store large amounts of attribute data, describing elements of the virtual environment, including discrete objects and characters. In a typical contemporary videogame, object assets may include: vertex data (multiple attributes—*e.g.* position, normal(s), texcoord(s)), texture data (multiple layers—*e.g.* albedo, normal, reflection mask), collision mesh (multiple versions of the prior data for LODs), animations and related audio files and particle effects, and physical data about the object's surface and volume. Although such data may not be organized in the form of a relational database, for reasons of simplicity and performance, many data tables adopt the simple row-column form of the standard attribute data matrix. However, because of the highly dynamic nature of VG environments, attribute data in VGs are more frequently stored in other data structures, such as trees, sparse voxel grids and hash tables.

In both VGs and GIS, attribute data are crucial, not only for data visualization and object selection, but also for more advanced forms of simulation, spatial analysis and reporting. Indeed, there is frequently little difference between VGs and GISs in either the type or quantity of non-graphic data they require. (For example, some 80,000 lines of spoken dialogue were specially written for *Grand Theft Auto IV*, which used 740 different voices.) However, one significant feature of VGs is that a great deal of game information is not attached to objects as standardized descriptive information. Unlike GIS, much of the information stored in VGs supports elements of gameplay not normally found in GIS. Thus, for example, the text in GTA IV are delivered in an intelligent fashion based on in-game events and player actions. *Half-Life 2*: *Episode 1*, for example, uses a fairly generic event-driven contextual system in which dialogue is situation, event, and action specific. This makes the world and its characters react to the player in a (seemingly) more intelligent fashion. Some lines of dialogue are always heard—usually associated with events which are guaranteed to occur—but some will only be heard if certain things happen.

A second difference resides in the uses that can be made of attribute data. In contrast to GIS, much VG data are only accessible to the games software, and not directly to the player. In general, GIS analysts actively extract data, while in videogame players are passively fed data. In both cases, algorithms are at work, and user decision making and interaction is required, but the degree of control of data-extraction and display is skewed towards analysts in GIS, and to the software in VGs. Thus, while in VGs data are mainly 'pushed' to the player, in GIS, data are mainly 'pulled' by the analyst.

However, there are some notable exceptions to this general rule. Titles across several game genres (*e.g.* role-playing adventure games, strategy games, racing games and sports games) not only maintain large attribute databases related to their virtual environments, objects and characters (and particularly for player avatars), but they also display some of these data on screen during gameplay, and update attribute data for the player's avatar and its behavior as the game proceeds. Stored attribute data is displayed in various ways during gameplay. In some car racing games (*e.g. Forza Motorsport 2*), detailed readouts of car settings and behavior can be displayed, and the player is able to adjust these during the game to modify the driving behavior of their vehicle. (In *Sega Splash! Golf*, for example, golfer statistics can be viewed in a player room.) Such data may provide players with feedback and motivation (as in many car racing and role-playing adventure games), or may aid dynamic decision-making (as in many strategy games).

In many adventure, strategy and simulation games, relevant text is provided on screen as a readout or mouseover display (akin to Windows tooltips), while many shooters provide head-up displays with summary information about the current state of the game. In *Civilization: Revolution*, visual modifiers (brief displays of summary information) are attached to objects on screen, while a street-name indicator appears at the top of the *Burnout Paradise* screen, reminiscent of GIS practice. Another genre of VG which routinely accumulates in-play data is the massively multi-player online game (MMOG). Such data are made available to game community websites which feed game statistics back to players so they can learn from others' gameplay performance. (A notable example is Bungie.net, which reports on online, multiplayer deathmatches in *Halo*.) These data are also used by game designers to help steer new feature development, and plan game marketing activities. There are lessons to be learnt here by GIS, where the actual problem-solving behavior of expert analysts might be of considerable value to less experienced users, or peers working in the same field. In addition, the accumulation of data on trainee actions during training courses could be used to make informed decisions on further learning activities.

Spatial data visualization

Few (if any) VGs adopt the kind of symbolic map representation that has benefited spatial data analysts for the best part of two hundred years, and are thus able to provide little guidance on effective 2D data visualization. (One of the few exceptions is *Defcon*, the multi-player world strategy game, which displays its symbolism on a cylindrically projected world map, see Figure 3.) However, because most contemporary VGs display realistically modelled scenes in 4D, they provide a suitable benchmark against which data visualization in virtual worlds modelled by GIS can be measured. Although few VGs would appear to adopt the strict rule-based approach to data visualization propounded by Bertin[24–28], it is nevertheless instructive to understand how VGs adopt the graphical sign system, albeit at an informal level, and to identify innovative applications in VGs that might be useful in a GIS context.

Perhaps the simplest example of the operation of data visualization rules is the use of categorical visual variables such as shape, color and texture to distinguish between qualitatively different groups of characters, objects or events. Most videogames pay careful attention to the visual design of entities so as to allow at-a-glance disambiguation between the many entity types present. The use of color-coding to relate groups of entities is a very common practice in videogame design. In the team-based online game *Team Fortress 2*, for example, players on opposing teams are differentiated through the use of simple color schemes (*e.g.* one team wears red clothing, the other blue), while in *Schizoid*, enemies of a particular kind are given distinct colors. Numerous combat games also distinguish friends from enemies by the style of clothing they adopt, many role-playing games (RPGs) distinguish male and female characters by their clothing styles, and in some shooter games the lethality level of monsters is identifiable from their 3D shapes. In *Black & White II*, the avatar changes its form based on behavior, so that it in turn looks good or evil, fat or skinny, weak or strong. Color is also widely used for signalling purposes. In *Mirror's Edge*, for example, features of the environment (such as ledges and machinery) that might be useful to the player in traversing the cityscape are depicted in primary colors, so they stand out from the mainly white and blue of the virtual world, and while safe paths are clearly illuminated, dangerous paths are shown in shadow. In the *Half-Life* series, interactive objects are colored red, and lighting is commonly used to attract attention to important objects or locations.

Figure 4 Lighting-enhanced rooftop scene with health indicators (*Left 4 Dead*)

Videogames also use gradational variations in visual variables to reflect continuous variations in numerical variables (*e.g.* counts or measurements). Normally, this applies to individual objects or characters in the game. Size, for example, is often related to the strength of enemies or the value of items in a game (as in *Blinx: The Time Sweeper*, in which gold ingot size implies value), and variations in lightness or hues along a color ramp is sometimes used to convey some internal state value (*e.g.* brightness of a head-up display (HUD) element relates to the health of the player's avatar, the health of an enemy is reflected in a 'health bar', typically floating in the world above the enemy's head, which changes in color from green (healthy) to yellow (damaged) to red (nearly destroyed). Games which use a perspective projection also provide implicit scalar visual feedback, by relating an object's size on screen to its distance from the viewpoint (in addition to the obvious relation between on-screen position and virtual 3D location).

Unlike most GIS, VGs also make use of time as a visual variable. A common example is found in shooter games, where colored bars of dynamically varying length represent the gradual decline or rapid replenishment of a player's health as in Figure 5. Another example is the pulsation of an object's size or color, where the frequency of the variation indicates the current threat level, *e.g.* when an object is likely to explode (*e.g.* in *Left 4 Dead*, pipe bombs beep with increasing frequency as they near detonation, and the player's view becomes increasingly dark and blurred as their health runs out when they are in need of rescue by another player).

Several VG practices may be of interest for data visualization in a geographical context. Some games eschew the normal per-symbol approach to data visualization, and adopt instead what might be termed ambient visualization. In *Brothers in Arms: Hell's Highway*, for example, a color intensity gradient is used to darken and redden the entire screen as the risk of the player being shot increases. A variation of this technique is used in *Half-Life 2*, where threats to the player from the side or rear are indicated by the appearance of edge-of-screen red bars. With the left bar appearing if the threat is from the left, the right bar if from the right, and both bars if from directly behind the player? Also, the redness of these bars is proportional to damage, and fades over time after the damage occurs. In the case of repeated damage from a given direction, the color will maintain a strong intensity due to the slow fade.

Another distinctive use of data visualization is to provide on-screen visual prompts derived from several relevant variables to assist the player in making spatial decisions, or undertaking complex actions. Examples include: the 3D trajectory line symbol in *Tiger Woods PGA Tour 08*; to assist the player in making a shot; and the quarterback's 'vision' cone in *Madden NFL 2006*, whose length and width is proportional to that player's level of expertise, is used to determine which receivers can be thrown the ball. The location of certain visual prompts in the virtual environment can often be of considerable importance to effective decision making by the player. A good example is the visual feedback provided by the balance bars included in early editions of the *Tony Hawks* skateboard games. Because these bars were placed next to the skater, it was found that found that they competed for the player's visual attention. Consequently, they were replaced in *Proving Ground* by a screen-wide colored arc, which shifted the visual prompt to the player's peripheral vision. (Similar thinking was used in the design of the edge-of-screen colored bars in *Half-Life 2*, as described earlier.)

A further lesson for spatial data visualization concerns how well 2D visual symbol coding translates to 3D. Although the approaches explored by Kraak[29, 30] do not appear to have been implemented in VGs, careful graphical design has established some working principles for effective 3D visualization. An example is found in the team-based, multi-player game *Team Fortress 2*, where the player takes on one of 9 distinct roles. Because it is important to be able instantly to identify another player's team and role, the avatar for each role has a unique silhouette, easily distinguishable from all the others, which retains its distinctiveness when viewed from various directions and under varying lighting conditions. A related technique is used to show objects even when occluded, as in the 'shadow Mario' used in *Mario Sunshine/Galaxy*, whose silhouette at least is always visible, even when the character is occluded. In general, automatic cameras and environmental design are concerned with mapping 3D worlds to effective 2D views at any given moment.

A final lesson for GIS concerns the way in which data visualization can often serve multiple roles in VGs. It is frequently the case that visual symbols not only represent information, but at the same time add emotional depth, invoke engagement and develop a sense of immersion for the player (see Figure 5). This suggests potentially fruitful lines of collaborative research between VG and GIS, involving such topics as denotative and connotative responses to data representations, the semiotic boundaries between monosemy and polysemy, and possible interactions between cognitive and affective responses.

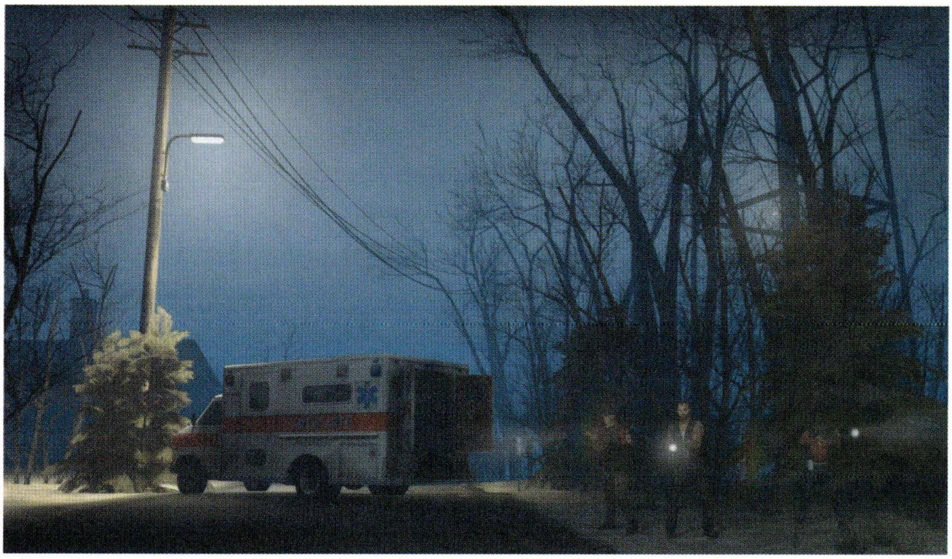

Figure 5 Atmospheric lighting (*Left 4 Dead*)

What Videogames Can Contribute to GIS

In this section, we consider areas in which VGs have surpassed GIS in their treatment of elements which may be relevant to spatial data explorers and analysts. Three areas will be discussed in which VGs can contribute in a significant way to GIS: dynamic process modelling, user interaction, and multi-sensory data representation. Space prevents discussion of several other significant aspects of VGs that may be relevant to GIS.

Dynamic process modelling

Videogames are change-rich environments. From their earliest appearance, videogames have provided dynamic experiences, with events continually unfolding on screen. Changes to the virtual environment are in part the result of dynamic simulations running within the software. However, they also result from user interactions with the virtual environment, including shifts in player position and viewpoint, but also user modifications to the virtual environment. Indeed, change tends to be continuous rather than episodic in most VGs, and permeates most aspects of the virtual environment, including parts that may currently be unseen by the player. The management of change therefore represents one of the key challenges to the games designer and player alike, and these challenges constitute what Poole[31] refers to as the kinetic narratives which are such a distinctive feature of videogames.

Not all change evident on screen is significant to the goal or purpose of the game; a great deal of dynamic behavior is modelled for purposes of ambient realism. Thus, for example, offices contain working computers and fax machines, homes have TVs which play ads, vehicles have SatNavs that emit spoken instructions, and in-car radios emit music and DJ banter. Some games mimic real environments by providing usable telephone booths in the street (as in *Proving Ground*) or mini-games that can be played in shopping mall arcades (as in *Sega Splash! Golf*). This dynamic modelling is taken to new levels in representing change initiated by player actions. Thus, for example, footprints appear when walking on wet sand; spray appears on a camera lens during a powerboat trip; newspapers flutter in the vortex created by a passing car; and plaster flies off a wall when hit by bullets. Some of this dynamic realism is more significant in terms of gameplay, as when a box breaks apart when hit by a crowbar, to reveal hidden ammunition.

In earlier games, the impact of change within the virtual environment was often temporary, with killed creatures fading away shortly after their death, and bullet holes disappearing or being generalized on vehicle bodies and building facades soon after they had been created. Nowadays, however, bodies will accumulate on the ground, and bullet holes will stay exactly where they first appeared. VGs thus not only represent events; they also record the accumulated effects of events, so that the evolving game environment becomes a dynamic palimpsest. What is often missing, taking a GIS perspective, are facilities to interrogate and analyse this change. However, in *Sim City IV*, the player can look at pollution or property-price color maps as the city evolves, and in *De Blob*, the player progressively covers more and more of a 3D environment with various colors of paint.

For their part, GIS initially drew their modelling approaches from static 2D maps, with digital spatial models typically representing a synchronic snapshot—the classic cross-sectional view. Over the past couple of decades, however, GIS technology has moved progressively from mimicking static 2D maps to providing digital representations of 4D data[32, 33]. Only comparatively recently have GIS begun to change, with the introduction of 3D viewing, brushing, and animation[34, 35]. Increasingly important is the visual representation of diachronic change, typically undertaken by creating sequences of temporally spaced snapshots (*e.g.* the spread of population across the UK from decennial census data; the diffusion across America of WalMart stores; or the tracks of hurricanes in the Caribbean). Objects have also begun to move and change on screen by coupling the software to real-time data feeds (*e.g.* from GPS or ground-based tracking technologies) or links to simulation model output.

Although most GIS can handle historical data, typically in the form of time-spaced snapshots (*e.g.* remotely sensed image sequences for a given area, or decennial population censuses for a particular country), the essential characteristic of dynamic representation in GIS is that it is mainly a data-driven process, where the data are the outcome of prior surveying or monitoring activity, or as a result of real-time data capture. Most do not display the output of dynamic process models. Thus, while most GIS are adept at handling spatio-temporal data, few model dynamic spatial processes. This is not to ignore the considerable success within the geoscientific community, in developing dynamic environmental and social models. These range from the geological and hydrological models of the Kansas Geological Survey during the 1960s and 1970s[36–38], through to the more recent geosimulation models of the 1990s[39]. There have been numerous attempts since the 1990s to link GIS to environmental process models (*e.g.* Ref. [40–42]) mainly through loose and tight coupling rather than embedding. Nevertheless, these models still occupy a parallel technological space to mainstream GIS, especially in the geosciences.

Videogames, unlike GIS, do not tend to store a great deal of time-varying data, other than music or recorded sounds. (Exceptions include character or object animation, which is often canned rather than simulated in real-time, which is discussed further below). Unlike GIS, most modern videogames, across several genres, embody numerous dynamic process models which not only enable various elements of the virtual environment to appear to behave realistically, but also permit the environment to respond in realistic ways to player interactions. Many car racing games (*e.g. Forza Motorsport 2*) combine a high degree of visual realism with a high degree of process realism. These simulation-heavy not only replicate circuits with considerable geographical and architectural fidelity, and highly realistic vehicles with every detail of the bodywork precisely modelled, but the behavior of the vehicles is accurately replicated in the player's driving experience. For example, specific vehicular sounds are replayed during the game, and dynamic vehicle handling characteristics are also faithfully modelled based on a wide range of actual performance variables, to give the player the experience of real vehicle driving. Another example of an embedded process model is found in *Hydrophobia*, in which a fluid dynamics model, encapsulated in a specialised software engine, manages the dynamic behavior of water bodies and its effect on objects. This model not only handles realistic water movement in complex spaces, and modifies the acoustics of the enclosed spaces it occupies, but it also models interactions between the water and the game protagonists.

The difference between VGs and GIS in this respect is crucial. While the outcome of a game's dynamic processes may appear to be realistic, and indeed may be generated by physically accurate models, they are synthetic outcomes, and will typically vary each time they are produced, usually in context-dependent ways. GIS also display the outcomes of real-world processes but, in contrast to VG outputs, this realism derives from the fact that what is being modelled usually consists of recorded occurrences. Here we find a contrast not only between the synthetic and the real, but also between the what-might-be and the what-was, between the possible and the actual, and between the fictional and the factual.

Dynamic realism in VGs is closely related to notions of real-world time, both chronological and perceptual. The representation of cyclic time, and changes associated with it, is relatively common. Several games track the circadian cycle, for example by moving the sun across the sky (*e.g.* the *Jak & Daxter* series), by revealing changes in urban air pollution (as in *Midnight Club*: *Los Angeles*), or by modelling inhabitants' daily routines (*e.g.* retiring to bed at night in *The Elder Scrolls III*: *Oblivion*). Others unfold their story through the seasons of the year (e.g. *Animal Crossing* and *Harvest Moon*), with some (*e.g.* the *Harvest Moon* series) modelling vegetation, crop and landscape changes through the seasons, while some games (*e.g. Grand Theft Auto 3* and *The Elder Scrolls III: Morrowmind*) also introduce varied weather during gameplay. Several epic adventure and role-playing games (*e.g. Day of the Tentacle* and *Eternal Darkness: Sanity's Requiem*) simulate change through far longer periods of time.

Process distortion

Not all games behave strictly in tune with chronological time or Euclidean space. For example, some games set in extensive virtual worlds (*e.g. The Elder Scrolls III: Morrowmind* and *Legend of Zelda*: *Ocarina of Time*) permit players to traverse long-distances between widely separated locations almost instantly, while others permit time to be frozen (*e.g. Timeshift* and *Blinx*: *The Time Sweeper*), sped up (*e.g. Frontier: First Encounters* and *Sim City IV*) or slowed down (*e.g. Proving Ground*, *Burnout* and *Max Payne*). (*Braid,* being a game entirely based around time manipulation, allows you to do all of these things and more.) In *Mirror's Edge*, the 'Runner Vision' tool slows down time during gameplay to assist the player in executing complex moves, and is similar in visual style to 'bullet time' adopted in Matrix-inspired games. In a GIS context, maps have long been distortable through map projection or topological restructuring (as in cartograms). As for temporal distortion, modern data visualization software permits analysts to vary the speed and duration of events in spatial animations and in dynamic brushing, while spatial simulation software has always enabled models to be run in other than real time. The key difference between VGs and GISs, in this respect, lies in the purpose of temporal and spatial distortion. In the former it enhances various elements of gameplay, while in the latter it supports effective analytical interpretation.

Even when VGs provide an accurate simulation of the world, they sometimes permit the player to select the degree of realism they wish to see operating. This is particularly common in car racing games. Thus, for players who enjoy the fidelity of precise simulation, vehicle handling realism might be set at a high level, while those who prefer a more thrill-seeking driving experience might set the vehicle's handling system at a lower and far more forgiving level. Although visual, aural and behavioral realism are the primary goals of this often costly modelling effort, other liberties are also taken to enhance playability at the expense of temporal realism. *Midnight Club: Los Angeles*, for example, permits its high-performance cars to be raced at speeds in excess of 200mph around the streets of Los Angeles. Another interesting deviation from behavioral realism occurs in those squad-based combat games (*e.g. Brothers in Arms: Hell's Highway*), in which the game storyline requires that if key squad members are wounded in battle, they can be revived to rejoin the action. In a similar fashion, the AI which directs non-player characters in numerous games genres may be varied between a level of optimal decision-making efficiency (where the human player has little chance of winning) to a much more flawed level where it makes human-like errors (which gives the player some chance of winning). These and many other examples suggest that while VGs may be increasingly adept at creating accurate models of real-world processes, the fidelity of such modelling will always need to serve the maximisation of player satisfaction, in terms of the balance to be struck between challenge and reward.

Dynamic world modification

Unlike GIS editing software, which is typically used to build and modify spatial data models prior to analysis, most VGs facilitate the dynamic modification of their game environments during gameplay, based on player actions and decisions. This requires sophisticated software not only to model multi-way interactions between player, environment and objects, but also to update spatial and attribute data models on the fly. Dynamic landscape modification takes several forms. Buildings and other items of infrastructure are blown up in many games as in Figure 6, while in *Fracture*, players can use various kinds of grenade and firearms to modify the land surface itself, in order to gain tactical battle advantage[43]. More constructive changes include the moving of objects (*e.g. Elebits*), altering the look and feel of the existing environment by pulling on the ground surface to change its shape (as in *Civilization* and *Age of Wonders*), by planting or cutting grass (as in *Viva Pinata*), or by extending the environment by adding land and infrastructure (as in *Sim City*). The virtual worlds of many videogames are becoming almost as malleable as the real world.

Terraforming, deformation, destruction and construction are not restricted to landscape elements. In car racing and shooter games, for example, it is common for discrete objects such as opponents, furniture, vehicles and roadside objects (such as lampposts or fire hydrants) to be damaged or destroyed. (The earlier but innovative *Red Faction* game had a geomod, or geographical modification, feature which enabled players to incrementally destroy infrastructure, while the more recent *Crysis* and *Mercenaries* 2: *World in Flames* have

Figure 6 Exploding farmhouse (*Half-Life 2*)

more extreme destruction capabilities.) Some games also permit damage to be undone. In *Burnout Paradise*, for example, drivers of damaged vehicles may visit a virtual repair shop to have them instantly repaired, while in *Grand Theft Auto*, layers of dirt accumulated on vehicles may be removed by passing through a virtual carwash.

Human behavior and social processes

Human behavior is also simulated with increasing realism in VGs. In games which involve competition between player and software, most titles incorporate some form of artificial intelligence (AI) to operate non-player characters (*e.g.* monsters or zombies) and competing drivers' vehicles in a realistic fashion. In many shooter games, for example, combatants may be given varying degrees of intelligence, which may be selected by the player to match their skill level. In car racing games, competing cars are also driven with increasingly sophisticated software. In at least one such game (*Forza Motorsport* 2), the player may select several AI drivers to drive for them.

Squad-based behavior is increasingly evident in combat games, and is increasingly modelled using sophisticated AI. In *Brothers in Arms: Hell's Highway*, and the World War I title, *To End All Wars*, the squad behavior mimics multi-level military decision-making behavior, down to squad and individual levels. Human behavioral modelling is taken further in the tactical shooter *BlackSite*, in which a squad morale system is used to influence the battlefield behavior of individual members of the player's team. (The player indicates what squad members are expected to do, but the current squad morale level affects how they do it.) Even more remarkable is the growing practice of developing and communicating the emotions of game characters, as in the trait and 'moodlet' system implemented in *The Sims 3*.

Many strategy and role-playing games involve large numbers of humans, with which the player can interact and/or control. At a relatively simple level, some games (*e.g. Assassin's*

Creed) provide realistic simulation of crowd behavior. At a more sophisticated level, are the strategic games, often set in cities (*e.g. Settlers*: *Rise of an Empire*), which have evolved through several iterations to present high-fidelity urban environments in which citizens are seen acting out their daily lives in transparent detail. AI is particularly important in such games (*e.g. The Sims 3*), which often involve large numbers of inhabitants, because it would be virtually impossible for the player to micro-manage the behavior of all of the individuals inhabiting the game. Some RPGs mimic society in other ways. *The Elder Scrolls III: Morrowmind*, for example, models a character-class system, in which attributes and behaviors are established for game characters (usually including the player) in a way that resonates with socially stratified societies in the real world. Although some of these class systems are caricatures of the real world, they do represent a form of social simulation that is likely to achieve greater realism as in-game AI is further refined. At a higher social level, some games attempt to model cultural differences, but this is currently done rather poorly. The crude national stereotyping in such games as *Battalion Wars 2*, for example, fares poorly when compared with the heavily researched national cultural classifications of Hamden-Turner and Trompenaars[44], Hofstede[45] and others.

However, some commentators (*e.g.* Ref. [46]) suggest that AI can become too clever for its own good, in that it can make a game too difficult to play against. Dumbing down the AI also has its limitations, however, because of the danger of mismatches appearing between the visual realism of game characters and the relative stupidity of their behavior. As a result, in games such as *Unreal Tournament* 3 and *Burnout*, the designers attempt to create AI that behaves in ways that humans would behave, and makes the same mistakes that humans would make.

There are several lessons here for GIS. One might be in terms of monitoring the behavior of spatial analysts during and across sessions to provide intelligent context-sensitive help (as in the case of *Left 4 Dead* described earlier). It is also worth noting that since the AI would be used entirely on behalf of the protagonist (*i.e.* the analyst) in the case of the GIS, there would be few of the problems associated with over-intelligent antagonists. In GIS, the AI would always be on the player's side; the problem to be solved would play the abiding role of combatant. A second lesson might be drawn from those racing games (*e.g. Forza Motorsport* 2) which provide both a highly realistic simulation mode and a more forgiving 'arcade' driving mode (also described earlier), with the latter permitting the player to accumulate considerable physical damage to their vehicles without incapacitating them. The principle that is useful in a GIS context is that the degree of simulation realism must always be fit for purpose. Thus, for example, while a highly realistic dynamic spatial model might be necessary for environmental planning purposes, a 'dumbed down' version might be more appropriate in an educational context.

User interfaces

While the effectiveness of a GIS may largely be measured in terms of the power and flexibility of its toolset, "a computer game, unlike most other computer applications, lives or dies by the effectiveness of its user interface design"[8]. Because several thousand games are published each year, and many introduce innovative interface ideas, the games scene represents a huge experimental laboratory for interface design. This provides a major resource that may be raided by GIS designers in developing more effective methods for spatial exploration, search and interrogation[47, 48] In this section, a number of VG interface ideas will be described, and their GIS potential indicated (See Ref. [49] for a fuller discussion). The discussion will focus primarily on software rather than on game hardware devices.

Viewpoint

One of the most important aspects of the videogame user interface is its viewpoint—*i.e.* the location of the player in relation to the virtual environment. Three viewpoints are especially evident: the first-person viewpoint, in which the player moves through a virtual environment using the eyes of the largely unseen protagonist, as in *Half-Life 2* (see Figure 1); the third-person viewpoint, in which the player is distanced from the game's protagonist, which is represented by an avatar which they control, as in *Team Fortress 2* (see Figure 7); and the god viewpoint, in which the player looks down on, or obliquely across, the field of play, directing events from a location that affords high situational awareness, as in *Sim City* (see Figure 8). (It should be noted that while the player's control of characters in *The Sims* has led to it being called a 'god-game', it does not adopt what is commonly referred to as the god viewpoint.) There are variations on these main viewpoints, such as the 'over-the-shoulder' approach of some third-person shooters, in which the player is partly immersed in

Figure 7 Stylised buttes in the Badlands (*Team Fortress 2*)

Figure 8 Incongruent urban landscape elements (*Sim City IV*)

the scene by following closely behind the seen protagonist, and the 'through-the-rear-window' viewpoint in some car racing games. Some games, such as the online *World of Warcraft*, allow the player to zoom in and out from the standard over-the-shoulder third-person viewpoint. The viewpoint serves multiple functions in VGs, not only providing a degree of immersion (as in the first-person viewpoint), or situational awareness (as in the other two viewpoints), but also providing opportunities for user identification with the seen protagonist (as in the third-person viewpoint).

Most games adopt a single viewpoint throughout. However, in recent years, some popular games (*e.g. Metroid*) have switched from one viewpoint to another in successive releases, and in other games, it is possible to switch between viewpoints during gameplay. In the first-person *Brothers in Arms* and *Ghost Recon*, the player can momentarily switch to a third-person camera to provide a tactical overview when the combatant has taken cover. The RPG game *The Elder Scrolls III*: *Morrowmind*, for example, offers both first- and third-person viewpoints, with the latter being useful when the player's avatar wishes to display its winnings. In some games, the viewpoint is switched automatically, either to reveal important events (as in *Civilization: Revolution*), to taunt opponents with amusing character animations or for replay purposes (as in *Team Fortress* 2, Figure 4), or to reveal the player's avatar when it is injured (as in *Left 4 Dead*). The temporary switch between viewpoints to make use of their particular advantages highlights the benefits of making both available in the same game.

In contrast, most GIS traditionally adopt the map-based, god viewpoint. Recent innovations in Web mapping, however, are introducing alternative viewpoints more akin to the first-person and third-person viewpoints found in videogames. (The elevated camera position in the terrain-following viewpoint in Google Earth, however, still maintains most of the elements of the god viewpoint.) As GIS introduce more 3D scenes into the analyst's working life, the need for additional and alternative viewpoints and interaction methods is likely to increase markedly[50]. One of the striking things about most GIS is that little attempt is made to use viewpoints to develop a sense of immersion, or to encourage user identification with a protagonist (the analyst does not normally appears as an avatar in GIS scenes). By and large, the analyst stays outside the scene, keeping his or her distance, and essentially playing god.

Viewpoint choice has major implications for GIS design. Detailed studies have been undertaken in a military context (*e.g.* Ref. [51–54]) which reveal that the egocentric and exocentric viewpoints are suited to different tasks and benefit users with different levels of experience. Since some of the tasks undertaken in these military studies (*e.g.* judgements of relative position, mobility assessment and line-of-sight visibility) are similar to those undertaken in GIS, it may be important for GIS software to enable users to switch as needed between the two types of viewpoint. A required first step might be to consider inserting an avatar into the scene, and using a third-person viewpoint rather than the conventional god-like viewpoint.

Camera position and movement

Camera control is an important means by which players are able visually to explore the game environment, and is critical for rapid decision making. In some games (*e.g.* the first-person shooter, and some car racing games), the camera equates to the player's eyes, and is locked to the player's movement through the game environment, which the player controls with mouse, keyboard or games controller. However, in games where both the camera and the protagonist's avatar have to be controlled simultaneously (*e.g.* those with a third-person viewpoint), this poses a heavy interaction load on players. To help resolve this problem, many such games take over camera control, leaving players to control the movement of the avatar. There is therefore a balance to be struck between player and software camera control, with some games (*e.g. Resident Evil 4*) offering higher levels of player control, and others (*e.g. Super Mario Galaxy* or *God of War*) offering lower levels of player control.

In some games (*e.g. Ratchet & Clank*), the player may opt either to control the camera or to cede control to the program's AI. A further distinction is made between, on the one hand, character-relative motion, in which the avatar moves relative to their current state in the game world (the classic example is the early *Resident Evil* series of games), and on the other hand camera-relative motion, in which the avatar moves relative to the viewpoint's current state in the game world (the recent *Resident Evil 4* finally switched to this approach). At root, this is a choice of how the user controls map to in-game state change, with the primary purpose of automatic camera control being to reduce the player's control burden, especially in fast-moving and stressful situations. However, in some games camera movement is also used for dramatic effect. In *Gears of War*, for example, the camera shakes as the player runs, which adds to the sense of immersion.

Fitness for purpose

A common feature of VGs is that specific interaction methods are designed to fit particular player actions and operations. For example, methods to control various locomotory actions by players are provided in a wide variety of games, and especially those adopting a third-person viewpoint. However, because of the god-like perspective adopted by the majority of GIS, their potential role in assisting spatial exploration and analysis has been largely unexploited. For example, the grip and climb systems available in several action games, including *Ico* and *Assassin's Creed*, could be adapted for the exploration of mountainous terrains in GIS.

Interface diversity is a significant feature of videogames, with the same action often controlled in different ways in different games. This can be illustrated by the process of targeting characters and objects, which supports a variety of purposes, including shooting, path planning, information extraction, selection, and general 'trigger' interactions. In *Metroid Prime 3*: *Corruption*, for example, a complex trigger facility allows players to lock onto an enemy, and then either fix their perspective on the target or look straight ahead. Players can also move the crosshair around the target extremely rapidly after they have locked onto it, which makes it extremely easy to handle multiple small targets within a clearly defined area. Finally, individual games may provide several options for weapon-targeting, as illustrated by *Grand Theft Auto IV* whose alternatives include: free aim; snap from target to target; target and modify; and lock-to-objects (*i.e.* the camera's anchor point locks to the object), all of which operate on both static and dynamic objects. Many of these innovative interface methods have potential applications in GIS.

Free and constrained exploration

Most early videogames constrained the player to movement in 2D space. The advent of full 3D in the 1980s (*e.g. Elite*) and 1990s (*e.g. Quake*) provided players with increasing freedom to explore and navigate in three dimensions. In parallel with this increased freedom emerged numerous ways of mapping player manipulations of their interface devices (mouse, keyboard, handheld controller, *etc.*) to a growing repertoire of exploratory movements in space (walking, running, climbing, jumping, flying, *etc.*). In time, extensive experimentation across thousands of game titles led to the emergence of a handful of classical interface mechanisms whereby spatial exploration and action could be undertaken. Among these were the first-person running and jumping (*Quake*), the third-person running and jumping (*Jak & Daxter*), the third-person per-scene viewpoints (*Alone In The Dark*), and the god's eye view (*Sim City*).

Although there has been a progressive move in VGs towards increasing freedom of spatial exploration, this freedom has often been deliberately constrained by games designers, chiefly because of the challenge to players of controlling free movement in three spatial dimensions. One form of constraint is found in those games (*e.g. Jak & Daxter, Ratchet & Clank* and *World of Warcraft*) which, although they present a highly realistic 3D virtual environment on screen, constrain the movement of the player's avatar mainly to the degrees of freedom of the ground surface. Another technique adopted in shooter games is to adopt an 'on rails' style during intense fighting episodes (as in *Resident Evil: The Umbrella Chronicles*), in which the camera follows a fixed path through a battle zone, and the player merely controls the direction it faces in order to shoot at large numbers of objects and assailants en route. In a similar manner, the circuit racer game largely prescribes where the player is able to drive during gameplay. Although they may have extensive rooms, the castle (as in *Wolfenstein 3D*) and the underground laboratory (as in *Half-Life*) offer similarly constrained spaces. The 'constrained navigation' approaches for moving across complex 3D terrains using 2DOF of movement[55] is of particular relevance in this context.

The recent emergence of non-linear, free-roaming games (championed by *GTA III*, and taken up by such titles as *Far Cry 2*, *Burnout Paradise* and *Tony Hawk's American Wasteland*) is beginning to challenge the hegemony of titles which keep the player within confined spaces or moving along constrained courses. This spatial opening up of games has led to players needing on-screen orientation and navigation aids. Although games have tended to avoid displaying on-screen aids, in order to sustain a sense of immersion, the emergence of 'open world' VGs set in extensive environments has led to the re-appearance of devices such as the inset map or minimap (*e.g. GTA San Andreas*, which has both an inset minimap and a fullscreen map view). *Metroid Prime* goes further than most games with its simplified, rotatable and zoomable 3D wireframe map. GIS have much to learn from videogames in this respect, especially as and when they adopt third-person user modes involving an analyst avatar.

Simplification

A notable feature of VG interfaces is the way in which often complex on-screen behavior is achieved by relatively simple interface actions on the part of the player. This process of interface simplification is enabled by the games software, which undertakes the necessary behavioral elaboration. In *Super Mario Galaxy*, for example, the protagonist's avatar is not directly controlled by the player (nor, indeed, is the camera, as discussed earlier). Moreover, not only are Mario's jump trajectories—especially between planetoids—software generated, but the use of modified gravity makes walking across planet surfaces or up walls without falling off as simple as walking across the ground. Indeed, this would be an entirely different game if players were required to take fine-grained behavioral control of their on-screen avatar, with no distortion of real-world physical laws to help them.

Many games (*e.g.* those involving driving, unarmed combat, skateboarding and parkour) require the player to control complex on-screen behavior, either of vehicles or their personal avatars. In most such games, these behaviors (*e.g.* the trick repertoire in successive editions of the *Tony Hawk's* skateboarding game, or the skyline leaping and running behavior of the hero in *Assassin's Creed*) are achieved by far less complex control manipulations, and in some cases only a single button press is required[56]. Much of this complex on-screen behavior is pre-animated, and in some cases (*e.g.* when the player wiggles their joystick around to control their avatar), several animations are cross-blended and phased in and out to produce plausible walking animation. In *Splinter Cell: Conviction*, not only are the player's relatively simple control actions (*e.g.* pressing a button) translated into more complex on-screen behavior (*e.g.* throwing an opponent across a room), but the software also intelligently analyses the current context, and elaborates the behavioral response to the player's control action in a way that is most appropriate for a given situation. A context-dependent approach is taken in several games, so that the action triggered by using a particular control depends on the current state of play. In *Mirror's Edge*, for example, instead of providing a 'grab' button, the software uses edge-detection routines to determine whether the player's avatar is close enough to a building ledge to invoke a gripping action. Similar techniques are used in *Fable II*, which generates moves for its melee fights that reflect their immediate context. It is in ways like these that the design of virtual environments in VGs is able to guide and encourage player behavior, in ways that are largely absent from GIS.

Although the advantages of interface simplification and behavioral elaboration are relatively obvious, there are also dangers to be avoided. One is the player frustration caused when the software takes autonomous decisions about the game context, especially when there are several potential actions available at a given game point, and the software chooses one that does not match the player's intention. Also, many commentators and players part company with games designers when elaborate cut scenes or heavy AI are used to animate spectacular action sequences (as found in the introductory sequences of many VGs, and in the destruction sequences in *Half-Life: Episode 2*). Because player interaction with the game is normally suspended while cut scenes are playing, the player's role is reduced to little more than that of a spectator.

Multi-sensory representation

In an earlier section, data visualization was shown to be an important ingredient of effective gameplay. However, not only is it an informal craft in most VGs, but the rule-based creation of visual symbols from data is intermixed with the design of graphical symbolism which serves other, less representational roles. Although the advantages of presenting spatial information to more than the visual human sense are well established[18, 19], most GIS still privilege the sense of sight. In plotting the future development of data representation, GIS designers could therefore do far worse than examine the track record of VGs, many of which have mastered the art of multi-sensory information presentation[57]. We will begin by examining how sound and haptics are used to convey information to players, and then consider the effective combination of multi-sensory information.

Sound

Sound, and especially music, is used mainly for emotional effect in videogames, as is exemplified by the mood established by the soundscapes of such games as *Defcon* and *Ico*. However, there are cases where music is also used to represent information. For example, it can indicate different states of play, including the presence of danger and a lack of remaining time, and can also provide notification of certain events. In *Left 4 Dead*, for example, one music track begins just before a 'horde' (of zombies) attacks, another starts up as a 'tank' (a super-powerful zombie) approaches (and continues playing until it is dead), and yet another gradually increases in volume and sound layers as the player approaches the hiding place of a 'witch' (another extremely dangerous zombie). In one scene of *Super Mario Galaxy*, the tempo of the music increases and decreases in step with the player's movement of a star ball. Such redundant representation, in which sounds reinforce visuals, is found in many games. In *Ratchet & Clank*, for example, sound encodings are used to convey the nature of the contrasting surfaces on which the protagonists are walking. In *Crysis*, however, the game deviates from the rather strict rules of data visualization, in that weapons which sound similar in the real world are given distinctive sounds in the game because of the importance of players distinguishing aurally between them. A similar form of aural exaggeration, this time for emotional effect, is used in the *Thief* games, in which the volume of the protagonist's footsteps on certain surfaces is enhanced to alert them to the fact that they might be heard by others.

It is very common for classes of entities to be identified by means of audio 'tags' in VGs. In *Half-Life 2*, for example, each type of enemy has a distinctive aural signature, and will emit easily recognized sounds when distinctive events occur. (These sounds are of crucial importance when those events are off screen or obfuscated by other on-screen events.) Similarly, audio tags are used to provide aural confirmation and identification as the player collects types of item (*e.g.* medical supplies or ammunition), or defeats types of enemy. Sound can also be used to reflect measured data, through variations in pitch or frequency (though not amplitude, which is generally used to encode distance-from-viewpoint). For example, puzzles in the *Zelda* series of games give timing information to the player in this way.

A further role for sound in VGs is in providing locational information. Variously named spatial audio, localized sound, specialised sound, positional audio or 3D sound, the ability to alert players aurally to the location of characters and events in 3D space is an important element of the soundscape of most modern VGs. 3D spatialization of audio implicitly provides scalar feedback by changing audio volume on the basis of a sound source's distance from the player or viewpoint. Most game designers have extensive soundscape-building tools, which means they are able to add sound to a virtual environment by identifying the location of sound emitters interactively and drawing boxes around spaces affected by individual sounds.

Haptics

Due to the widespread availability of interface devices (including joysticks and game controllers) that emit vibrations or output force feedback, many games are able to harness the haptic senses (the vibrotactile, which is mainly located in the skin, and the kinaesthetic, which is mainly located in the joints) to transmit information to the player. (The kinaesthetic sense has the unique property of both receiving and issuing force or stress, and is thus uniquely able to engage in active exploration of environments.) Many games (*e.g. Ratchet & Clank*: *Tools of Destruction*, *Motorstorm*, and *Heavenly Sword*) adopt rumble, *i.e.* vibrotactile feedback, to identify events occurring in the virtual environment, with the 'profile' of a rumble (that is, how its strength varies over time) commonly used to differentiate between different kinds of event. Controller rumble is also used to generate a sensation which indicates that the player's avatar is in contact with certain surfaces in the virtual environment. Scalar feedback can also be provided by the controller, as when the amplitude of a sound or the frequency of an aural pulse is used to convey proximity to otherwise secret (*e.g.* underground) items.

Force-feedback devices are not normally used to provide categorical information to players, but are commonly used to emit the values of continuously varying parameters. In some flight simulator games, for example, force-feedback joysticks are used to mirror the forces felt through the control column in real aircraft, and force-feedback steering wheels available for some driving games convey the resistance of the car's wheels to turning based on physical simulation of the vehicle in contact with the road.

Combining sensory outputs

Multi-sensory output has two broad roles: sensory substitution, in which one sensory modality is replaced by another, and sensory combination, in which one sensory output is used to reinforce another through redundant data coding, or to encode multiple data variables[57]. A great deal of sensory fusion plays a redundant encoding role. In *Zelda: Twilight Princess*, for example, the Wii controller combines both sound and tactile feedback (by emitting a 'twang' sound and rumbling) when the player fires a bow and arrow. Similarly, in *Super Mario Galaxy*, the Wii controller rumbles when the player's avatar lands from a great height. Audio cues can help players focus on a critical visual element, for example, by teaching the player to associate a distinctive audio cue with a given enemy action (and its related animation, such as raising an arm high prior to striking the player). In several games, spatial distance and audio volume are often used together to help separate the camera from the player avatar, or conversely to make them feel more tightly bound to them. An interesting (but rather rare) effect found in some third-person games (in which the player controls an avatar), is where the visual and audio information is decoupled, in that audio effects are spatialized as if heard from the avatar's location (*i.e.* as if a virtual microphone was positioned there) rather than from the camera position.

The multi-sensory output of multivariate information is perhaps less frequently used than redundant encoding of individual variables. However, in some car racing games (*e.g. Forza*

Motorspot 2), the driving experience is fully multi-sensory in a multivariate sense, with a combination of visual, aural and tactile (*e.g.* rumble) feedback being continually provided. The non-visual information is highly nuanced, conveying a variety of engine, exhaust and tyre noises with both a high degree of realism and emotional impact.Some other games also use multi-sensory output to reflect multivariate information. For example, *Pikmin* has a hundred or so little creatures running around following the player's avatar, and color is used to represent the type of pikmin and sound (in addition to animation) is used to represent their current state (*i.e.* whether they are working, running, endangered, hurt, *etc*). It should be borne in mind that the integration of multiple sensory outputs in VGs combines with tightly integrated storylines and compelling game mechanics to transcend mere information presentation, enabling the player to participate in an immersive experience. While this might give VGs the capacity to support marketing activities, it might not mesh in an obvious way with the typical requirements of the spatial analyst.

Conclusions

In this chapter, we have examined two kinds of spatially intelligent software: the videogame and the GIS, and have compared several of their key characteristics and capabilities. One of the major conclusions drawn from our analysis is that some of the differences between videogames and GISs are more apparent than real, and that the former have a surprising number of similarities with the latter. A second conclusion is that VGs constitute a public laboratory in which numerous innovative ideas have been road-tested with large numbers of highly demanding users. The results of these experiments are freely accessible to GIS designers, and can serve to short-circuit the improvement of existing features as well as the introduction of additional functionality.

Videogame contributions

We have evaluated VGs against key GIS capabilities to reveal their strengths and weaknesses, and have identified several areas in which VGs could make potentially significant contributions to tasks normally undertaken with GIS. Of course, this is not to suggest that particular videogames could be used off the shelf as fully-fledged GIS, but it is to suggest that some widely exhibited functionalities in VG technology could contribute in significant ways to the future development of GIS.

It has been argued that in three specific areas (dynamic process modelling, interfaces and multi-sensory representation), VGs exhibit more developed capabilities than GIS, and have much to offer the spatial analyst. Firstly, VGs may be characterized as dynamic, process-based simulation systems, in which the interactions between players and modelled processes provide much of the software's functionality and appeal. In contrast, GIS are essentially spatial data handling environments, which for the most part display spatio-temporal data rather than model spatio-temporal processes. In this context, VGs arguably have more in common with dynamic, process-based geospatial simulations than with GIS. Nevertheless, given numerous past attempts to enable dynamic modelling in a GIS framework, VG achievements in this field are well worth examining.

Secondly, it has been shown how VG interfaces and interaction styles are highly diverse, and may be selected according to the needs of individual players and the tasks they are required to perform. The way in which VG interfaces are designed for exploring 3D virtual environments offers GIS designers many ideas for providing analysts with additional interactive approaches for exploring their own geographical worlds. Finally, we have shown that while GIS are generally more advanced in rule-based data visualization, the leadership roles are reversed when the representation of data to other sensory modalities is concerned. We have demonstrated ways in which VGs might help to rectify a longstanding gap in GIS functionality.

Both VGs and GIS may be broadly interpreted as being concerned with solving problems in a spatial context. While it is readily apparent that GIS tools for spatial modelling, display and analysis provide support for environmental and social problem solving and decision making, this role may not be so evident with VGs. Nevertheless, games present players with considerable perceptual, motor and cognitive challenges in spatial environments, and innumerable tools with which to solve spatial problems, at several levels of challenge and complexity. VG approaches and tools may therefore be relevant for undertaking tasks equivalent to those found in a GIS context. For example, given the highly developed capability of VGs in assisting players to navigate, explore and interrogate virtual environments, many of the innovative techniques developed for 3D navigation and interaction might also be useful in digital GIS worlds. In this context, it is instructive to note Google's current positioning of its own GIS-related functionality in term of: 'Explore, Search and Discover'. However, VGs offer considerably more than quasi-3D map browsers in application areas such as virtual tourism and virtual heritage exposition[58] which require the use of highly realistic 4D technology. The fact that an increasing number of today's youth generation is already well versed in using VG technology should give GIS developers considerable food for thought.

Charting the future

A not insignificant question raised by this chapter concerns the means by which VGs might contribute to GIS in the future. Several possible scenarios may be outlined whereby spatial analysts might be able to acquire more game-like tools. The first would involve analysts modifying selected VGs to undertake GIS-related activities. There are already indications of this happening in several other application areas, including: scientific visualization[59], marketing[60], education[61], scientific research[62], military training[63,64], and environmental planning[65]. A second route might see developers using VG design toolkits (*e.g.* VG middleware) for rapidly prototyping and deploying games technology for GIS applications. This approach is perhaps most attractive for 3D data visualization applications[66]. A third scenario sees games developers broadening their target markets to include spatial exploration applications. There are several indications (*e.g.* in serious games, passive games and some casual games) that the industry might have an appetite for this. A fourth scenario might be based on interoperable collaboration between GIS and VGs, creating powerful software federations that capitalise on the individual strengths of each technology[5]. A final scenario would be for the developers of GIS and other spatial software to adopt ideas read-tested in VGs in their own software, to deliver added value to users. An example of this approach may be seen in the adoption of MMOG game mechanics in business software designed for trans-national organizations[67].

It remains an open question as to whether 'entertaining GIS' will become as popular or as relevant as 'serious games'. Equally uncertain is whether there will be increasing convergence of videogame and GIS technologies in the future. It remains to be seen, for example, whether continuing innovations in videogame technology will contribute to major improvements in GIS, whether videogames will chip away at the GIS consumer base with products that are increasingly able to model worlds for insight as well as for entertainment, or whether a visionary product might appear that encapsulates the best that both technologies currently have to offer. Amongst this exciting uncertainty, one conclusion emerges with clear-cut outlines: those of us who use technology to understand and manage the world we inhabit are facing a future of ever-increasing possibilities, with technological options becoming increasingly available from complementary sources. Without succumbing entirely to the hubris of a former political leader, it may truly be said that never in the history of spatial science have we had it so good.

Notes

The titles of the videogames included as acronyms in Figure 2 are:

AC (*Assassin's Creed*), BiAHH (*Brothers in Arms: Hell's Highway*), BP (*Burnout Paradise*), BR (*Braid*), CIV (*Civilization*), CoD4 (*Call Of Duty 4*), CoDWaW (*Call of Duty: World at War*), CR (*Crysis*), DC (*Defcon*), EC (*Echochrome*), FR (*Fracture*), F2 (*Fable II*), FF (*Final Fantasy*), FM2 (*Forza Motorsport 2*), GA (*Getaway*), GoW (*God of War*), GTAVC (*Grand Theft Auto: Vice City*), HGL (*Hellgate: London*), J&D (*Jak & Daxter*), KD (*Katamari Damacy*), LBP (*Little Big Planet*), MFS (*Microsoft Flight Simulator*), PM (*Pacman*), RCGC (*Ratchet & Clank: Going Commando*), SC4 (*Sim City IV*), SMG (*Super Mario Galaxy*), SP (*Spore*), STH (*Sonic The Hedgehog*), TS (*The Settlers*), and WoW (*World of Warcraft*).

Acknowledgements The *authors* wish to thank the organizers and attendees at the Virtual Geographical Environments conference, held in Hong Kong in January 2008, at which some of the ideas discussed in this chapter were first presented. Thanks are also due to EA Games, Introversion Software Ltd. and Valve Software for permission to reproduce images from their games.

References

[1] Marble D F and Peuquet D J. Geographic Information Systems and Remote Sensing. In: Colwell R

N (Editor). Manual of Remote Sensing, second edition, Vol. 1, ASPRS/ACSM. Falls Church, VA, 1983: 923–958.

[2] Fisher P F and Lindenberg R. On Distinctions Among Cartography, Remote Sensing, and Geographic Information Systems. Photogrammetric Engineering and Remote Sensing, 1989, 55: 1431–1434.

[3] Duecker K J. Geographic Information Systems and Computer-Aided Mapping. Journal of the American Planning Association, 1987, 53: 383–390.

[4] Cowe D J. GIS versus CAD versus DBMS: What are the Differences? Photogrammetric Engineering and Remote Sensing, 1988, 54: 1551–1555.

[5] Shepherd I D H. Build your own GIS? Land and Minerals Surveying, 1990, 8: 176–183.

[6] Picard M. Video Games and Their Relationship with Other Media. In: Wolf M J P (Editor). The Video Game Explosion: A History from Pong to PlayStation and Beyond. Westport, Conn: Greenwood Press, 2008: 293–300.

[7] Murray J. Hamlet on The Holodeck: The Future of Narrative in Cyberspace. Cambridge, MA: The MIT Press, 1997.

[8] Aarseth E. Allegories of Space: The Question of Spatiality in Computer Games. In: von Borries F, Walz S P, and Böttger M (Editors). Space, Time, Play: Computer Games, Architecture and Urbanism: The Next Level. Zurich: BirkHäuser, 2007: 44–55.

[9] Wolf M J P. The Medium of the Video Game. Austin, TX: University of Texas Press,2002.

[10] Anon. The Making of Grand Theft Auto. Edge, 2008, April: 61–75.

[11] Anon. (2006) Never a City so Real? http://ask.metafilter.com/46273/Never-a-city-so-real [2008-10-18].

[12] Parish Y I H and Müller P. Procedural Modelling of cities. In: Proceedings ACM SIGGRAPH. Los Angeles, CA: ACM Press, 2001: 301–308.

[13] Hudson-Smith A and Evans S. Virtual Cities: from CAD to GIS. In: Longley P A and Batty M (Editors). Advanced Spatial Analysis: The CASA Book of GIS. Redlands, CA: ESRI Press, 2003: 41–60.

[14] Taylor F. Paragliding simulator with Google Earth plug-in. gearthblog.com/blog/archives/ 3d_models/ [2008-10-23].

[15] Biffo Mr. Down with Physics. Edge, 2007, November: 124.

[16] Crawford C. The Art of Computer Game Design. Berkeley, CA: McGraw-Hill, 1984.

[17] Anon. Conflicting Stories. Fun or Real: What Do War Games Want To Be? Edge, 2008, February: 42.

[18] Shepherd I D H. Meeting the Design Challenge of Multi-Sensory GIS. In: Proceedings AGI'94. London: Association for Geographical Information, 1994: 12.1.1–12.1.9.

[19] Shepherd I D H. Multi-Sensory GIS: Mapping Out the Research Frontier. In: Waugh T (Editor). Proceedings of the 6th International Symposium on Spatial Data Handling. London: Taylor & Francis, 1995: 356–390.

[20] Godchild M F and Goal S (Editors). Accuracy of Spatial Databases. London: Taylor & Francis, 1989.

[21] MacEachren A M. Visualizing Uncertain Information. Cartographic Perspectives, 1992, 13: 10–19.

[22] Flowerdew R. Spatial Data Integration. In: Maguire D J, Goodchild M F, and Rhind D (Editors).

Geographic Information Systems: Principles and Applications, Vol. 1., Harlow, UK: Longman Scientific and Technical, 1997: 375–387.

[23] Shepherd I D H. Information Integration. In: Maguire D J, Goodchild M F, and Rhind D (Editors). Geographic Information Systems: Principles and Applications, Vol. 1. Harlow, UK: Longman Scientific and Technical, 1997: 337–360.

[24] Bertin J. Semiologie Graphique: Les Diagrammes, Les Reseaux, Les Cartes, Mouton. France : Paris-La Haye, 1973.

[25] Bertin J. Theory of Communication and Theory of 'The Graphic'. International Yearbook of Cartography, 1978, 18: 118–126.

[26] Bertin J. Visual Perception and Cartographic Transcription. World Cartography, 1979, 15: 17–27.

[27] Bertin J. Graphics and Graphic Information-Processing. Berlin and New York: Walter de Gruyter, 1981.

[28] Bertin J. A New Look at Cartography. In: Taylor D R F (Editor). Graphic Communication and Design in Contemporary Cartography. Chichester, UK: John Wiley, 1983.

[29] Kraak M-J. Computer-Assisted Cartographical Three-dimensional Imaging Technique. Delft, Holland: Delft University Press, 1988.

[30] Kraak M-J. Computer-Assisted Cartographical 3D Imaging Techniques. In: Raper J (Editor). Three-Dimensional Applications in Geographical Information Systems. London: Taylor & Francis, 1989: 99–114.

[31] Poole S. Trigger Happy: The Inner Life of Videogames. London; Fourth Estate, 2000.

[32] Peuquet D J. Making Space for Time: Issues in Space-Time Representation. Geoinformatica, 2001, 5:11–32.

[33] Peuquet D J. Representations of Space and Time. New York: Guilford, 2002.

[34] Dorling D. Stretching Space and Splicing Time: From Cartographic Animation to Interactive Visualization. Cartography & GIS, 1992, 19: 215–227.

[35] Roberts J C and Wright M A E. Towards Ubiquitous Brushing For Information Visualization. Information Visualization, 2006, 5: 151–156.

[36] Harbaugh J W and Bonham-Carter GF. Computer Simulation in Geology. New York: John Wiley, 1970.

[37] Merriam D F. Computer Applications in Geology—Two Decades of Progress. Proceedings of the Geological Association, 1980, 91: 53–58.

[38] Bonham-Carter G F. Geographic Information Systems for Geoscientists: Modelling with GIS. Oxford, UK: Pergamon Press, 1994.

[39] Benenson I and Torrens P. Geosimulation: Automata-based Modeling of Urban Phenomena. Chichester, UK: John Wiley & Sons, 2004.

[40] Goodchild M F, Parks B O, Steyaert L T (Editors). Environmental Modelling with GIS. New York: Oxford University Press, 1993.

[41] Goodchild M F, Steyaert L T, Parks B O, Johnston C, Maidment D, Crane M, Glendinning S (Editors). GIS And Environmental Modelling: Progress and Research Issues. GIS World Books. Fort Collins, CO., 1996.

[42] Steyaert L T and Goodchild M F. Integrating Geographic Information Systems and Environmental Simulation Models: A Status Review. In: Michener W K, Brunt J W, and Stafford S G (Editors). Environmental Information Management and Analysis: Ecosystem to Global Scales. Bristol, PA: Taylor & Francis, 1994: 333–356.

[43] Smith R. Toeing the Disassembly Line. Edge, 2007, Christmas: 122.

[44] Hamden-Turner C and Trompenaars F. Riding the Waves of Culture: Understanding Diversity in Global Business. New York: McGraw-Hill, 1997.

[45] Hofstede G. Culture's Consequences: Comparing Values, Behaviours, Institutions and Organizations across Nations. Beverly Hills, CA: Sage Publications, 2003.

[46] Anon. Artificial Idiocy. Edge, 2008, June: 74–79.

[47] Germanchis T, Pettit C, Cartwright W. Using Computer Gaming Technology to Explore Human Wayfinding and Navigation Abilities Within a Built Environment. In: Proceedings of the 21st International Cartographic Conference. Coruña, Spain: International Cartographic Association, 2005.

[48] Cartwright W E. Exploring Games and Gameplay as a Means of Accessing and Using Geographical Information.Human IT, 2006, 8: 28–67.

[49] Shepherd I D H and Bleasdale-Shepherd I D. Towards Effective Interaction in 3D Data Visualizations: What Can We Learn From Videogames Technology? In: International Conference on Virtual Geographic Environments. Hong Kong, January 7-8, 2008.

[50] Shepherd I D H. Travails in The Third Dimension: A Critical Review of 3D Computer Graphics in Geographic Data Visualization. In: Dodge M. and Turner M (Editors). Geographic Visualization: Concepts, Tools and Applications. Chichester, UK: John Wiley, 2008: 199–222.

[51] Wickens C D. Frame of Reference for Navigation. In: Gopher D and Koriat A (Editors). Attention and Performance, 1999, 16: 113–143.

[52] Banks R and Wickens C D. Commanders' Display of Terrain Information: Manipulations of Display Dimensionality and Frame of Reference to Support Battlefield Visualization. In: Technical Report ARL-97-12/FED-LAB-97-2. Aberdeen, MD: US Army Research Laboratory, 1997.

[53] McCormick E P, Wickens C D, Banks R, Yeh M. Frame of Reference Effects on Scientific Visualization Subtasks. Human Factors, 1998, 40: 443–451.

[54] Thomas L C, Wickens C D, Merlo J. Immersion and Battlefield Visualization: Frame of Reference Effects on Navigation Tasks and Cognitive Tunnelling. In: Technical Report ARL-99-3/FED-LAB-99-2. Aberdeen, MD: US Army Research Laboratory, 1999.

[55] Hanson A and Wernert E. Constrained 3D Navigation With 2D Controllers. IEEE Visualization'99. San Francisco, CA, 1999: 175–182.

[56] Anon. Leap of Faith. Edge, 2008, August: 54–61.

[57] Shepherd I D H and Bleasdale-Shepherd I D. Multi-sensory Data Realization in Virtual Worlds: Abstraction or Pragmatism. In: the International Conference on Virtual Geographic Environments. Hong Kong, January 7-8, 2008.

[58] Champion E. Applying Game Design Theory to Virtual Heritage Environments. In: Proceedings of the First International Conference on Computer Graphics and Interactive Techniques in Australasia and South East Asia. Melbourne, Australia, 2003: 273–274.

[59] Rhyne T-M. Computer Games and Scientific Visualization. Communications of the ACM, 2002, 45: 40–44.

[60] Miliano V. Unreality: Application of a 3D Game Engine to Enhance the Design, Visualization and Presentation of Commercial Real Estate. VSMM99, http://www.unrealty.net/vsmm99/[1999].

[61] Woods S. Loading the Dice: The Challenge of Serious Videogames. In: Game Studies: The International Journal of Videogame Research, 4(1). http://www.gamestudies.org/0401/woods/[2007-11-19].

[62] Lewis M and Jacobson J. Game Engines in Scientific Research. Communications of the ACM, 2002, 45: 31.

[63] Riddell R. Doom Goes to War. Wired Magazine, 1997, April: 114–118.

[64] Raybourn E M and Bos N. Design and Evaluation Challenges of Serious Games. In: Proceedings CHI '05. New York: ACM Press, 2005: 2049–2050.

[65] Guy B, Bidwell N J, Musumeci P. Gameplan: Serious Gaming for Place Making. In: Proceedings of the Second Australian Conference on Interactive Entertainment, http://eprints.jcu.edu.au/view/year/2005.html [2005].

[66] Shepherd I D H. It's only a Game: Using Interactive Graphics Middleware to Visualise Historical Data. Society of Cartographers Bulletin, 2002, 36: 51–55.

[67] Kim A J. Putting the Fun in Functional: Applying Game Mechanics to Functional Software. http://www.slideshare.net/amyjokim/putting-the-fun-in-functional [2008-9-23].

Subject Index